5th International Conference on Applied Physics, Simulation and Computing (APSAC 2021)

AA007329

Journal of Physics: Conference Series Volume 2162

Salerno, Italy
3 – 5 September 2021

ISBN: 978-1-7138-5133-2
ISSN: 1742-6588

Printed from e-media with permission by:

Curran Associates, Inc.
57 Morehouse Lane
Red Hook, NY 12571

Some format issues inherent in the e-media version may also appear in this print version.

Institute of Physics
Dirac House, Temple Back
Bristol BS1 6BE UK

Phone: 44 1 17 929 7481
Fax: 44 1 17 920 0979

techtracking@iop.org

Additional copies of this publication are available from:

Curran Associates, Inc.
57 Morehouse Lane
Red Hook, NY 12571 USA
Phone: 845-758-0400
Fax: 845-758-2633
Email: curran@proceedings.com
Web: www.proceedings.com

TABLE OF CONTENTS

Preface

Peer Review Declaration

Natural Radioactivity and Mineralogical Composition of Different Particle Size Fractions of a River Sediment from Calabria, Southern Italy: A Case Study 1
F Caridi, G Acri, G Belmusto, V Crupi, G Faggio, R Grillo, G Messina, G Paladini, V Venuti, D Majolino

The Radioactivity Distribution and Radiation Hazard in Honey Samples from Italian Large Retailers 8
F Caridi, V Venuti, G Paladini, V Crupi, G Belmusto, D Majolino

Quantum (hyper)computation Through Universal Quantum Gates in Water Coherent Domains 16
Luigi Maxmilian Caligiuri

A "user-Friendly" Phantom to Conduct Quality Controls on MRgFUS Device 40
G Acri, F Caridi, B Testagrossa, S Gurgone, C Anfuso, G Paladini, G Vermiglio, C Greco, V Venuti

Early-Age Heat Flow and Hydration in Cementitious Composites: Numerical Modelling Based on Temperature Measurements 49
E Martinelli, M Pepe, C Lima

Application of the Harmonica Index for Noise Assessment in Different Spatial Contexts 57
G Graziuso, A B Francavilla, S Mancini, C Guarnaccia

Predicting Daily Water Tank Level Fluctuations by Using ARIMA Model. a Case Study 67
S Mancini, A B Francavilla, A Longobardi, G Viccione, C Guarnaccia

Physical Agents Quality Index (PAQI): Application in a University Campus 78
G Graziuso, A B Francavilla, S Mancini, M Guida, C Guarnaccia

An Application of ARIMA Modelling to Air Pollution Concentrations During Covid Pandemic in Italy 86
S Mancini, A Francavilla, G Graziuso, C Guarnaccia

Morphological and Chemical Characterization of Atmospheric Particles PM10 in an Urban Site in Leon, Guanajuato, Mexico 96
J G Cerón Bretón, R M Cerón Bretón, A A Espinosa Guzmán, R C Lara Severino, M L Espinosa Fuentes, S E Carranco Lozada, E Ramírez Lara, C Guarnaccia, V G Bautista Nolasco

Machine Learning Techniques Applied to Road Health Status Recognition Through Tyre Cavity Noise Analysis 116
Gloria Schiaffino, Lara Ginevra Del Pizzo, Stefano Silvestri, Francesco Bianco, Gaetano Licitra, Filippo Giammaria Praticò

Suitability of 3D Printing Materials for Printing Anthropomorphic Phantoms: A Simulation Study 127
Nikolay Dukov, Elica Encheva, Kristina Bliznakova

Computational Simulations and Assessment of Two Approaches for X-Ray Phase Contrast Imaging 135
Kristina Bliznakova, Zhivko Bliznakov, Nikolay Dukov

Open-Source Software Tools for Strategic Noise Mapping: A Case Study 145
G Graziuso, A B Francavilla, S Mancini, C Guarnaccia

Sustainable Building Materials Based on Hemp Shives and Geopolymer Paste 154
L Vitola, M Vilnitis, I Pundiene, D Bajare

Visualization of the Plasma Frequency by Means of a Particle Simulation Using a Normalized Periodic Model 162
Joschka Grunwald, Paolo Mercorelli

Conception and Analysis of Cascaded Dual Kalman Filters as Virtual Sensors for Mastication Activity of Stomatognathic Craniomandibular System 173
Bogena Carlotta, Sirarpi Hovsepyan, Paolo Mercorelli

A Physical Ground for the Scaling of Peak Ground Acceleration (PGA) with the Integral of Squared Velocity (I2
Vp and Its Potential for Earthquake Early Warning 185
F Vallianatos, I Spingos

Vision-Based Deep Learning Algorithm for Detecting Potholes 194
K Gajjar, T van Niekerk, Thomas Wilm, P Mercorelli

Three-Dimensional CFD Modelling of Urban Flood Forces on Buildings: A Case Study 205
G Viccione, C Izzo

Traffic Noise Prediction Using Machine Learning and Monte Carlo Data Augmentation: A Case Study on the Patiala City in India 216
Daljeet Singh, Priyal Kaler, Ishita Lyall, Aekamjot Singh, H S Pannu

Modular Enumeration - A New Method for Solving Discrete Programming Problems 226
Vitaly O Groppen

Calculation of Depressurization Coordinate in Underground and Offshore Gas Pipelines 235
G I Kurbatova, V A Klemeshev, V B Filippov

Biometric Identification Based on the Multidimensional Transient Functions of the Human Oculo-Motor System 249
Vitaliy Pavlenko, Tetiana Shamanina, Vladysla Chori

Fast and Accurate Control of Gates for Quantum Hypercomputation Based on Coherent Domains of Water 258
Luigi Maxmilian Caligiuri

Author Index

Preface of the 5th International Conference on Applied Physics, Simulation and Computing (APSAC 2021), Salerno, Italy, September 3-5, 2021

The 5th International Conference on Applied Physics, Simulation and Computing (APSAC 2021) was held online from September 3rd to September 5th, 2021 for safety reasons that are related to the sanitary emergency conditions that every country has taken in order to fight against coronavirus and for the preservation of the public health.

Virtual conference rooms were arranged, on the Microsoft Teams platform, in which participants could attend live or recorded presentations, interacting with questions and answers.

In line with the tradition of this conference, several professors, researchers, scholars and students shared their latest research results on the topics of the conference.

The direction of the event was at the University of Salerno. The Applied Physics Laboratory of the Department of Civil Engineering, University of Salerno, was the headquarter of the local organizing committee, that handled the schedule of the presentations and chaired the sessions. The conference program included regular and special sessions divided by topics in order to increase the quality of the discussions.

In addition, the conference hosted two reputable events, the 2nd International Workshop on "MOdelling, SImulation and DAta Analysis in Engineering and Physics Applications" (MOSIDA21) and the 2nd edition of the International Symposium on “Sustainable Construction” (SustaCon21).

There were more than 100 participants, considering lecturers, presenters, co-authors and attendants, coming from many academic and scientific institutions from all over the world. The range of topics increased, bringing this year novelties and new researches in the focus of the events.

I would like to thank the APSAC Conference Organizers and the secretariat for supporting all the phases of the conference as well as the University of Salerno and the Department of Civil Engineering for supporting the events. Special thanks to my colleagues and friends, Badea Lepadatescu, as co-chair of the conference, Martins Vilnitis and Michele Guida, as promoters and co-directors of the SustaCon symposium. Thanks to my collaborators Antonella Francavilla, Simona Mancini and Gabriella Graziuso for serving as co-chair and local organizing committee members. Thanks to all the lecturers and presenters where, beside the high number of participants, the quality of the presentations made this conference a great success. The final, greatest, thank goes to all the participants that trusted us and decided to invest resources and time to be part of the event.

It can be concluded that this valuable online event has fostered by sharing the research results and building of scientific networks among professors, lecturers, scholars and students in the field of Applied Physics, Simulation and Computing, as well as cross collaborations between companies and academic institutions in order to transmit the scientific research to the development and to the applications of new technologies.

Claudio Guarnaccia
Department of Civil Engineering,
University of Salerno,
via Giovanni Paolo II 132,
84084, Fisciano,
SA, Italy,
E-mail: cguarnaccia@unisa.it

Journal of Physics: Conference Series **2162** (2022) 011001 doi:10.1088/1742-6596/2162/1/011001

Peer Review Declaration Form Guidance

On the following page you will find the declaration form.

- Please answer each question.
- You should submit the form along with the rest of your submission files.
- The deadline is the submission date written in your publishing agreement.

All conference organisers/editors are required to declare details about their peer review.

We will published the information you provide as part of your proceedings.

Peer review declaration

All papers published in this volume of Journal of Physics: Conference Series have been peer reviewed through processes administered by the Editors. Reviews were conducted by expert referees to the professional and scientific standards expected of a proceedings journal published by IOP Publishing.

- **Type of peer review:**
  Single-blind

- **Conference submission management system:**
  Software system for tracking and management of electronic submissions

- **Number of submissions received:**
  95

- **Number of submissions sent for review:**
  95

- **Number of submissions accepted:**
  32

- **Acceptance Rate (Number of Submissions Accepted / Number of Submissions Received X 100):**
  34%

- **Average number of reviews per paper:**
  3

- **Total number of reviewers involved:**
  115

- **Any additional info on review process:**
  -

- **Contact person for queries:**
  Name : Badea Lepadatescu
  Affiliation: Transylvania University of Brasov, Brasov, Romania
  Email : lepadatescu@unitbv.ro, apsac.conference@gmail.com

Natural radioactivity and mineralogical composition of different particle size fractions of a river sediment from Calabria, southern Italy: a case study

F Caridi[1], G Acri[2], G Belmusto[3], V Crupi[4], G Faggio[5], R Grillo[5], G Messina[5], G Paladini[1,a], V Venuti[1] and D Majolino[1]

[1] Dipartimento di Scienze Matematiche e Informatiche, Scienze Fisiche e Scienze Della Terra, Università degli Studi di Messina, V.le F. Stagno D'Alcontres, 31-98166 Messina, Italy
[2] Dipartimento di Scienze Biomediche, Odontoiatriche, e Delle Immagini Morfologiche e Funzionali, Università degli Studi di Messina, c/o A.O.U. Policlinico "G. Martino" Via Consolare Valeria, 1-98125 Messina, Italy
[3] Agenzia Regionale Protezione Ambientale Calabria (ARPACal), Dipartimento di Reggio Calabria, Via Troncovito SNC, 89135 Reggio Calabria, Italy
[4] Dipartimento di Scienze Chimiche, Biologiche, Farmaceutiche e Ambientali, Università degli Studi di Messina, V.le F. Stagno D'Alcontres, 31-98166 Messina, Italy
[5] Dipartimento di Ingegneria dell'Informazione, delle Infrastrutture e dell'Energia Sostenibile (DIIES), Università "Mediterranea", Loc. Feo di Vito, 89122 Reggio Calabria, Italy

*E-mail: giuseppe.paladini@unime.it

Abstract. The natural radionuclides (^{238}U, ^{232}Th and ^{40}K) content and the mineralogical characteristics of three different particle size fractions of a sediment sample from the Amendolea river, Calabria region, southern Italy, were investigated.
High Purity Germanium (HPGe) gamma spectrometry was employed for the radioactivity measurements and specific activity data were reported. X-Ray Diffraction (XRD) measurements were carried out to identify the mineral contents of each particle size fraction.
This case study was developed with the aim of evaluating the specific activities of natural radionuclides and their relation to specific minerals. Data obtained can provide useful information in view of a deep understanding of the adsorption of radioactive elements and the influence of the sediment granulometry on the radionuclides specific activities.

1. Introduction
Natural radioactivity provides the greatest contribution to the dose received by the population. From this point of view, important radionuclides are those belonging to the radioactive families ^{238}U, ^{232}Th and ^{235}U, together with the radioisotope of potassium, ^{40}K [1]. Together with their product of decay, they are commonly found in different concentrations in environmental matrices, especially in water, rocks and soils [2–5].

Among the environmental matrices, river sediments are of course particularly interesting, due to their mineralogical composition, particle size distribution and because they are reliable long-term indicators of river pollution by radionuclides [6]. Furthermore, knowing the distribution of the latter in terms of activity concentration in river sediments is extremely important, as it allows us to evaluate any temporal variations of specific activities within the river [7].

On the other side, geochemical studies of river sediments are also particularly useful, as the geomorphology of the investigated site greatly influences the chemical composition of this type of sediment, as well as the mineralogical composition reflects the geological history of the transport and the sorting process [8–11].

In this article, a case study was taken into account; in particular, natural radioactivity and minerals composition of three different particle size fractions of a sediment sample of the Amendolea river, Calabria region, southern Italy, were investigated with the aim to evaluate the influence of sediment particle sizes on natural radionuclides concentrations and the influence of the mineralogical composition on the ^{226}Ra, ^{232}Th and ^{40}K activity concentration. River sediments can be employed as building materials and thus the knowledge of their granulometric distribution, as well as the radioactivity concentration in each particle size fraction and the mineralogy of the investigated sediment, strictly correlated with its natural radioactive content, is extremely useful from a radiological point of view [12].

The schematic block diagram of figure 1 summarizes what was performed in the present study.

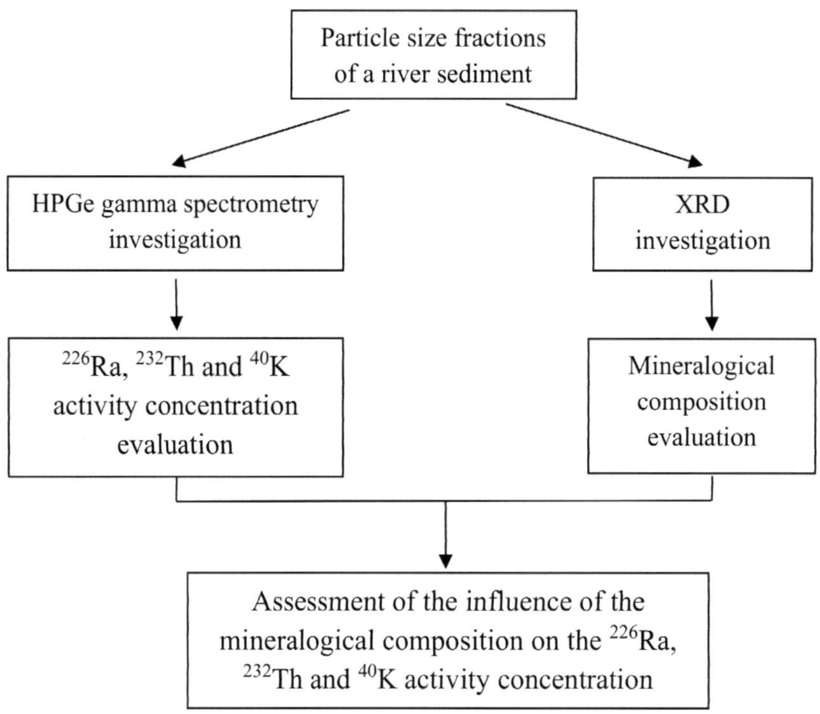

Figure 1. Schematic block diagram.

2. Materials and Methods

2.1 Sample collection and treatment

The investigated river sediment sample, around 1 kg, was collected in the sampling point indicated in figure 2 (GPS coordinates: latitude 37.9283; longitude 15.8878).

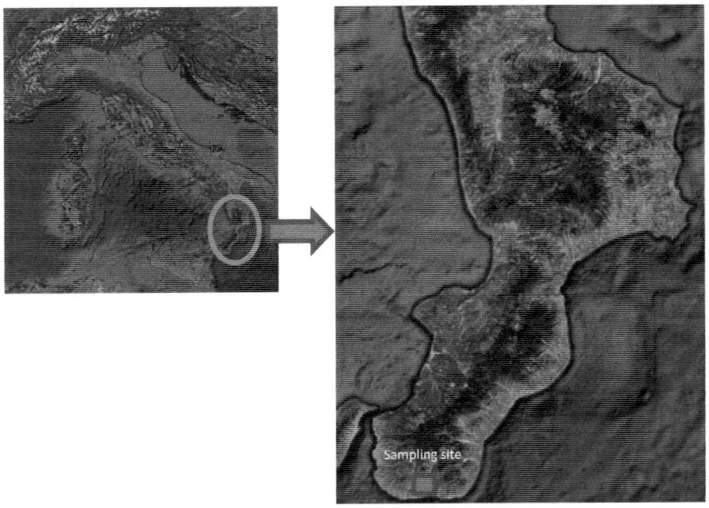

Figure 2. Map of the sampling site.

The sampling procedure is described in [10].

In laboratory, the sample was dried at 105° C in an oven and sieved to obtain the various particle size fractions: pelite (< 65 μm), fine sand (65 μm - 250 μm) and coarse sand (250 μm - 1 mm). Each size fraction was packed in a hermetically sealed 20 mL polyethylene plastic vial. After 40 days, the secular radioactive equilibrium between ^{226}Ra and their daughter products was attained and samples were ready for gamma spectrometry counting [13].

2.2 Radioactivity analysis

For the gamma spectrometry analysis, the pelite, fine sand and coarse sand fractions were counted for 70000 s. The ^{226}Ra and ^{232}Th activity concentrations were obtained by the 295.21 keV and 351.92 keV ^{214}Pb, the 1120.29 keV ^{214}Bi and the 911.21 keV and 968.97 keV ^{228}Ac γ-lines, respectively. For the ^{40}K specific activity evaluation, its γ-line at 1460.8 keV was employed.

The experimental setup was composed by an Ortec HPGe positive biased detector (GEM), with FWHM of 1.85 keV, peak to Compton ratio of 64:1 and relative efficiency of 40 % at 1.33 MeV (^{60}Co), and integrated digital electronics.

Eckert and Zigler Nuclitec GmgH traceable multinuclide radioactive standard, number AK-5901, was employed for efficiency and energy calibrations [14].

The ANGLE 4 code was employed for the efficiency transfer factors calculations [15].

The Gamma Vision (Ortec) software was used for data acquisition and analysis [16].

The activity concentration of each identified radionuclide was calculated using the following formula [17]:

$$C = \frac{N_E}{\varepsilon_E t \gamma_d M} \qquad (1)$$

where N_E, ε_E and γ_d account for the net area, efficiency and yield of a photopeak at energy E, respectively, M is the mass sample (kg) and t is the live time (s).

The quality of the gamma spectrometry experimental results was certified by the Italian Accreditation Body (ACCREDIA) [18].

2.3 X-Ray Diffraction (XRD) analysis

A Malvern Panalytical Empyrean Diffractometer, equipped with a solid-state detector, PIXcel, was employed for the XRD tests on about 1 g of finely powdered sample for each investigated fraction [19].

Acquisition conditions were 40 kV and 40 mA. XRD spectra were recorded from 2 to 54 degrees, with step size of 0.007 degrees and counting time of 20 seconds.

The background correction of the Cu Ka2 component with a digital filter and the identification of the crystalline mineral components through the ICDD JCPDS database were also performed [20].

3. Results and Discussion

3.1 Radioactivity analysis

The specific activity of detected radionuclides, ^{226}Ra, ^{232}Th and ^{40}K, in the investigated sediment sample, is shown in figure 3 for pelite, fine sand and coarse sand.

Obtained values are strictly correlated with the sampling point. It is, in fact, well-established that the radionuclides activity concentration in river sediments varies from location to location, considering that the river bottom can exhibit large variations in chemical and mineralogical properties and rare-earth elements [21].

As we can see from an inspection of the figure, experimental values are 1561 Bq/kg d.w. for ^{226}Ra, 691 Bq/kg d.w. for ^{232}Th and 4461 Bq/kg d.w. for ^{40}K, for particle size < 65 μm (pelite); 41 Bq/kg d.w. for ^{226}Ra, 69 Bq/kg d.w. for ^{232}Th and 1260 Bq/kg d.w. for ^{40}K, for particle size 65–250 μm (fine sand); 31 Bq/kg d.w. for ^{226}Ra, 13 Bq/kg d.w. for ^{232}Th and 1156 Bq/kg d.w. for ^{40}K, for particle size 250 μm–1 mm (coarse sand).

The adsorption of radioactive elements is then highest in the pelite fraction and it decreases from fine to coarse-grained materials [22].

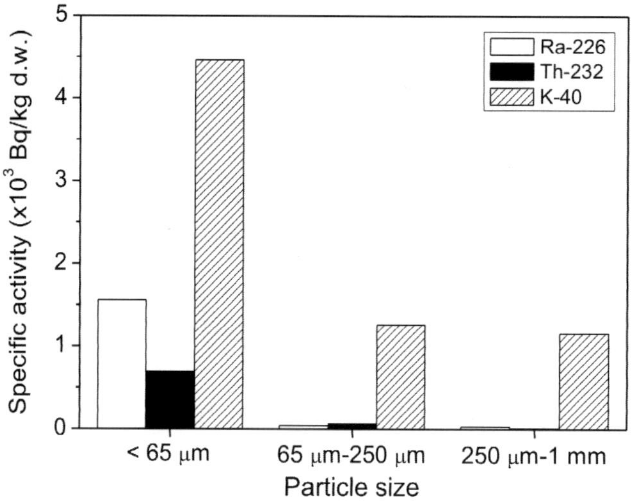

Figure 3. The activity concentration of ^{226}Ra, ^{232}Th and ^{40}K in the various particle size fractions of the investigated river sediment.

Moreover, the decrease in the specific activity values for ^{226}Ra, ^{232}Th and ^{40}K with increasing particle size, seems to indicate that the three particle size fractions are different from a mineralogical point of view and, consequently, have a different radionuclides content. The assessment of the influence of the mineralogical composition on the activity concentration of detected radioisotopes was then performed through the XRD analysis of the three particle size fractions of the investigated sample.

3.2 Mineralogical analysis

XRD spectra of the three particle size fractions of the investigated river sediment are shown in figure 4.

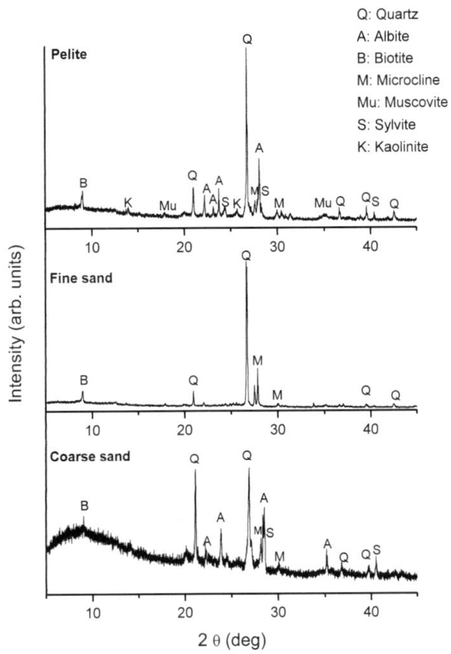

Figure 4. XRD patterns of the three particle size fractions.

From this figure we can notice that for each investigated size fraction, quartz (SiO_2) is the most abundant mineral. Biotite, $K(Mg;Fe)(AlSi_3O_{10})(OH)_2$, and microcline feldspar, $KAlSi_3O_8$, are also present in all samples. Kaolinite mineral, $Al_2Si_2O_5(OH)_4$, was identified only in the pelite fraction, whereas albite, $(Na,Ca)Al(Si,Al)_3O_8$, and sylvite (KCl) mineral components were detected only in the pelite and coarse sand samples. Finally, the muscovite mineral, $KAl_2(Si_3Al)O_{10}(OH,F)_2$, was found in the pelite.

Worth of note, kaolinite is detected only for the pelite fraction, not for the fine and coarse sand ones. Now, according to the literature [23], the kaolinite mineral is shown to be responsible for ^{232}Th and ^{40}K concentrations and its highest dispersion is produced by particles of 25 - 35 μm diameter [24]. In the light of this we can hypothesize, then, that kaolinite plays the major role, more than microcline feldspar and quartz, in the observed trend of the specific activity of these radionuclides as detected by radioactivity measurements in the investigated particle size fractions. No specific correlation between some particular mineralogical phase and specific activity of ^{226}Ra was found in literature. Nevertheless, being ^{226}Ra in secular equilibrium with ^{238}U, its decreasing radioactivity concentration with increasing particle could be related to the uranium concentration. An investigation of the elemental composition of each particle size fraction, by means of spectroscopic techniques (like as X-Ray Fluorescence and ICP-MS), with the aim to quantify the uranium content in each of them and thus to verify a possible decrease in concentration by increasing the particle size, can be helpful in clarifying this aspect. It will be planned in the next development of the research.

4. Conclusions

In the present paper, the ^{226}Ra, ^{232}Th and ^{40}K specific activity for three different particle size fractions of a sediment sample of Amendolea River (Calabria region, southern Italy), taken as case study, was quantified with HPGe gamma spectrometry, in order to investigate the influence of the granulometry on the radionuclides specific activity. The highest concentrations of detected radionuclides were found in the pelite fraction, because the adsorption of radioactive elements has a strong dependence on the sediment nature and decreases from fine to coarse-grained materials.

At the same time, XRD tests were performed to identify the mineralogical composition of each particle size fraction, with the aim of correlating the radioactivity results to specific minerals. From the results, the main role of kaolinite in increasing the radioactivity concentration for ^{232}Th and ^{40}K in pelite fraction with respect to, in the order, fine and coarse sand was hypothesized. As far as ^{226}Ra is concerned, being this radioisotope in secular equilibrium with ^{238}U, the observed decreasing level of its activity concentration with increasing particle size was supposed to be related to the uranium concentration in each of them.

The obtained data can be used as a baseline for future investigations about radioactivity background levels in river sediments samples of the investigated region and they can provide useful information to better understand the adsorption of radioactive elements and the variation of the radionuclides specific activity with the sediment granulometry.

References

[1] United Nations Scientific Committee on the Effects of Atomic Radiation. 2000 *Sources and Effects of Ionizing Radiation: Report to the General Assembly, with Scientific Annexes* vol I
[2] Navas A, Soto J and Machín J 2002 238U, 226Ra, 210Pb, 232Th and 40K activities in soil profiles of the Flysch sector (Central Spanish Pyrenees) *Appl. Radiat. Isot.* **57** 579–89
[3] Caridi F, D'agostino M, Belvedere A, Marguccio S and Belmusto G 2016 Radon radioactivity in groundwater from the Calabria region, south of Italy *J. Instrum.* **11**
[4] Caridi F, Messina M and D'Agostino M 2017 An investigation about natural radioactivity, hydrochemistry, and metal pollution in groundwater from Calabrian selected areas, southern Italy *Environ. Earth Sci.* **76** 668
[5] Bayramli G 2020 The environmental problems of Azerbaijan and the search for solutions *WSEAS Trans. Environ. Dev.* **16** 423–33
[6] Kücükömeroglu B, Kurnaz A, Keser R, Korkmaz F, Okumusoglu N T, Karahan G, Sen C and Cevik U 2008 Radioactivity in sediments and gross alpha–beta activities in surface water of Fırtına River, Turkey *Environ. Geol.* **55** 1483–91
[7] Madruga M J 2008 Environmental radioactivity monitoring in Portugal *Appl. Radiat. Isot.* **66** 1639–43
[8] Caridi F, Marguccio S, D'Agostino M, Belvedere A and Belmusto G 2016 Natural radioactivity and metal contamination of river sediments in the Calabria region, south of Italy *Eur. Phys. J. Plus* **131** 1–10
[9] Caridi F, Marguccio S, Belvedere A, Belmusto G, Marcianò G, Sabatino G and Mottese A 2016 Natural radioactivity and elemental composition of beach sands in the Calabria region, south of Italy *Environ. Earth Sci.* **75**
[10] Caridi F, Bella M Di, Sabatino G, Belmusto G, Fede M R, Romano D, Italiano F and Mottese A F 2021 Assessment of Natural Radioactivity and Radiological Risks in River Sediments from Calabria (Southern Italy) *Appl. Sci.* **11** 1729
[11] Atzori P, Ferla P, Paglionico A, Piccarreta G and Rottura A 1984 Remnants of the Hercynian orogen along the "Calabrian-Peloritan arc", southern Italy: a review *J. Geol. Soc. London.* **141** 137–45
[12] Caridi F, Testagrossa B and Acri G 2021 Elemental composition and natural radioactivity of refractory materials *Environ. Earth Sci.* **80** 170
[13] Sima O and Dovlete C 1997 Matrix effects in the activity measurement of environmental samples—implementation of specific corrections in a gamma-ray spectrometry analysis program *Appl. Radiat. Isot.* **48** 59–69
[14] Web site 2020 Eckert & Ziegler Calibration Sources

https://www.ezag.com/home/products/isotope_product
[15] Ortec 2016 Angle 4 Software User Manual
[16] Caridi F, D'Agostino M, Belvedere A, Marguccio S, Belmusto G and Gatto M F 2016 Diagnostics techniques and dosimetric evaluations for environmental radioactivity investigations *J. Instrum.* **11**
[17] Caridi F, Marguccio S, Durante G, Trozzo R, Fullone F, Belvedere A, D'Agostino M and Belmusto G 2017 Natural radioactivity measurements and dosimetric evaluations in soil samples with a high content of NORM *Eur. Phys. J. Plus* **132** 56
[18] ACCREDIA 2021 https://www.accredia.it/
[19] Malvern Panalytical. 2013 Empyrean diffractometer user manual
[20] Caridi F, Santangelo S, Faggio G, Gnisci A, Messina G and Belmusto G 2019 Compositional and Mineralogical Analysis of Marine Sediments from Calabrian Selected Areas, Southern Italy *Int. J. Environ. Res.* **13** 571–80
[21] Blanco Rodríguez P, Tome F, Lozano J and Pérez-Fernández M 2008 Influence of soil texture on the distribution and availability of U-238, Th-230, and Ra-226 in soils *J. Environ. Radioact.* **99** 1247–54
[22] McCubbin D, Leonard K S, Maher B A and Hamilton E I 2000 Association of 210Po(210Pb), 239+240Pu and 241Am with different mineral fractions of a beach sand at Seascale, Cumbria, UK *Sci. Total Environ.* **254** 1—15
[23] Ramasamy V, Suresh G, Meenakshisundaram V and Ponnusamy V 2011 Horizontal and vertical characterization of radionuclides and minerals in river sediments *Appl. Radiat. Isot.* **69** 184–95
[24] Yahaya S, Jikan S S, Badarulzaman N A and Adamu A D 2017 Chemical Composition and Particle Size Analysis of Kaolin *Path Sci.* **3** 1001–4

The radioactivity distribution and radiation hazard in honey samples from Italian large retailers

F Caridi[1], V Venuti[1,a], G Paladini[1], V Crupi[2], G Belmusto[3] and D Majolino[1]

[1] Dipartimento di Scienze Matematiche e Informatiche, Scienze Fisiche e Scienze Della Terra, Università degli Studi di Messina, V.le F. Stagno D'Alcontres, 31-98166 Messina, Italy
[2] Dipartimento di Scienze Chimiche, Biologiche, Farmaceutiche e Ambientali, Università degli Studi di Messina, V.le F. Stagno D'Alcontres, 31-98166 Messina, Italy
[3] Agenzia Regionale Protezione Ambientale Calabria (ARPACal), Dipartimento di Reggio Calabria, Via Troncovito SNC, 89135 Reggio Calabria, Italy

*E-mail: vvenuti@unime.it

Abstract. The natural (^{40}K) and anthropogenic (^{137}Cs) radioactivity concentration in twenty-five honey samples of five different typologies (*acacia, chestnut, wildflower, linden* and *honeydew*), coming from Italian large retailers, was investigated by High Purity Germanium (HPGe) gamma spectrometry.
The study was carried out in order to estimate the background levels of detected radionuclides in various honey samples, as well as to assess the dose levels due to their ingestion, taking into account the average yearly direct consumption in Italy and assuming this requirement satisfied from a single kind of honey.
Experimental results were compared with the total natural radioactivity value (external + internal) for humans, in order to evaluate any possible radiological health risk.

1. Introduction

Honey is one of the most complex foods produced naturally [1]. It is made from the nectar collected by bee workers from various flowers and spread in the surrounding areas where the bees are raised [2].

From a general point of view, its chemical composition and predominant simple sugars content [3], even considering the expected variability related to its botanical and geographical origin [4], make honey a precious energy source, with valuable nutritional and healing properties. As a matter of fact, it contains a mixture of different carbohydrates, high sugars, proteins, aminoacids, vitamins and minerals [5], and is also used as an ingredient in foodstuffs because of its flavor, color and sweetness [6]. Honey has, in addition, healing and antibacterial properties [7,8].

As a foodstuff used also for beneficial purposes, it must be free of any of unpleasant contents [9,10]. In particular, although, as well established, radionuclides can pose a health risk for humans, their presence in honey has not been the subject of a systematic investigation to date.

As apiculture is popular in Italy, i.e. every year the country produces approximately 20000 tons of honey [11], authors decided to investigate the levels of radionuclides in honey samples.

With this aim in mind, twenty-five honey samples, of five different typologies (*acacia, chestnut, wildflower, linden* and *honeydew*), coming from Italian large retailers, were analyzed in the present study, as far as natural (^{40}K) and anthropogenic (^{137}Cs) radioactivity concentration is concerned. In particular, as far as natural specific activity is concerned, our attention was focused only on ^{40}K since natural radioactivity in food mainly comes from this radioisotope, uranium and thorium daughter products being usually present in traces [12].

The assessment of the dose levels due to honey consumption, together with a comparison with the total natural radioactivity value (external + internal) for humans (2.4 mSv y^{-1}), were also performed in order to evaluate any possible radiological health risks for the population.

The main advantages of the presented research methodology are due to the possibility of evaluating:

- any possible anthropogenic radioactivity contamination of the investigated samples;
- the content of the most significant natural radioisotope in food (^{40}K);
- if the honey consumption is safe from the radiological point of view and if it does not adversely affect the human health.

2. Materials and Methods

2.1 Samples description

Investigated samples were divided into 5 groups (G#, # = 1, 2, 3, 4, 5), based on their typology, according to what reported in Table 1.

Group ID	Typology	Number of samples
G1	*Acacia*	5
G2	*Chestnut*	5
G3	*Wildflower*	5
G4	*Linden*	5
G5	*Honeydew*	5

Table 1. Investigated groups, together with their identification code (ID), typology, and number of samples for each group.

Acacia honey is obtained from the *robinia pseudoacacia*, a plant belonging to the *mimosaceae* family. It is from its flowers, when pollinated by bees, that this type of honey is obtained, with a vaguely floral scent and sweet taste. *Acacia* is then a monofloral honey, very light yellow in color, transparent and with a liquid consistency. It has a very low crystallization rate and can remain in the liquid state for a very long time, thanks to its high concentration of fructose compared to glucose. By virtue of its organoleptic and gustatory characteristics, as well as its excellent solubility, *acacia* honey is excellent as a sweetener for drinks (tea, herbal teas, milk, yogurt, coffee, vegetable milk, etc.). It does not excessively alter the taste and aroma of the products; on the contrary, it generally exudes its main characteristics [13].

Chestnut honey is a monofloral one, produced by bees that live near chestnut trees and use their flowers practically exclusively. That of the chestnut tree (*Castanea*) is a botanical genus that belongs to the *Fagaceae* family.

It is dark amber or brown in color, vaguely translucent and with a fluid but very viscous consistency. Second only to *acacia* honey, *chestnut* one has a low crystallization rate and can remain liquid for a long time. Its taste is sweet with a strong bitter aftertaste; aroma and flavor are decisive, with hints typical of the aromas left by the botanical species of origin. In Italy, *chestnut* honey is among the most used to accompany cheeses with an intense flavor [14].

Wildflower honey is produced by the natural mixing of the nectar collected by bees inside the hive. There is, therefore, no preeminent floral type, but is possible to find flavors and characteristics of various types of flowers.

Depending on the color it can be divided into two types:

- *wildflower* honey with a light color, almost transparent: its flavor, very delicate and sweet, makes it pleasant for everyone and suitable for multiple uses and recipes;

- *wildflower* honey with a slightly darker and amber color: it has a stronger taste and can vaguely recall honeydew. It is typical of mountain areas, where there are numerous chestnut trees and berries.

At nutritional level, *wildflower* honey is more complete than monofloral ones, presenting all the main characteristics of honey: it is very good for children, pregnant women and the elderly being an important energy food for those who are growing or for those who need to reinvigorate their strength; it also promotes bone development and is capable of improving cognitive functions [15].

Linden honey is obtained from the flowers of the plant of the same name, that belongs to the *Tiliaceae* family and can be found both in its "wild" version (*Tilia Cordata*) and as an ornamental tree (*Tilia Platyphyllos* and *Tilia americana*). *Linden* honey is presented in a rather light, amber color, with reflections tending to yellow-green in its purest varieties. In addition to the normal anti-inflammatory, anti-bacterial and anti-oxidant action common to other types of honey, *linden* honey has a real anti-spasmodic effect, acting as a calming of the nervous system, and is often used to combat anxiety and the resulting disorders [16].

Honeydew honey originates from a sugary substance secreted by some small insects on tree sap, such as spruce, pine, oak, lime and maple. This type of honey is much less sweet than the others and has a bitter aftertaste. Its color ranges from dark brown to almost black; the dark color is due to the richness of antioxidants, especially polyphenols. This type of honey has the greatest number of mineral salts and trace elements. It contains many vitamins and numerous essential amino acids, it has a viscous consistency, but it never crystallizes completely. Due to its strong antibacterial properties, it is particularly recommended for relieving diseases involving the respiratory system. It is an excellent source of energy for athletes and can be used to supplement a lack of minerals and trace minerals [17].

2.2 Radioactivity measurements

For High Purity Germanium (HPGe) gamma spectrometry evaluations, investigated samples were directly inserted in Marinelli containers of 1 L capacity, without any preparation [18], and counted for 70000 s.

The 661.66 keV and 1460.8 keV γ-ray lines of ^{137}Cs and ^{40}K, respectively, were used to quantify their specific activity.

Two Ortec HPGe detectors and integrated digital electronics were employed for the analysis. The first is a negative biased detector (GMX), with full width at half maximum (FWHM) of 1.94 keV, peak to Compton ratio of 65:1 and relative efficiency of 37.5 % at 1.33 MeV (^{60}Co). GMX is cooled by the Ortec recycler condensing liquid nitrogen cooling Mobius system [19]. The second one is a positive biased detector (GEM), with FWHM of 1.85 keV, peak to Compton ratio of 64:1 and relative efficiency of 40 % at 1.33 MeV (^{60}Co). GEM is cooled by liquid nitrogen [20].

The Gamma Vision (Ortec) software was used for data acquisition and analysis [21,22]. The specific activity (Bq kg^{-1}) of ^{137}Cs and ^{40}K was calculated as follows [23]:

$$C = \frac{N_E}{\varepsilon_E t \gamma_d M} \quad (1)$$

where N_E, ε_E and γ_d account for the net area, efficiency and yield of a photopeak at energy E, respectively, M is the mass sample (kg) and t is the live time (s).

The data uncertainty (coverage factor k=2) takes into account the contribution of the counting, calibration source, efficiency calibration, background subtraction, γ-branching ratio and internal absorption estimation [24].

The reliability of the experimental results was certified by the Italian Accreditation Body (ACCREDIA) on the basis of the quality controls performed according to the UNI 11665:2017 [25].

2.3 Radiological health risk assessment

The assessment of the radiological health risk was performed as the effective dose for honey samples ingestion [26]:

(2)

$$D_{ing}\ (Sv\ y^{-1}) = h_{ing,K-40} \times J_{ing,K-40}$$

where $h_{ing,\ K\text{-}40}$ indicates the ^{40}K conversion coefficient of effective dose, estimated [27] equal to 4.2 x 10^{-8} Sv Bq^{-1} for the age category between 1 and 5 years old, 2.1 x 10^{-8} Sv Bq^{-1} between 5 and 10 years old, 1.3 x 10^{-8} Sv Bq^{-1} between 10 and 15 years old, 7.6 x 10^{-9} Sv Bq^{-1} between 15 and 17 years old, and 6.2 x 10^{-9} Sv Bq^{-1} for adults, respectively, and $J_{ing,\ K\text{-}40}$ accounts for the intake of ^{40}K (Bq $year^{-1}$) and it is obtained by multiplying the yearly honey consumption (kg) for the experimentally measured activity concentration of the investigated radionuclide (Bq kg^{-1}) [28]. The assessment of the dose levels due to honey consumption is a critical point in order to evaluate if it is safe from the radiological point of view and does not adversely affect human health.

3. Results and Discussion

3.1 Radionuclides concentration

The specific activities of ^{40}K and ^{137}Cs for all the analyzed honey samples are reported in Table 2. For each group, the mean values are also reported, together with the standard deviation.

Group ID	**Specific activity**	
	^{40}K (Bq kg^{-1})	**^{137}Cs (Bq kg^{-1})**
G1	5 ± 1	< 0.15
	15 ± 2	< 0.17
	7 ± 3	< 0.19
	18 ± 3	< 0.18
	34 ± 5	< 0.16
Mean value ± standard deviation	*16 ± 3*	*< 0.17*
G2	66 ± 10	< 0.19
	108 ± 12	< 0.21
	71 ± 10	< 0.17
	84 ± 10	< 0.18
	74 ±10	< 0.17
Mean value ± standard deviation	*81 ± 10*	*< 0.18*
G3	40 ± 8	< 0.13
	56 ± 10	< 0.12
	30 ± 6	< 0.12
	47 ± 7	< 0.14
	51 ± 9	< 0.16
Mean value ± standard deviation	*56 ± 8*	*< 0.13*
G4	43 ± 7	< 0.19
	65 ± 9	< 0.18
	15 ± 5	< 0.13
	43 ± 7	< 0.14
	39 ± 7	< 0.17
Mean value ± standard deviation	*41 ± 7*	*< 0.16*
G5	114 ± 19	< 0.22
	109 ± 17	< 0.19
	116 ± 20	< 0.23
	106 ± 16	< 0.21
	111 ± 14	< 0.17

Mean value ± standard deviation	*111 ± 17*	*< 0.20*

Table 2. Specific activities (Bq kg^{-1}) of ^{40}K and ^{137}Cs in the analyzed honey samples.

The variation of the ^{40}K content from honey to honey is mainly attributed to the differences in botanical structure, as well as in the mineral composition of the soil in which the plants are cultivated. Other factors responsible for the observed variation are preferential absorbability of the plant and climatic conditions [29].

The ^{137}Cs activity concentration was in particular quantified in order to investigate about any possible artificial contamination. In all cases, it was found to be lower than the minimum detectable activity, giving evidence of the absence of any residual contamination from anthropogenic radioactivity.

All these experimental results are in good agreement, being of the same order of magnitude and also very similar, with those reported in the database of the "Italian Institute for the Environmental Protection and Research" (ISPRA) [30].

It should be noted that the specific activity of ^{40}K, measured for each sample, indicates the amount of radioactivity, and not the radiological hazard to individuals. In order to evaluate it, additional factors have to be taken into account, as reported in the following.

3.2 Potential health hazards resulting from honey consumption

In addition to the environmental concern regarding honey element composition, the analysis of the quality of honey is of great relevance, given the increasing global trends in total honey production and the fact that the European Union is the world's largest consumer of this foodstuff.

With regard to national and supranational regulations related to honey, the Directive 2001/110/EC refers to some properties related to its composition, without however making any mention of the radionuclide content [31].

The regulation EC 733/2008, which sets the maximum levels for certain contaminants in foodstuffs, does not address apiary products [32].

Figure 1 reports the annual effective dose due to the ingestion of the investigated honey samples, as calculated by the Equation (2). Obtained results take into account the average yearly honey consumption per person in Italy (500 g) [33] under the a priori hypothesis that this need can be satisfied by a single variety of honey [34].

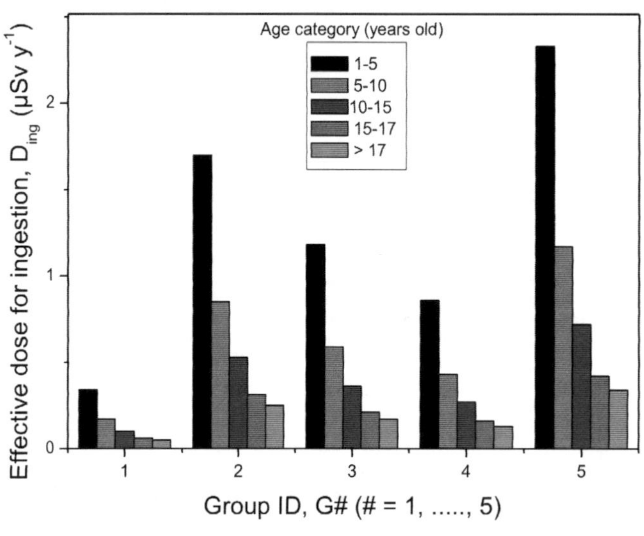

Figure 1. The calculated annual effective dose for honey ingestion.

The consumption of this foodstuff for infants (lower than 1 year old) is not recommended by the World Health Organization [35], for this reason this age group was not considered in the analysis.
The annual effective dose due to the honey ingestion was found to be:

- 0.34 μSv y^{-1}, 0.17 μSv y^{-1}, 0.10 μSv y^{-1}, 0.06 μSv y^{-1} and 0.05 μSv y^{-1} for the age category between 1 and 5 years old, between 5 and 10 years old, between 10 and 15 years old, between 15 and 17 years old and higher than 17 years old (adults), respectively, for the acacia honey consumption;
- 1.70 μSv y^{-1}, 0.85 μSv y^{-1}, 0.53 μSv y^{-1}, 0.31 μSv y^{-1} and 0.25 μSv y^{-1} for the age category between 1 and 5 years old, between 5 and 10 years old, between 10 and 15 years old, between 15 and 17 years old and higher than 17 years old (adults), respectively, for the chestnut honey consumption;
- 1.18 μSv y^{-1}, 0.59 μSv y^{-1}, 0.36 μSv y^{-1}, 0.21 μSv y^{-1} and 0.17 μSv y^{-1} for the age category between 1 and 5 years old, between 5 and 10 years old, between 10 and 15 years old, between 15 and 17 years old and higher than 17 years old (adults), respectively, for the wildflower honey consumption;
- 0.86 μSv y^{-1}, 0.43 μSv y^{-1}, 0.27 μSv y^{-1}, 0.16 μSv y^{-1} and 0.13 μSv y^{-1} for the age category between 1 and 5 years old, between 5 and 10 years old, between 10 and 15 years old, between 15 and 17 years old and higher than 17 years old (adults), respectively, for the linden honey consumption;
- 2.33 μSv y^{-1}, 1.17 μSv y^{-1}, 0.72 μSv y^{-1}, 0.42 μSv y^{-1} and 0.34 μSv y^{-1} for the age category between 1 and 5 years old, between 5 and 10 years old, between 10 and 15 years old, between 15 and 17 years old and higher than 17 years old (adults), respectively, for the honeydew honey consumption;

Worth of note, the aforementioned doses are about three order of magnitude lower than the total (external + internal) natural radioactivity value for humans, equal to 2.4 mSv y^{-1} [36]. Then, the calculated effective doses reasonably appear to be too low to induce significant health hazards [37].

4. Conclusions

In the present article, the activity concentration of natural and anthropogenic radionuclides (^{40}K and ^{137}Cs, respectively) was measured for honey samples of five different typologies, collected from Italian large retailers.

Samples were analyzed using HPGe gamma spectrometry and results indicate that there is a variation of the ^{40}K content in the investigated samples, mainly due to the differences in botanical structure, as well as in the mineral composition of the soil in which the plants are cultivated. The ^{137}Cs activity concentration was found to be lower than the minimum detectable activity in all cases, thus excluding a contamination due to anthropic radionuclides.

In order to evaluate the radiological hazard effects for the population, the effective dose for honey ingestion was assessed. The calculated values were compared with the total natural radioactivity value for humans, resulting about three orders of magnitude lower. It is therefore possible to conclude that radionuclide intoxication resulting in human adverse effects is not a concern.

References

[1] Pisani A, Protano G and Riccobono F 2008 Minor and trace elements in different honey types produced in Siena County (Italy) *Food Chem. - FOOD CHEM* **107** 1553–60
[2] Tuzen M, Silici S, Mendil D and Soylak M 2007 Trace Element Levels in Honeys from Different Regions of Turkey *Food Chem.* **103** 325–30
[3] H.-D. Belitz, W. Grosch P S 2005 *Food Chemistry, Third Edition* vol 382
[4] Caridi F, Marguccio S, Belvedere A, D`Agostino M and Belmusto G 2018 The Natural Radioactivity in Food: A Comparison Between Different Feeding Regimes *Curr. Nutr. Food Sci.* **15** 493–9
[5] Buldini P, Cavalli S, Mevoli A and Sharma J 2001 Ion chromatographic and voltammetric determination of heavy and transition metals in honey *Food Chem.* **73** 487–95

[6] Caridi F, D'Agostino M and Belvedere A 2020 Radioactivity in calabrian (Southern Italy) wild boar meat *Appl. Sci.* **10** 3580
[7] Akbari B, Gharanfoli F, Khayyat M H, Khashyarmanesh Z, Rezaee R and Karimi G 2012 Determination of heavy metals in different honey brands from Iranian markets *Food Addit. Contam. Part B* **5** 105–11
[8] Bizerra F, Da Silva Junior P I and Hayashi M 2012 Exploring the antibacterial properties of honey and its potential *Front. Microbiol.* **3** 398
[9] Caridi F, Messina M, Belvedere A, D'Agostino M, Marguccio S, Settineri L and Belmusto G 2019 Food salt characterization in terms of radioactivity and metals contamination *Appl. Sci.* **9**
[10] Caridi F, Pappaterra D, Belmusto G, Messina M, Belvedere A, D'Agostino M and Settineri L 2019 Radioactivity and heavy metals concentration in Italian (Calabrian) DOC wines *Appl. Sci.* **9** 1–9
[11] Benvenuti M, Fraschetti L, Gubiani M and Masci A 2009 Schema di riferimento per la programmazione delle iniziative nel settore apistico
[12] Caridi F, Acri G, Belvedere A, Crupi V, Agostino M D, Marguccio S, Messina M, Paladini G, Venuti V and Majolino D 2021 Evaluation of the Radiological and Chemical Risk for Public Health from Flour Sample Investigation *Appl. Sci.* **11** 3646
[13] Muhammad A, Odunola O, Ibrahim M A, Sallau A, Erukainure O, Aimola I and Malami I 2015 Potential biological activity of acacia honey *Front. Biosci. (Elite Ed).* **8** 351–7
[14] Stoic M, Yoshiyama M and Kimura K 2016 Potential Antibacterial Activity of Chestnut Honey against Paenibacillus larvae *J. Apic.* **31** 351–8
[15] Yadzir Z H, Misnan R, Abdullah N, Arip M and Murad S 2011 Identification of major allergens of wildflower honey *Southeast Asian J. Trop. Med. Public Health* **42** 370–5
[16] Olah N-K, Dărăban A, Iacob A, Chișe E, Turcuș V, Ramona Flavia C, Petrescu S and Dehelean G 2016 Preliminary analysis of some linden honey originary from Arad county *Stud. Univ. Vasile Goldis Arad, Ser. Stiint. Vietii* **26** 339–42
[17] Escuredo O, Fernández-González M and M. Carmen S-C 2012 Differentiation of Blossom Honey and Honeydew Honey from Northwest Spain *Agriculture* **2** 25–37
[18] Caridi F, Testagrossa B and Acri G 2021 Elemental composition and natural radioactivity of refractory materials *Environ. Earth Sci.* **80** 170
[19] Caridi F, Bella M Di, Sabatino G, Belmusto G, Fede M R, Romano D, Italiano F and Mottese A F 2021 Assessment of Natural Radioactivity and Radiological Risks in River Sediments from Calabria (Southern Italy) *Appl. Sci.* **11** 1729
[20] Caridi F, D'agostino M, Belvedere A, Marguccio S and Belmusto G 2016 Radon radioactivity in groundwater from the Calabria region, south of Italy *J. Instrum.* **11**
[21] Caridi F, Messina M and D'Agostino M 2017 An investigation about natural radioactivity, hydrochemistry, and metal pollution in groundwater from Calabrian selected areas, southern Italy *Environ. Earth Sci.* **76** 668
[22] Caridi F, D'Agostino M, Messina M, Marcianò G, Grioli L, Belvedere A, Marguccio S and Belmusto G 2017 Lichens as environmental risk detectors *Eur. Phys. J. Plus* **132**
[23] Caridi F, Marguccio S, Durante G, Trozzo R, Fullone F, Belvedere A, D'Agostino M and Belmusto G 2017 Natural radioactivity measurements and dosimetric evaluations in soil samples with a high content of NORM *Eur. Phys. J. Plus* **132** 56
[24] Caridi F, D'Agostino M, Marguccio S, Belvedere A, Belmusto G, Marcianò G, Sabatino G and Mottese A 2016 Radioactivity, granulometric and elemental analysis of river sediments samples from the coast of Calabria, south of Italy *Eur. Phys. J. Plus* **131**
[25] ACCREDIA 2021 https://www.accredia.it/
[26] Fadlalla H E 2005 Radioactivity levels of basic foodstuffs anddose estimates in Sudan *Thesis Master Degree, Univ. Khartoum* 60
[27] ICRP 2012 ICRP publication 119: compendium of dose coefficients based on ICRP publication 60 *Ann. ICRP* 130
[28] Caridi F, D'Agostino M, Belvedere A, Marguccio S, Belmusto G and Gatto M F 2016

Diagnostics techniques and dosimetric evaluations for environmental radioactivity investigations *J. Instrum.* **11**
[29] Vanhanen L P, Emmertz A and Savage G P 2011 Mineral analysis of mono-floral New Zealand honey. *Food Chem.* **128** 236–40
[30] Ispra web site https://www.isprambiente.gov.it/it
[31] EU Directive relating to honey http://www.fao.org/faolex/results/details/en/c/LEX
[32] EC 733/2008 https://eur-lex.europa.eu/legal-content/IT/LSU/?ur
[33] Ismea 2020 Tendenze miele 2020
[34] Stockigt J R 1977 Potassium Metabolism *Anaesth. Intensive Care* **5** 317–25
[35] World Health Organization Infant and young child feeding https://www.who.int/news-room/fact-sheets/detail/i
[36] EC (European Commission) 1999 Radiation Protection 112: Radiological protection principles concerning the natural radioactivity of building materials *Eur. Comm.* 1–16
[37] Bratu I and Georgescu C 2005 Chemical contamination of bee honey-identifying sensor of environment pollution *J. Cent. Eur. Agric.* **6**

Quantum (hyper)computation through universal quantum gates in water coherent domains

Luigi Maxmilian Caligiuri
Foundation of Physics Research Center (FoPRC), Cosenza, ITALY
E-mail: luigimaxmilian.caligiuri@foprc.org

Abstract. According to quantum electrodynamics (QED), liquid water is composed by two phases, the first one characterized by a coherent state in which all the molecules oscillate in phase, the other by a non-coherent state in which they are all non-correlated. In the coherent state, the phase-correlated oscillations take place within macroscopic spatial regions called "coherent domains" (CD), admitting a spectrum of excited energy levels and generating, at their borders, an evanescent coherent e.m. field. When two or more excited CDs are sufficiently close to each other, the overlapping between their evanescent fields gives rise to a novel type of interaction due to the mutual exchange of virtual photons by quantum tunnel effect. Furthermore, when such water coherent domains are enclosed with waveguides consisting of suitable materials and design, these effects are stabilized and enhanced, allowing for the realization of an extended network of interacting coherent domains. In this paper, we'll discuss how to exploit this dynamics to perform quantum computations by setting-up a set of universal quantum gates and calculate their operational time as a function the main parameters of the proposed physical model. We show this model can represent a basic architecture for a novel kind of quantum hyper-computer, characterized by a very high computational speed and able to overcame some of the main issues currently affecting the quantum computational frameworks so far proposed.

1. Introduction

The fundamental unit of any quantum computer is the logical quantum bit (or simply qubit), namely a mathematical entity, associated to a unit vector state in a two-dimensional complex space. In order to build a quantum computer, it is needed to realize a physical implementation of such logical qubits, called "physical" qubits. A qubit can be built by considering many different two-levels physical systems having certain specific requisites [1] and different proposal in this direction has been advanced so far [2].

Apart its theoretical interest, the importance of realizing quantum computation lies in the common belief it would offer several key advantages with respect its classical counterpart and in the deep impact it would have, for this reason, on a lot of scientific and technological fields such as computer science, cryptography, simulation of very complex systems, artificial intelligence, quantum information, finance and many others.

Nevertheless, despite this enormous potential, many challenges affect its physical implementation so limiting the quantum systems suitable for this scope to, for example, trapped ions, superconductors, quantum dots, molecular spins and optical cavities, each being characterized by specific strengths and weaknesses [3].

A common issue affecting all the above implementations is the decoherence that compromises the stability of the quantum state, namely the effect of environmental noise (due to the interaction of the quantum system with its surrounding) able to destroy the phase information required for quantum computation. On the other hand, the reading of computational results needs for a measurement of the quantum state of the system (namely its interaction with a classical system) thus requiring a delicate and effective compromise between stability and measurability. A failure to achieve such compromise could heavily limit the scalability of any quantum computational system so preventing it from being actually useful.

A key parameter used to evaluate the "merit" of a particular quantum computational scheme is the ratio $\lambda^{-1} = \tau_Q/\tau_{op}$ [4] representing the longest possible quantum calculation, where τ_Q is the "decoherence time" (i.e. the time for which a system remains quantum coherent) and τ_{op} the operation time (namely the time required by the system to perform elementary unitary operations involving at least two qubits). The value of λ^{-1} is larger when the greater is τ_Q and the smaller is τ_{op}. The latter is a measure of the calculation speed, defined as the maximum number of logical operations for unit time achievable by the system. The attainment of very high values of computational speed is then a top priority to achieve for any quantum computational system.

In this paper we discuss a completely novel proposal regarding how to exploit the QED coherent dynamics occurring in liquid water to perform quantum computations.

According to the theory of QED coherence in condensed matter [5], liquid water exhibits a two – phases behavior, due to the presence of a coherent and non-coherent fractions of molecules. The coherent fraction is composed by arrays of so-called "coherent domains" (CDs), namely macroscopic quantum regions in which all the molecules oscillate in phase with each other and with a self-trapped electromagnetic field, while the non-coherent one is made by an ensemble of uncorrelated molecules (vapor phase) filling the interstices between coherent domains. Inside a coherent domain, the electromagnetic field, generated by coherent dynamics, spreads across its boundaries in the form of evanescent field whose tail extends far from it, being therefore able to overlap the analogous evanescent fields produced by the other CDs in the surroundings.

We have yet shown [6-10] this overlapping gives rise to a quantum-type interaction between CDs, we have called "evanescent tunneling coupling" interaction, mediated by the exchange, through tunnel effect, of evanescent (virtual) photons between close CDs, provided the involved CDs have a spectrum of excited states as always occurs in the case of coherent liquid water [9,11].

The resulting transition probability between quantum states of the coupled interacting coherent domains can be described by a "Rabi -like equation" [9]. This suggests the fascinating possibility to use this still unexplored "evanescent tunneling coupling" interaction to achieve quantum computation in liquid water since, as also discussed in this paper, such interaction can be used to realize every type of one- and two-qubit gates and, consequently, every type of quantum computation at all [10].

This possibility is further strengthened by enclosing the water CDs with waveguides made of metamaterials (MTMs) [6,10,12]. This has two important consequences: a) it ensures the stability of coherent fraction of water against the disruptive effect of thermal collisions at room temperature [9] and b) it determines an important amplification of the evanescent e.m. field crossing the boundaries of a given CD in turn allowing the possibility for each of them "to communicate", through the evanescent tunneling-coupling mechanism, with other and farther CDs, so generating an extended network of superfast mutually interacting coherent domains [6-8,10].

On the other hand, the alleged extraordinary computational power of quantum computing with respect its classical counterpart is limited by several factors that meaningfully reduce the achievable calculation speed with respect to the largest theoretically reachable one. According to some theoretical results [13,14], the "ultimate" limit of computation would be theoretically affected by the energy spread ΔE of the physical system performing computation. Furthermore, as known, quantum computational efficiency is heavily reduced by the loss of coherence of the system due to its interaction with the environment.

In this paper we'll show how the use of water as physical substrate for quantum computation could offer several meaningful advantages with respect the present-day state of the art in quantum computation as, firstly, its stability against the environmental "decoherence" and a very high computational speed and density.

As regards as the calculation speed, that can be define as the maximum number of logical operations the system can perform per unit of time, if we could "indefinitely" increase the value of ΔE we would, in principle, "indefinitely" boost the calculation speed to its theoretical limit.

As we'll prove in this paper, the time required to set-up the elementary quantum gates based on the evanescent tunneling-coupling interaction decreases when the energy spread ΔE increases. A very distinguishing feature of the proposed model is the presence of an energy gap per molecule characterizing water in its coherent state (with respect to the non coherent one). In fact, due to the very high number of molecules contained in a single coherent domain, the energy storable in its excited energy levels could be very high, so that the resulting spectrum would have virtually no upper bound [11]. Such energy gap could be further increased by the occurrence of so-called "supercoherence", namely the onset of a coherence between coherent domains [12,15], that would make available to each quantum logic gate a greater quantity of energy (due to a wider range of possible excited energy levels for each CD).

Supercoherent dynamics could then further boost the computational power of the proposed quantum system based on water CDs [10] by considering, for example, a system composed by an extended network of correlated water CDs, oscillating in phase with each other.

In this kind of system, a high number N of tuned oscillating elementary systems (water CDs) would work in parallel achieving a computational time inversely proportional to N. This reminds, in principle, the concept of quantum "hyper-computing" or "accelerated" quantum computation (namely the execution of a virtually countable infinite number of computational steps within a finite time interval) already investigated in some previous publications [6,7,8,16].

In this paper, we'll particularly focus on a possible scheme to employ coherent dynamics of liquid water to realize a universal set of quantum logic gates, also giving a theoretical estimation of the operational time for each of them. Finally, we'll also discuss some aspects of a possible novel architecture for quantum hypercomputer system that uses liquid water as physical substrate and tunneling interaction between water coherent domains.

2. A brief overview of QED coherence in liquid water

In 1973 mathematical physicists K. Hepp and E. Lieb [17] found that the Dicke Hamiltonian, the first conceptual physical model of the laser, under suitable conditions (almost always verified in the condensed matter) about density and temperature of the system, undergoes a spontaneous phase transition (called "Superradiant Phase Transition" or SPT) to a new state in which a large classic coherent electromagnetic field, oscillating in tune with the atomic transitions between the ground state and a particular excited state, spontaneously emerges and gets trapped inside the atomic or molecular ensemble.

Later, G. Preparata [5], within the framework of Quantum Field Theory (QFT), shown such SPT to be a general phenomenon occurring in a large class of physical systems whose stationary (long-time) dynamics automatically selects the couple of levels, driving the transition itself, characterized by an energy gap given by

$$E = \hbar\omega_0 = \frac{hc}{\lambda} \tag{1}$$

where λ is the wavelength of light, c the speed of light in vacuum and $\hbar = h/2\pi$ is the reduced Planck constant (from this point on, unless differently said, we assume the "natural units" $\hbar = c = 1$). The need to consider the QFT approach to describe such dynamics is rooted in the assumption that the Lagrangian function of a physical system (and then its equations of motion) must be invariant under local transformation of the phase $\phi(\vec{x},t)$ of the matter field $\psi(\vec{x},t)$ due to the fluctuations of its components. Such invariance is achieved by introducing a gauge field $A_\mu(\vec{x},t)$ in the Lagrangian so that the space-time derivative D_μ of the matter field is given by

$$D_\mu\psi = i\partial_\mu\psi - A_\mu\psi \tag{2}$$

so that the Lagrangian is invariant under the gauge transformation

$$A_\mu \to A'_\mu = A_\mu - \partial_\mu X \tag{3}$$

in which $X(\vec{x},t)$ is an arbitrary function of space-time. When we consider the space-time scales of atoms and molecules, the gauge field $A_\mu(\vec{x},t)$ is just the usual electromagnetic field so that the description of the system dynamics is just QED.

For a system composed by N elementary units (atoms and /or molecules) there exist a critical value of density $\bar{\rho} = (N/V)_c$, that depends on the "electric polarizability" of the elementary species, and a temperature T_0 such that, when $\rho > \bar{\rho}$ and $T < T_0$) for which all the matter components are phase correlated among them by means of the action of an electromagnetic field oscillating in tune with them too, confined within a defined spatial region, called "Coherence Domain" (CD), associated to the wavelength of the tuning electromagnetic field, given by equation (1). More specifically, we consider a two-levels quantum system, whose matter field is described by the two wavefunctions $\psi_1(\vec{x},t)$ and $\psi_2(\vec{x},t)$, respectively associated to the two levels, subjected to the normalization condition

$$|\psi_1|^2 + |\psi_2|^2 = 1 \tag{4}$$

and an electromagnetic filed described by a scalar field $A(\vec{x},t)$. If now we restrict ourselves to a region whose spatial dimension is of order of λ given by eq. (1) and introduce the adimensional time $\tau = \omega_0 t$, the Eulero – Lagrange equations describing the two - levels matter system interacting with the e.m. field are given by [5]

$$\begin{array}{c} i\frac{\partial}{\partial\tau}\psi_1(\tau) = gA^*(\tau)\psi_2(\tau) \\ i\frac{\partial}{\partial\tau}\psi_2(\tau) = gA(\tau)\psi_1(\tau) \\ -\frac{1}{2}\ddot{A}(\tau) + i\dot{A}(\tau) - \mu A(\tau) = g\psi_1^*(\tau)\psi_2(\tau) \end{array} \tag{5}$$

with

$$g \equiv 2\pi\left(\omega_p/\omega_0\right) f_{01}^{\frac{1}{2}} \tag{6}$$

where ω_p is electron "plasma frequency" (m_e is the electron mass)

$$\omega_p = \left(e/\sqrt{m_e}\right)(N/V)^{1/2} \tag{7}$$

the term f_{01} is the oscillator strength for the electronic transition from the ground state ψ_1 to the excited state ψ_2 and μ is the photon "mass" term [5]. We search for the solution of the system (5) around $\tau = 0$, where we assume the initial conditions

$$A(0) \sim 0,\ \psi_1(0) \sim 1,\ \psi_2(0) \sim 0 \tag{8}$$

By differentiating the third of Eqs. (5) and substituting it into the second one, we obtain the following equation

$$-\frac{1}{2}\frac{\partial^3}{\partial\tau^3}A + \frac{\partial^2}{\partial\tau^2}A + i\mu\frac{\partial}{\partial\tau}A + gA^2 = 0 \tag{9}$$

We now look for solutions of Eq. (10) in the form

$$A = \exp(ip\tau) \tag{10}$$

that, inserted in Eq. (10), gives

$$\frac{p^3}{2} - p^2 - \mu p + g^2 = 0 \tag{11}$$

Equation (11) describes two completely different dynamical regimes for our system according as [5] $g > g_{crit}$ or $g < g_{crit}$, where

$$g_{crit} = \left[\frac{8}{27} + \frac{2}{3}\mu + \left(\frac{4}{9} + \frac{2}{3}\mu\right)^{3/2}\right]^{1/2} \tag{12}$$

In the first case, Eq. (11) admits three real solutions and, by Eq. (10), the system performs zero-point periodic oscillations around a state characterized by the conditions (8). In the second case, Eq. (11) admits two complex conjugate $(p_1, p_2 = p_1^*)$ and one real p_3 roots. This means the field A has a "runaway" solution like

$$A \sim \exp[\mathrm{Im}(p)\,\tau] \tag{13}$$

where p is the complex solution of Eq. (11) for which $\mathrm{Im}(p) > 0$. Such a solution is characterized by an exponentially growing amplitude from its "perturbative" point-zero value corresponding to vacuum fluctuations towards a non-vanishing value that forces a similar increase of the matter field component ψ_2. The system then becomes instable and evolves, in a very short time [5], towards a limit cycle defined by $(0 < \gamma < \pi/2)$

$$\begin{gathered}\psi_1(\tau) = \cos\gamma\exp[i\theta_1(\tau)]\\ \psi_2(\tau) = \sin\gamma\exp[i\theta_2(\tau)]\\ A(\tau) = A_0\exp[i\varphi(\tau)]\end{gathered} \tag{14}$$

characterized by the **phase-locking** constrain [5]

$$\frac{\partial\varphi}{\partial\tau} = \frac{\partial\theta_1}{\partial\tau} - \frac{\partial\theta_2}{\partial\tau} \tag{15}$$

Equations (14) and (15) imply the new state of the system is just the coherent state above described, called "coherent ground state" (CGS), characterized by the coherent common oscillation of matter and e.m. fields all assuming non-vanishing amplitudes.

One of the most important consequences of such dynamical evolution is the shift of the frequency of the coherent e.m. field, that becomes renormalized by the coherent interaction to a value $\omega_{coh} < \omega$ [5] given by

$$\omega_{coh} = \omega_0 - \frac{\partial\varphi}{\partial\tau} < 0 \tag{16}$$

Such frequency shift, not corresponding to a similar change in wavelength λ , ensures the coherent e.m. field to be trapped inside the CD, a process in all similar to that of total reflection of light between two media having different refraction indices.

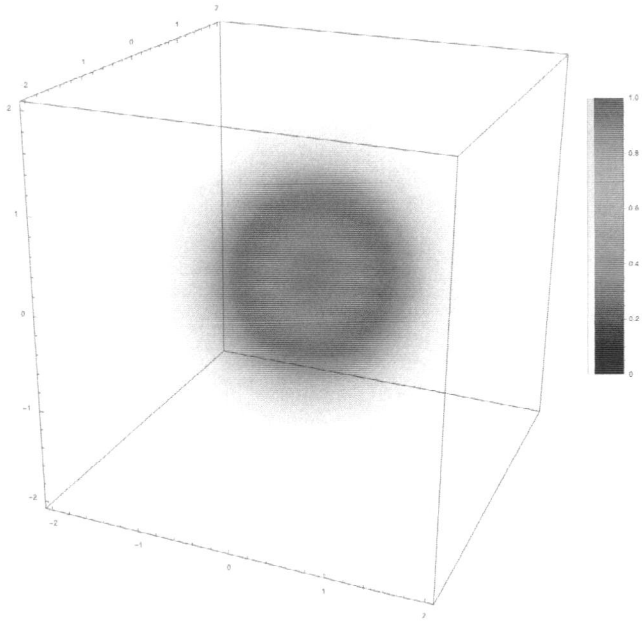

Figure 1. Schematic view of a spherical CD and the coherent evanescent e.m. field escaping from it.

In particular, it can be shown [5] the e.m. field, in the case of a spherical CD, can be described by the following equations

$$A(\vec{x}, \tau) = A(0)\,\frac{\sin(\omega_0 \tau)}{\omega_0 \tau} \qquad (r < r_0) \tag{17}$$

$$A(r) \simeq \frac{A(0)}{\sqrt{2}}\,\frac{\exp\left[-r\sqrt{\omega_0^2 - \omega_{coh}^2}\,(r - r_0)\right]}{\omega_0 r} \qquad (r > r_0) \tag{18}$$

At the interface between CD, whose radius is given in general by [5]

$$R_{CD} \cong \frac{3}{8}\lambda \tag{19}$$

and the surrounding medium, an evanescent e.m. field is then produced (fig. 1).

This evanescent field is characterized by a pulsation ω_{coh} and it crosses the CD's border, falling off at a rate of the order of λ [5] (fig. 2) as given by Eq.(18).

According to the coherent dynamics, the phase agreement between matter and co-resonating e.m. field defines a new macroscopic quantum state (corresponding to a high number N of elementary components) characterized by a well-defined value of phase Φ which defines the rhythm of oscillation of whole the system. From a quantum viewpoint this means the state vector of the system to be an eigenstate of a quantum phase operator [5, 18].

The transition from the non-coherent state to the coherent one can be considered as a "condensation" in which the system releases outwards a given energy, previously "borrowed"

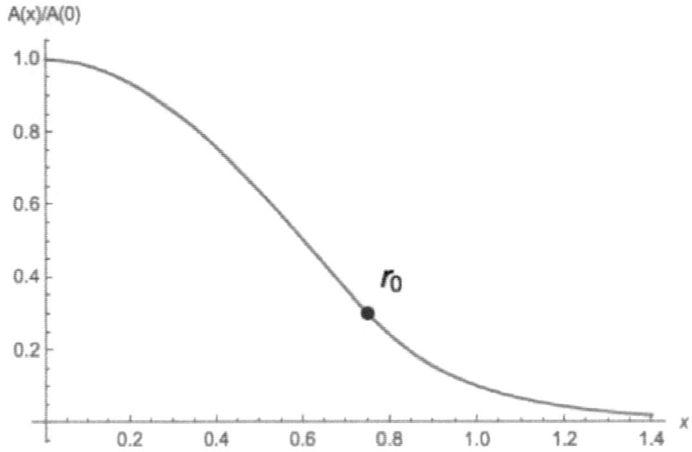

Figure 2. Radial profile of evanescent e.m. field $A(r)$ of a CD, with $r_0 = R_{CD}$. $A(0)$ is the e.m. field intensity at the center of CD.

from quantum vacuum zero-point oscillations. Every coherent domain is then characterized by an energy gap per molecule $\Delta E/N$ that makes it more stable with respect the non-coherent state. The value of this gap depends on the couple of levels involved in the coherent oscillations [5]. For absolute temperatures different than $T = 0$, thermal collisions could transfer to molecules an energy higher than the energy gap so pushing them out of tune. In the general case we can identify a coherent and a non - coherent fraction of the system $F_c(T)$ and $F_{nc}(T)$, respectively indicating the number of molecules belonging, at each instant, to the coherent and non - coherent phases satisfying, at a given temperature, the following constraint:

$$F_c(T) + F_{nc}(T) = 1 \tag{20}$$

For every $T \neq 0$, the system oscillates between the coherent and the non-coherent state, according to whether the coherent electrodynamic attraction is overtaken or not by the thermal collisions transferring to the CD the energy E_{ther} . In the case of liquid water, already studied in several previous works [5,15,19-23], the molecules are characterized by a very rich spectrum of electronic excitations and the coherent oscillations occurs between a ground state and a $5d$ excited level characterized by the transition energy $E_{exc} = 12.06\, eV$ that lies just below the ionization threshold. The water CD can be then considered as a very huge reservoir of quasi-free electrons (a water CD can include, at the density of liquid water, a number of molecules of the order of $n \simeq 2 \cdot 10^5$, corresponding to a number of quasi-free electrons of about $n_e \simeq 2 \cdot 10^4$) that are very easily excitable in the form of cold electron "vortices", as far as this energy supply is lower than the energy gap per molecule $\Delta E/N \simeq 0.26\ eV$ associated to the coherent state. These vortices form the metastable coherent excited spectrum of water CD [11] whose energy depends on the quantized value of their angular momentum and on the external magnetic field. A very important feature of such excited states is that, because of the coherence, energy supplied from the outside is absorbed by the CD as a whole and such vortices cannot be excited nor decay thermally since they are coherent and have a quantized magnetic moment that allows them to align to an external static magnetic field. For this reason, they are characterized by a very long lifetime.

Furthermore, being the spectrum $\{E_n\}$ a collective property of the entire CD, it has practically no upper limit since, although every molecule in the coherent state cannot absorb an energy higher than E_{coh} , all the N resonant molecules can account for a total energy of excitation of the order of NE_{coh} a very high quantity (including frequencies up to visible and HV spectrum) since, at room temperature, $N/V \sim 10^{22}\,cm^{-3}$. Finally, we remark another important feature of coherent water when it is close to a surface. In this case the disruptive effect of thermal collisions is compensated by the attraction of the water molecules to the wall [6,9,24,25]. In this way the coherent fraction of water is stabilized to a value close to the unity so that it can be considered as fully coherent [25] even at room temperature.

3. Tunneling-coupling interaction between coherent domains of water

Our idea to exploit coherent dynamics occurring in liquid water in order to perform quantum computation is based on the chance the water CDs could interact each other [10,11,26]. This is possible if the CDs has a spectrum of excited states and can release outwards the energy stocked in its cold excited vortices. A conceivable chemo-electrodynamical mechanism has been proposed so far [10, 27] to explain the emission of energy from excited CDs.

On the other hand, we have proven [6-9,15,26] the existence of a purely quantum process allowing the water CDs to interact each other through the exchange of virtual photons by quantum tunneling effect.

Such interaction would occur when the evanescent coherent fields produced at the boundaries of two or more CDs (sufficiently close each other) partially or totally overlap, allowing for the exchange of virtual photons among them as schematically shown in fig. 3.

From the standpoint of classical physics, evanescent waves usually originate by the "frustrated total internal reflection" (FTIR) of a light beam at the boundary between two media characterized by different values of refraction index. In this case a "transmitted" wave component is nevertheless present in the second medium, namely just the so-called "evanescent" wave. From the quantum standpoint, the tunneling of evanescent modes of e.m. waves is just interpreted as the tunneling of virtual photons through the corresponding potential barrier. Such tunneling photons are characterized by a negative square mass in the ordinary metric [6,7]. The "quantum-tunneling" interaction above described can be also considered as the result of the process of excitation-deexcitation of the CD energy levels, that determines the tunneling of virtual photons between them (since the CDs are stable in their fundamental state) as the result of the exchange of virtual photons between the quasi-free electrons belonging to the coherent vortices associated to the couple of involved CDs.

From a QFT viewpoint, this process can be described, assuming a not too strong coupling, by the following Hamiltonian (we limit ourselves to the photon dynamics) [10,28]:

$$H = \sum_{\vec{k}} \left[\omega_{\vec{k}} \left(a^{\dagger}_{\vec{k},1} a_{\vec{k},1} + a^{\dagger}_{\vec{k},2} a_{\vec{k},2} \right) + \Gamma \left(a^{\dagger}_{\vec{k},1} a_{\vec{k},1} + a^{\dagger}_{\vec{k},2} a_{\vec{k},2} \right) \right] \tag{21}$$

where $a_{\vec{k},i}$ ($a^{\dagger}_{\vec{k},i}$) is the annihilation (creation) operator for the mode $\vec{k}$ in the i-th CD ($i = 1, 2$) acting on the vacuum state $|0\rangle$ and Γ is a real parameter quantifying the coupling strength. We then assume:

$$|1\rangle \rightarrow |\psi_1\rangle = a^{\dagger}_1 |0\rangle \tag{22}$$

$$|2\rangle \rightarrow |\psi_2\rangle = a^{\dagger}_2 |0\rangle \tag{23}$$

and Γ is approximatively independent of the value of $\vec{k}$. From a macroscopic viewpoint, the exchange of a virtual photon can be described as the time evolution of a two – levels system (described by the energy eigenstates $|1\rangle$ and $|2\rangle$ respectively associated to the values E_1 and E_2) whose Hamiltonian has the form:

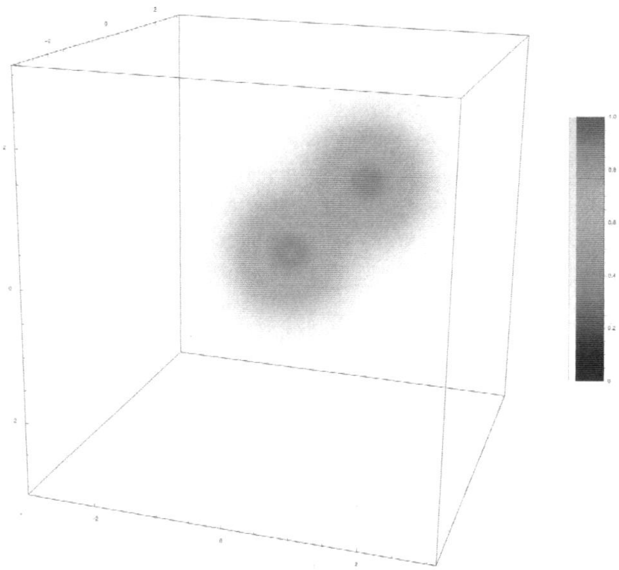

Figure 3. Overlapping between the evanescent fields associated to two nearby CDs.

$$H = E_1 |1\rangle \langle 1| + E_2 |2\rangle \langle 2| + \Gamma (|1\rangle \langle 1| + |2\rangle \langle 2|) \tag{24}$$

where, again, the parameter $\Gamma \in R$ represents the "strength" of the interaction, giving the coupling of the two states, namely the "amplitude" of the tunneling of virtual photons. The transition probability between the two eigenstates $|1\rangle$ and $|2\rangle$ is then given by a "Rabi equation" [9,26]:

$$P_{12}(t) = \frac{\Gamma^2}{\Gamma^2 + (E_1 - E_2)^2} \sin^2\left(\sqrt{\frac{(E_1 - E_2)^2 + 4\Gamma^2}{4}} t\right) \tag{25}$$

that describes quantum oscillations in which a photon wave packet of a given wave number is coupled back and forth between two CDs.

In particular, we note the probability of transition between the two states admits the maximum value:

$$P_{\max} = \frac{\Gamma^2}{\Gamma^2 + (E_1 - E_2)^2} \tag{26}$$

that is equal to 1, namely the excitation is completely transferred from one state to the other (from a CD to the other) when $E_1 = E_2$, whatever the value of $\Gamma \neq 0$. In this case the two CDs are energy degenerate.

The time evolution of such a coupled system, originating from the Hamiltonian (24), is described by the unitary evolution operator [4,9]:

$$U(t) = \exp\left[-i\frac{(E_1+E_2)}{2}t\right]\left[\cos(|\vec{n}|\,t)\,I - i\sin(|\vec{n}|\,t)\,\frac{\vec{n}}{|\vec{n}|}\cdot\vec{\sigma}\right] \tag{27}$$

where $\vec{\sigma} = (\sigma_x, \sigma_y, \sigma_z)$ are the Pauli matrices, I is the 2×2 identity matrix and the three - dimensional vector $\vec{n}$ has the components:

$$n_1 = \Gamma,\; n_2 = 0,\; n_3 = \frac{1}{2}(E_1 - E_2) \tag{28}$$

Mathematically, the evolution operator expressed by Eq. (27) is a "rotation" operator of angle θ in the quantum states space:

$$U = \exp(i\alpha)\exp\left(-i\theta\hat{u}\cdot\frac{\vec{\sigma}}{2}\right) \tag{29}$$

in which, by comparing Eqs (27) and (29):

$$\alpha = -\frac{E_1+E_2}{2}t;\quad \theta = 2\,|\vec{n}|\,t;\quad \hat{u} = \frac{\vec{n}}{|\vec{n}|} \tag{30}$$

The time-evolution operator of the system composed by two tunneling - interacting CDs can be then expressed as the product of a rotation of angle θ and an overall phase shift of angle α. As a consequence of the above reasoning, two interacting CDs cannot be further considered as independent entities but as interconnected parts of the same system.

3.1. Tubular coherent domains of water and their interaction

When two CDs are sufficiently close to each other (or when the evanescent field amplitude is sufficiently high even at higher inter-distances), their respective e.m. evanescent fields can overlap. The width of the resulting overlapping zone, through which tunneling of virtual photons takes place, will depend on the distance between the two CDs and on the spreading of the two evanescent fields outside the respective CDs, whose "extension", described by their tales, is of order of $\lambda = 2\pi/\omega_0$ (fig. 3). For a water CD, it has been calculated [6] $d_p \simeq 10^{-7}\,m$, so that, in order to the evanescent fields overlapping to take place, a couple of water CDs should have their mutual intercentrum distance satisfying the condition:

$$d_{CD} \leq 2(R_{CD} + d_p) \simeq \frac{11}{4}\lambda \cong 137.5\;nm \tag{31}$$

On the other hand, in order to the tunneling coupling interaction to take place effectively, once the coherent state has been achieved, it is needed to ensure: a) the stabilization of the coherent fraction of liquid water against disruptive thermal fluctuations (so that $F_c(T) \sim 1$) b) a suitable amplification of the evanescent fields generated by each coherent domain. The latter is a necessary condition since, although the condition (31) is satisfied, the evanescent field intensity could be too low for ensure a sufficient coupling between them. On the other hand, such amplification could also allow for the evanescent coupling to spread, for a given field intensity $A(0)$, even at distances $d >> d_{CD}$.

As already proven [6,9,10,16,24] both the conditions a) and b) can be assured by enclosing water inside the inner volume of (even symmetrical) waveguides whose walls are made up of suitably designed metamaterials (MTMs) (fig. 4), a kind of materials characterized by a purely imaginary value of refraction index. We can also consider the water molecules contained in such waveguide as a whole tubular coherent domain, according to a process we could define as CD "compactification" (whose dynamics and consequences will be studied in detail in a forthcoming paper), at least with respect to the generation of the evanescent field inside the contained CDs [5].

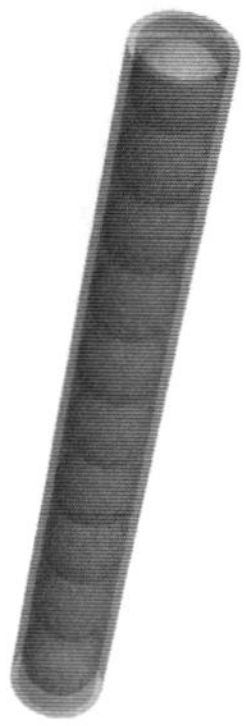

Figure 4. Tubular structure of water CD inside a waveguide made up of metamaterial.

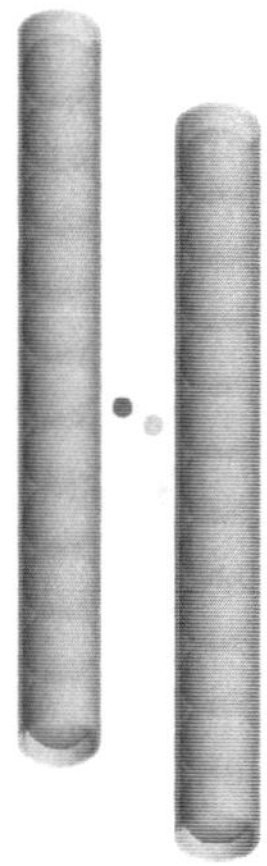

Figure 5. Evanescent coupling by virtual photons (green and red spots) exchange between two waveguides filled with water CDs.

The evanescent field amplification induced by MTMs determines an important increase of the evanescent penetration depth d_p (at least up to four time the original value) allowing a single CD to interact with other ones even if placed far away from it, so forming a spatially extended network of mutually interacting CDs that could be used, as we'll see, to realize superfast parallel quantum computations.

In our theoretical framework the propagation of a photon in a waveguide (namely a tubular water CD) is likened to the presence of a photonic spatial mode, due to tunneling, in one or the other waveguide as sketched in fig. 5.

The previous discussion can be easily generalized to the case of many interacting water CDs, by restricting our analysis to single excited states in which the i-th CD is described by its coherent ground state $|\varphi_{ig}\rangle$ and by an excited state $|\varphi_{ie}\rangle$ so that the overall quantum ground state is given by:

$$|\varphi_0\rangle = \prod_i |\varphi_{ig}\rangle \tag{32}$$

while all possible excited states are described by the state vector:

$$|\varphi_k\rangle = |\varphi_{ke}\rangle \prod_{i \neq k} |\varphi_{ig}\rangle \tag{33}$$

In this picture, the excited state $|\varphi_k\rangle$ is the state in which the k-th tubular CD is in the excited state while all the other are in the coherent ground state. The set $\{|\varphi_k\rangle\}$ is a basis for the space of excited states and satisfies the general orthogonality relation:

$$\langle \phi_{i,k} | \phi_{j,l} \rangle = \delta_{i,j}\delta_{k,l} \tag{34}$$

with $k, l = e, g$ (where "e" stands for "excited" and "g" for "ground"). This basis is often referred to as "site basis" since each state, when expressed in this basis, can be identified by the specific CD where the excitation is localized.

The Hamiltonian of a system composed by N interacting CDs can be then written, in this basis, as:

$$H = \sum_{i=1}^{N} \varepsilon_i |i\rangle \langle i| + \sum_{i \neq j}^{N} \Gamma_{ij} |i\rangle \langle j| \tag{35}$$

where ε_i is the relative excitation energy of the i-th tubular CD, namely $\varepsilon_i = E_i - \omega_{coh}$, and Γ_{ij} is the "strength" of the tunneling interaction between the i-th and j-th CDs. In our system configuration, the values of Γ_{ij} will definitively depend on the mutual distance between tubular units as well as on the structural features (that in turn control the evanescent field amplification) of MTMs tubular waveguides, whose detailed design and features will be studied in a separate work.

4. The mapping of unitary quantum transformations on interacting water CDs and the realization of universal quantum computation

Quantum computation relies on the manipulation of quantum information by exploiting dynamic transformation of quantum states of a physical system.

Each of these transformations is described by unitary linear operator (a matrix when expressed in a suitable basis) defined in the complex vector space associated with the quantum state space. In the framework of classical computation, there exists a small set of gates (like, for example, AND, OR, NOT gates) that can be used to compute an arbitrary classical function and such set is then said to be "universal".

A similar result holds in quantum computation framework where a set of gates can be found to be used to realize universal quantum computation, since it can be shown [4] that any arbitrary unitary transformation can be implemented from a set of primitive transformations, including the two-qubits CNOT (controlled-NOT) gate in addition to three one-qubit gates.

A suitable set of gates that also allows a fault-tolerant construction is [4]:

(1) the Hadamard gate H (to be not confused with the Hamiltonian);
(2) the $\pi/8$ gate T;
(3) the "phase" gate S;
(4) the CNOT gate.

To each of these gates is associated a unitary operator whose matrix representation in the computational basis that is given in table 1. The physical realization of quantum computation algorithms implies the mapping of logical qubits to physical states of the considered quantum system (namely the "physical" qubits), and then the implementation of a suitable set of unitary operators, acting on these states and realizing the universal quantum gates, each corresponding to a unitary time evolution of the system, as described by its Hamiltonian.

Generally, for a given physical quantum system, this mapping is not unique and, also, different mappings could give rise to different values of computational speed and overall fidelity.

As we'll see, the physical implementation of such unitary transformations generally sets some constraints on the parameters of physical system. The possibility to engineering the CDs interaction in order to use them for practical quantum computation relies on the ability to set and control such parameters.

Encoding quantum operations into the dynamics of physical system then implies the ability to state a correspondence between quantum unitary transformations, the system's Hamiltonian and the physical (geometrical and/or structural) properties of the system itself.

For a closed quantum system, the time evolution is driven by a set of unitary transformations that are functions of the system's Hamiltonian H according to:

$$U(t) = e^{-\frac{i}{\hbar}Ht} \tag{36}$$

We can then "invert" Eq. (36) and search for the Hamiltonian able to originate, after a time interval τ_G, a desired unitary transformation τ_G, namely:

$$H_Q = \frac{i\hbar}{\tau_G}\ln U_Q \tag{37}$$

By using the site basis defined by Eqs (32) and (33), the logical qubit states are mapped to the physical states $|i\rangle$ where i indicates the i-th CD that, in this representation, is directly associated with a state of the qubit register.

4.1. Realization of one-qubit gates

According to the above mapping, an N-dimensional quantum state needs 2^N CDs to be represented by a physical circuit made of water CDs. In our model, a single quantum qubit gate, acting on a two-dimensional quantum space, needs two interacting CDs (labelled by 1 and 2) to be realized.

Then we can write, using the notation of Eqs (32) and (33):

$$\begin{aligned} |0\rangle &\rightarrow |\phi_1\rangle = |1_e 2_g\rangle \\ |1\rangle &\rightarrow |\phi_2\rangle = |1_g 2_e\rangle \end{aligned} \tag{38}$$

in which, for further clarity, we have explicitly labeled the single CD (as 1 and 2) and its energy state (ground or excited). For a two-qubits gate, the related quantum space is four dimensional, so we must consider four CDs and adopt the following mapping of the (computational) basis states:

$$\begin{aligned} |00\rangle &\rightarrow |\phi_1\rangle = |1_e 2_g 3_g 4_g\rangle \\ |01\rangle &\rightarrow |\phi_2\rangle = |1_g 2_e 3_g 4_g\rangle \\ |10\rangle &\rightarrow |\phi_3\rangle = |1_g 2_g 3_e 4_g\rangle \\ |11\rangle &\rightarrow |\phi_4\rangle = |1_g 2_g 3_g 4_e\rangle \end{aligned} \tag{39}$$

It is worthy to be observed how the adopted mapping can be directly related, within our model, to the underlying system dynamics, since it can be associated to the tunneling exchange of virtual photons between two water-filled waveguides (that is a couple of interacting tubular

CDs of water) by adopting the dual-rail representation for the photon. According to such representation, the two logic states $|01\rangle$ and $|10\rangle$ respectively represent the state in which a photon is absent and present as spatial mode in the i-th waveguide.

We are now in position to study how to realize the elementary quantum transformations needed to perform universal quantum computation.

The simplest gate to be implemented is the T gate whose unitary matrix is given in table 1. The corresponding Hamiltonian is given, using Eq. (38), by the matrix:

$$H_T(\tau) = \frac{\pi\bar{h}}{4\tau}\begin{pmatrix} 0 & 0 \\ 0 & -1 \end{pmatrix} \tag{40}$$

The effect of T gate is to change the relative phase between $|0\rangle$ and $|1\rangle$ by the amount $e^{i\pi/4}$; this transformation then doesn't involve any exchange between the basis states and then no interaction between the corresponding CDs. This is confirmed in the Hamiltonian (40) in which the off-diagonal terms, associated to the terms $\Gamma_{12} = \Gamma_{21} = \Gamma$ are zero. It's easy to obtain the expression of the time τ_T required by the system evolution to realize the T gate. By remembering the difference in the diagonal term of the Hamiltonian corresponds to the difference between the relative energy of excitation of the two states $|0\rangle$ and $|1\rangle$ we have:

$$\frac{\pi\bar{h}}{4\tau_T} = \Delta\varepsilon_{01} = E_1 - E_2 \qquad \Rightarrow \tau_T = \frac{\pi\hbar}{4(E_1 - E_2)} \tag{41}$$

We would obtain the same result, through a more formal and general approach, by considering the Eqs (27)-(30) and recalling that (see Table 1), in terms of Pauli matrices:

$$U_T = \exp\left(i\frac{\pi}{8}\right)\exp\left(-i\frac{\pi}{8}\sigma_z\right) \tag{42}$$

so, by equating Eqs. (27) and (42) and using Eq. (30) we find [10,26]:

$$u_z = 1; \quad u_x = u_y = 0; \quad \theta = \frac{\pi}{4}; \quad \alpha = \frac{\pi}{8} \tag{43}$$

By inserting these values in the Eq. (30) and using Eq. (28) we find:

$$\frac{\frac{E_1 - E_2}{2}}{\sqrt{\Gamma^2 + \frac{(E_1 - E_2)^2}{4}}} = 1 \tag{44}$$

and

$$t = \frac{\pi\bar{h}}{8|\vec{n}|} = \frac{\pi\bar{h}}{8}\frac{1}{\sqrt{\Gamma^2 + \frac{(E_1 - E_2)^2}{4}}} \tag{45}$$

From Eq. (44) we derive $\Gamma = 0$ that, inserted in the Eq. (45) just gives the Eq. (41). The latter indicates the time required to perform T gate is inversely proportional to the difference of the excited energy levels of the two CDs. Equation (41) and the requirement $\Gamma = 0$ show the T gate can be implemented, in our model, by considering a couple of non-interacting ($\Gamma = 0$) tubular CDs, characterized by non-degenerate excited energy states ($E_1 \neq E_2$). The Hamiltonian corresponding to S gate (whose unitary matrix is given in table 1) is given by:

$$H_S(\tau) = \frac{\pi\bar{h}}{2\tau}\begin{pmatrix} 0 & 0 \\ 0 & -1 \end{pmatrix} \tag{46}$$

The effect of S gate is to apply the T gate twice since $S = T^2$. So, like the T gate, we have in this case $\Gamma_{12} = \Gamma_{21} = \Gamma = 0$ and the time required to implement the gate is obtained as above:

$$\tau_S = \frac{\pi \hbar}{2\left(E_1 - E_2\right)} \tag{47}$$

with $E_1 \neq E_2$. The physical implementation of such gate is then realized in the same way as the T gate. The values of the parameters $\hat{u}$, θ and α are found to be [10]:

$$u_z = 1; \quad u_x = u_y = 0; \quad \theta = \frac{\pi}{2}; \quad \alpha = \frac{\pi}{4} \tag{48}$$

As known, the application of the NOT gate to a qubit $|\phi\rangle = \alpha|0\rangle + \beta|1\rangle$ interchanges the basis vectors so that we have $U_{NOT}|\phi\rangle = \alpha|1\rangle + \beta|0\rangle$ Its unitary matrix, in computational basis, is given in table 1.

Table 1. Parameters to implement quantum gates and related operational times.

gate	unitary matrix
Hadamard	$\frac{1}{\sqrt{2}}\begin{pmatrix} 1 & 1 \\ 1 & -1 \end{pmatrix}$
T	$\exp\left(i\pi/8\right)\begin{pmatrix} \exp\left(-i\pi/8\right) & 0 \\ 0 & \exp\left(i\pi/8\right) \end{pmatrix}$
S	$\begin{pmatrix} 1 & 0 \\ 0 & i \end{pmatrix}$
NOT	$\begin{pmatrix} 0 & 1 \\ 1 & 0 \end{pmatrix}$
CNOT	$\begin{pmatrix} 1 & 0 & 0 & 0 \\ 0 & 0 & 1 & 0 \\ 0 & 1 & 0 & 0 \\ 0 & 0 & 0 & 1 \end{pmatrix}$

The corresponding Hamiltonian is:

$$H_{NOT}\left(\tau\right) = \frac{\pi \hbar}{2\tau}\begin{pmatrix} -1 & 1 \\ 1 & -1 \end{pmatrix} \tag{49}$$

We note U_{NOT} can be expressed, in terms of Pauli matrices as:

$$U_T = \sigma_x \tag{50}$$

the same analysis of NOT gate performed as above leads us to the constraints [10]:

$$u_x = 1; \quad u_y = u_z = 0; \quad \theta = \pi; \quad \alpha = \frac{\pi}{2} \tag{51}$$

By inserting these values in Eq. (30) and using Eq. (28) we find:

$$\frac{\Gamma}{\sqrt{\Gamma^2 + \frac{\left(E_1 - E_2\right)^2}{4}}} = 1 \tag{52}$$

and

$$t = \frac{\pi\hbar}{2\,|\vec{n}|} = \frac{\pi\hbar}{2}\frac{1}{\sqrt{\Gamma^2 + \frac{(E_1 - E_2)^2}{4}}} \tag{53}$$

Equation (52) implies $E_1 = E_2$ while, from Eqs. (51) and (53) we obtain:

$$\tau_{NOT} = \frac{\pi\hbar}{2\Gamma} \tag{54}$$

as can be also confirmed by the analysis of the Hamiltonian (49). We note τ_{NOT} is equal to half of the "Rabi time" as calculated through Eq. (26) so that the physical implementation of such gate can be achieved by considering a couple of two interacting tubular CDs ($\Gamma_{12} = \Gamma_{21} = \Gamma \neq 0$) characterized by degenerate excited energy levels ($E_1 = E_2$) as required in order to ensure a fully transfer of excitation between them (corresponding to the transition $|0\rangle \rightleftarrows |1\rangle$). The Hadamard gate, often referred to as "square root of NOT gate", turns $|0\rangle$ into $(|0\rangle + |1\rangle)\big/\sqrt{2}$ and $|1\rangle$ into $(|0\rangle - |1\rangle)\big/\sqrt{2}$ in a sense "halfway" between the states $|0\rangle$ and $|1\rangle$ [4]. Its associated unitary matrix is given in table 1 and the corresponding Hamiltonian is:

$$H_{HAD}(\tau) = \frac{\pi\hbar}{2\sqrt{2}\tau}\begin{pmatrix} 1-\sqrt{2} & 1 \\ 1 & -1-\sqrt{2} \end{pmatrix} \tag{55}$$

The matrix U_{HAD} can be expressed, in terms of Pauli matrices as:

$$U_{HAD} = \frac{\sigma_x + \sigma_z}{\sqrt{2}} \tag{56}$$

this implies [10]

$$u_x = \frac{1}{\sqrt{2}}; \quad u_y = 0; \quad u_z = \frac{1}{\sqrt{2}}; \quad \theta = \pi; \quad \alpha = \frac{\pi}{2} \tag{57}$$

By inserting these values in Eq. (30) and using Eq. (28) we find:

$$\frac{\Gamma}{\sqrt{\Gamma^2 + \frac{(E_1 - E_2)^2}{4}}} = \frac{1}{\sqrt{2}}; \qquad \frac{E_1 - E_2}{2\sqrt{\Gamma^2 + \frac{(E_1 - E_2)^2}{4}}} = \frac{1}{\sqrt{2}} \tag{58}$$

leading to the constraint

$$\frac{E_1 - E_2}{2\Gamma} = 1 \tag{59}$$

By using Eq. (59) in Eq. (30) we find:

$$\tau_{HAD} = \frac{\pi\hbar}{2\sqrt{2}\Gamma} = \frac{\pi\hbar}{\sqrt{2}\,(E_1 - E_2)} \tag{60}$$

The Hadamard gate can be physically realized by considering two non-degenerated interacting tubular CDs evolving for a time given by Eq. (60). It's important to note Eq. (59) sets a constraint on the circuital realization of such port in terms of geometrical and/or engineering design of the involved tubular CDs.

Table 2. Parameters to implement one-qubit gates and related operational times.

Gate	α	θ	u_x	u_y	u_z	Δt
Hadamard	$-\pi/2$	π	$1/\sqrt{2}$	0	$1/\sqrt{2}$	$\pi\hbar/\sqrt{2}\Delta E$
T	$-\pi/8$	$\pi/4$	0	0	1	$\pi\hbar/4\Delta E$
S	$-\pi/4$	$\pi/2$	0	0	1	$\pi\hbar/2\Delta E$

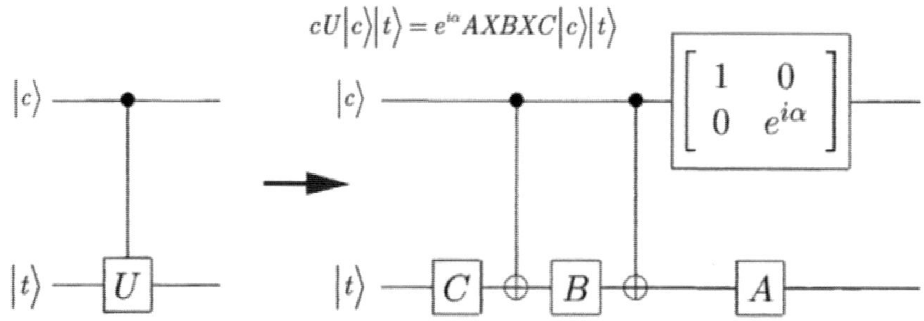

Figure 6. Circuit representation of controlled unitary operation on single qubit.

4.2. Controlled operations

The above discussion clearly indicates how, by a suitable selection of the physical parameters E_1, E_2, Γ and by letting the system evolving for an appropriate time Δt, an ensemble of interacting water CDs can realize, in principle, any generic quantum gate. Such dynamics is also able to reproduce any controlled unitary operations cU (such as, for example, CNOT), so allowing the realization of any quantum computing circuits. A generic controlled unitary operation U on a single qubit can be realized by considering the circuit shown in fig. 6, by decomposing U as:

$$U = e^{i\alpha} AXBXC \tag{61}$$

where $X = \hat{\sigma}_X$ and

$$\begin{aligned} A &\equiv R_z(\beta) R_y\left(-\tfrac{\gamma}{2}\right) \\ B &\equiv R_y\left(-\tfrac{\gamma}{2}\right) R_z\left(-\tfrac{\delta+\beta}{2}\right) \\ C &\equiv R_z\left(\tfrac{\delta-\beta}{2}\right) \end{aligned} \tag{62}$$

with the unitary operators A, B, C acting on a single qubit such that $ABC = I$ and for some angles β, δ and γ. To prove how this allows us, within our model of quantum computation, to physically realize controlled operations on qubits, it suffices to note any unitary transformation can be expressed in terms of rotations in the space of quantum states that, in turn, can be simulated through interacting water coherent domains for a suitable choice of the transformation parameters (see table 2). The relations between the parameters in Eq. (62) and the angle θ in table 1 has been calculated in [10]. The previous discussion can be easily generalized to realize multiple - qubits conditioned unitary operations [4,10].

4.3. Realization of two-qubits gates

In order to show how to implement a two-qubits gate, we apply the above framework to the example of the CNOT gate, due to its fundamental importance in quantum computing. The unitary matrix describing CNOT is given in table 1 and its corresponding Hamiltonian can be written as:

$$H_{CNOT}(\tau) = \frac{\pi^{-}h}{2\tau}\begin{pmatrix} 0 & 0 & 0 & 0 \\ 0 & 0 & 0 & 0 \\ 0 & 0 & -1 & 1 \\ 0 & 0 & 1 & -1 \end{pmatrix} \tag{63}$$

The CNOT gate acts on a two-qubits state $|ct\rangle$, where the first qubit is the "control" and the second one is the "target", in such a way that if $c = 0$ the target qubit is unchanged while, if $c = 1$, it is subject to the NOT operation. According to our model, CNOT gate can be physically constructed by considering four tubular CDs according to the following scheme. We note the upper left quarter of U_{CNOT}, acting on upper part of the two-qubit state, is the identity transformation while the lower right quarter, acting on the lower part of the two-qubit state, is just the NOT gate already considered. So, the CNOT gate can be physically realized by considering a couple of degenerate interacting CDs to perform the NOT gate, and another couple of non-degenerate non-interacting CDs to realize the control. In this case the time required by realize the CNOT operation is simply given by:

$$\tau_{CNOT} = \frac{\pi^{-}h}{2\Gamma} \tag{64}$$

5. Quantum (hyper)computation by means of water coherent domains

Any feasible quantum computing scheme should satisfy some general requirements [4]:

(1) the physical realization of well-defined qubits;
(2) the initialization of the system in a well-defined initial state, such, for example, the qubit $|0\rangle$;
(3) the set-up and "programming" of a set of universal quantum gates;
(4) the efficient storing and retrieving of information;
(5) a sufficiently long decoherence time in order to ensure the quantum computation to finish.

A general treatment of how to address the above requirements within our model is the subject of previous and forthcoming publications [10,26]. Here we'll focus specifically on the point c), limiting ourselves to recall some key previous results related to the other aspects only when needed. According to the quantum coherent dynamics of water, the wavefunction describing a coherent domain, $\Psi = \psi \exp(i\Theta)$, represents an eigenstate $|\Theta\rangle$ of the quantum phase operator (whose eigenvalue is related to the coherent frequency of oscillation including the angular frequency of the excited vortices) [9,10,21,27].

So, any coherent state corresponds to an eigenstate of quantum phase operator and then is always a well-defined quantum state. On the other hand, we have shown the systems composed by tubular CDs can be used as circuits able to hard code the action of specific elementary quantum unitary operations. The actual feasibility of a quantum computer based on the water coherent domains dynamics demands for the "programmability" of such circuits in order to perform sequential and / or parallel operations. This also implies our ability to control and "adjust" the strength of the evanescent photon tunneling interaction between CDs, macroscopically measured by the value of Γ. We can assume such value depends:

(1) on the reciprocal distance between interacting CDs (or, more generally, their geometrical configuration) since, for a given couple of tubular CDs, Γ is inversely proportional to their spatial separation;

(2) for a given spatial separation, on the intensities of the evanescent coherent fields produced by them, in turn related to the specific design of the enclosing shell made up of MTMs.

Many different configurations of two or more tubular CDs could reproduce the same quantum gate, because of the "freedom" in the choice of ΔE and Γ. Each of these configurations is characterized, for a given specific quantum gate they belong to, by a different computational time, depending on the values of ΔE or Γ. The value of quantum phase Θ exactly defines the quantum state $|\Psi\rangle$ of the water tubular CDs (ground or excited) since a change in Θ would imply a change in the value of the coherent oscillation frequency and vice versa. The "identification" of the quantum state $|\Psi\rangle$ with its phase state $|\Theta\rangle$, together with the precise arrangement of the tubular CDs in a given circuit, allows us:

(1) to set the quantum state of each tubular CD, so initializing the quantum qubits register;
(2) to "program" and "control" the quantum gates made by water CDs [10,26];
(3) to achieve a "selective" excitation and a measurement scheme of the phase of such coherent domains capable to address each of them and the qubits they are associated to in order to realize the quantum circuits and operations.

The last point is particularly important in controlled operations in which the setting of the control qubit is crucial. For example, in the case of CNOT gate, the four state $\{|00\rangle, |01\rangle, |10\rangle, |11\rangle\}$ have to be selectively addressed, initialized and measured. In the mapping we have defined (as results from the Hamiltonian H_{CNOT}) the two sets $A = \{|00\rangle, |01\rangle\}$ and $B = \{|10\rangle, |11\rangle\}$, corresponding to different values of the control qubit, are characterized by different transition energies $\hbar\omega_A$ and $\hbar\omega_B$ (the set A is implemented by a couple of non-interacting CDs while the set B by interacting ones) so that they can be distinguished and selected by using, for example, a resonating approach [10].

With regard to the performance of a quantum computational system, as we have seen a rough indication is given by the value of $\lambda^{-1} = N_{op}$ that is directly proportional to the decoherence time τ_Q of the system. As already shown [9,24,25], the decoherence time of coherent quantum state of liquid water can be extraordinary long (virtually even "indefinitely" long) provided that we are able to ensure the condition of interfacial water inside tubular CDs. If this condition is satisfied, as we assume for water inside waveguides made by MTMs, then τ_Q could be of many orders of magnitude greater than τ_{op} so overcoming the critical issue related to environmental decoherence that affect all the other quantum computational schemes so far proposed. Equations (41), (47) and (60) also show computing time can be reduced by increasing ΔE (for T, S and Hadamard gates respectively), whereas for the NOT and CNOT gates, according to Eqs. (54) and (64), a reduction of computational time corresponds to increase the interaction strength Γ.

The energy "accumulated" in an excited water CDs, in the form of rotational energy of its cold vortices, can be very high, leading to a value of ΔE up to about $10^5\,eV$ and a corresponding operational time as small as:

$$\tau_G \sim 10^{-39}\,s \tag{65}$$

for T, S and Hadamard gates. In order to estimate the transformation times required to implement the NOT and CNOT gates, we note the value of Γ (which can be increased by a suitable geometrical and structural design of the tubular MTMs shells) is related, for a given coupling configuration of the tubular CDs, to the tunneling time Δt_{tun} of the evanescent photon from one CD to the other [6,8,9,10,16,26], in the following way:

$$\Gamma \propto \frac{1}{\Delta t_{tun}} \tag{66}$$

On the other hand, it is just such tunneling process that allows the excitation to transfer between a couple of interacting CDs so we can assume:

$$\Delta t_{tun} = \frac{\tau_{Rabi}}{2} \tag{67}$$

where τ_{Rabi} is the "Rabi" time, namely the time required by the system of coupled CDs to perform a complete oscillation between its excited states, that is exaclty twice the time to implement the NOT and CNOT gate, leading us to:

$$\tau_{CNOT} = \tau_{NOT} \cong \Delta t_{tun} \tag{68}$$

The tunneling time is related, in turn, to the energy of the tunneling photon [6,10], that is:

$$\Delta t_{tun} \leq \frac{2\pi^{-}h}{\omega_{coh}} \tag{69}$$

where ω_{coh} is the frequency of the coherent evanescent field associatd to the water CD. For a single exctited water CDs we can extimate an upper bound for Δt_{tun} as [10]:

$$\Delta t_{tun} \leq \frac{2\pi^{-}h}{N_{CD}\Delta E_{mol}} \tag{70}$$

where N_{CD} is the number of water molecules belonging to the the CD and ΔE_{mol} is the energy gap per molecule carachterizing the coherent state. Equation (68) gives for our system [10]:

$$\tau_{CNOT} = \tau_{NOT} \sim 10^{-20}\, s \tag{71}$$

In general, it can be assumed the overall operational time of a computational system depends on the:

(1) overall gates operational time $\sum\limits_{G} \tau_G$;
(2) time required by a circuit to "communicate" with another τ_C;
(3) decoherence time of the system τ_{dec};
(4) time of writing and retrieval of information τ_{RW};
(5) implementation of parallel and/or serial computational schemes.

For a generic system, several theoretical upper bounds for computational speed (expressed as the number of logical operations performed per unit of time n_{op}) related to energy has been proposed so far, even if the computation if performed reversibly. More specifically, for a quantum system characterized by an energy uncertainty ΔE, the time $T_{\perp}$ required to evolve to an orthogonal quantum state obeys the bound [13,14]:

$$T_{\perp} \geq \frac{\pi\hbar}{2\Delta E} \tag{72}$$

that gives the corresponding upper limit for the computational speed n_{op}

$$n_{op} \equiv \frac{1}{T_{\perp}} \leq \frac{2\Delta E}{\pi^{-}h} \tag{73}$$

For a two-levels system, the energy uncertainty of the system can be written as $\Delta E = |E_2 - E_1|$ so that the minimum computational speed is directly proportional to ΔE. The identification of $T_{\perp}^{-1}$ with a maximum computationl clock speed seems quite reasonable for some reasons. In quantum circuits each logic gate brings the system to a near-orthogonal state so we can tipically assume $T_{\perp}$ as the minimum time required, for example, to flip a qubit from $|0\rangle$ to $|1\rangle$ (and vice versa), like in the NOT operation. In our model of computation, the value of $n_{op,G} = 1/\tau_G$ increases, for every quantum gate, with the value of relative excitation energy $\Delta E = E_1 - E_2$, that can also represent the energy uncertainty of the quantum state of the system composed by a couple of interacting CDs. Then we can write, for a given quantum gate:

$$n_{op,G} = \alpha_G \frac{\Delta E}{\pi^- h} \tag{74}$$

where the coefficient α_G is specific for every quantum gate and can be easily obtained from the values of $\Delta t = 1/n_{op}$ in table 2 and agrees with the general bound (56). More specifically, for the NOT gate, that realizes the flip $|0\rangle \Leftrightarrow |1\rangle$, the computational speed can be extimated by Eq. (70) that, compared with Eq. (73), gives the constraint:

$$\frac{N_{CD}\Delta E}{2\pi^- h} \leq n_{op} \leq \frac{2\Delta E}{\pi^- h} \tag{75}$$

One of the most important reason of interest in quantum computing is the common believe it could be intrinsically much more powerful than classical computation. The model of quantum computation based on water coherent domains could be, in principle, even much more powerful of the quantum computation schemes so far proposed.

The discussed model in fact, apart to be able to solve efficiently the critical issue of environmental decoherence, could ensure, if properly implemented, a precise and superfast (theoretically almost instantaneous as already discussed in some previous publications [6,10,26]) set-up and addressing of the quantum states used in computation. A similar mechanism could be also used to perform writing and reading of quantum information as we'll show in a forthcoming publication.

But one of the most important and interesting features of the proposed model of computation concerns its scalability to a large number of quantum circuits composed by many interacting CDs that are able to meaningful increase the computational speed.

According to the commonly accepted picture, the overall speed of a computational system wouldn't depend on the computational architecture, namely if the logical operations are performed in serial or in parallel [10,26].

This can be noticed, for example, by generalizing the Eq. (73) to N logical gates acting in parallel:

$$n_{op}(N) \leq \frac{2}{\pi^- h} \sum_{l=1}^{N} \Delta E_l(N) \tag{76}$$

if ΔE_{tot} is the total energy available to the system then we can suppose $\Delta E_i = \Delta E_{tot}/N$ and write:

$$n_{op}(N) \leq \frac{2}{\pi^- h} \sum_{l=1}^{N} \frac{\Delta E_{tot}}{N} = \frac{2N}{\pi^- h} \frac{\Delta E_{tot}}{N} = \frac{2\Delta E_{tot}}{\pi^- h} \tag{77}$$

that is independent on the number N of the involved water CDs.

Yet, as we have seen, the onset of the QED coherence is characterized by an energy gap that allows the coherent state to have an energy smaller than the non-coherent one. A further increase of the coherence of the system, as occurring in the so-called "supercoherence" [10,12,26,27], would further increase the stability of the system and the related energy gap. For the proposed water system instead, the use of a greater number of logic gates (associated to an increasing number of water CDs) would be accompanied by a higher energy ΔE available for

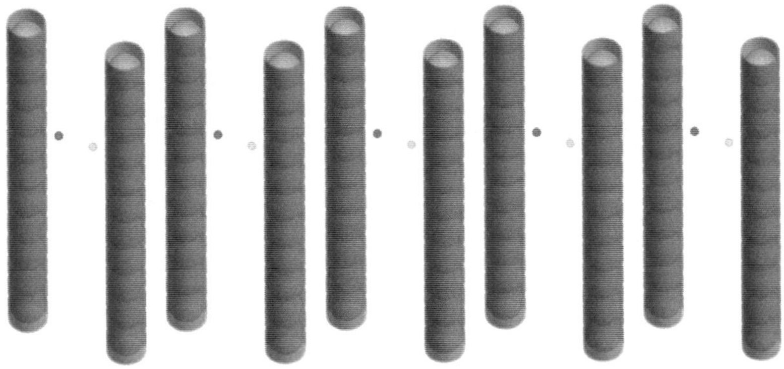

Figure 7. "Synchronized" evanescent coupling and virtual photon exchange between waveguides filled water CDs acting in parallel.

each gate (due to a wider range of possible excited energy levels for each CD associated to the increased number of elementary systems involved in the coherent dynamics) with a consequent reduction of the related computational time for each gate. The underlying dynamics of the proposed computational model then opens a very remarkable perspective to further boost its computational power.

The consideration of an extended network of correlated (through the tunneling and/or phase interaction) water CDs oscillating in phase with each other (fig. 7), is then able to reduce, in principle, the overall parallel computational time, in a way proportional to the number N of the tuned oscillating elementary systems. Such result, that would be impossible to achieve for a closed system, is a natural consequence of the quantum coherent dynamics of liquid water as an open system locally exchanging energy with the fluctuations of quantum vacuum.

So, in the Eq. (77), when N increases, each term of the sum increases as well and the time required to execute each step proportionally reduces, so that, if $N_1 \gg N_2$ we recover, in principle, the concept of "hyper-computing", already analyzed, in a preliminary form, in several previous publications [6-8,9,10,12,16], for which the time required to perform a given number of operations decreases or, at least, remains constant whatever the number of operations.

Finally, in an actual quantum computer architecture, it is very important to consider the issue of memorization and readout of quantum information and the magnitude of the time interval required to perform such operations. This is strictly related to the specific processes used to implement memorization and readout of quantum information that, in our model, are linked to the mutual phase interaction between water CDs and to the interaction between them and their surrounding electromagnetic-thermodynamics environment which they "probe" through e.m. vector potentials. A more detailed study of the role of such interactions in the storage and retrieval of quantum information is not trivial and will discussed in a forthcoming work.

6. Conclusions

Starting from the predictions of QED coherence in water, the realization of quantum circuits, based on the dynamics of energy excitations of coherent water domains, as well as the proposal of a novel architecture of quantum hypercomputing system have been discussed. We have shown such dynamics is able to perform universal quantum gates for one and two-qubits and then, in principle, any type of quantum computation.

The operational speed of the quantum gates so obtained can reach very high values close to the ultimate limit imposed by the fundamental principles of quantum mechanics. This result can be achieved by suitably "adjusting" the energy levels of the water coherent domains (corresponding to its excited levels), by composing the quantum circuits and controlling the parameters ruling the interaction between them.

Apart from the very high computational speed reachable by each elementary universal gate and the precise control and setting of the quantum states representing the qubits, further very amazing features of the proposed model concern the possibility to overcome the environmental decoherence and to realize "supercoherent" quantum computing. The latter refers to the possibility of implementing an extended network of quantum gates acting in parallel, whose computational speed would be proportional to number of the matter components that join the common coherent oscillation, so achieving a much higher operational speed than of a single quantum gate. "Supercoherence" then opens the door to a sort of "quantum hypercomputing" able to boost the computational speed well beyond the ultimate limit so far assumed in classical and quantum computation.

Despite some features of the proposed model (like, for example, storing and retrieving of the quantum information and the study of other interaction mechanisms apart the tunneling of virtual photons already considered) would obviously need further closer examinations (a lot of which already strongly in progress), we argue it could open very fascinating perspectives towards a new conception of quantum computing using water molecules as physical and computational substrate as well as very intriguing technological applications requiring a very high computational power and stability as, for example, the simulation of very complex quantum systems, environmental and health forecasting, large-scale and strong artificial intelligence, encryption / decryption of data and cybersecurity, big data analysis, financial predictions and many others.

7. References

[1] Di Vincenzo D P 2000 The Physical Implementation of Quantum Computation *Fortschr. Phys.* **48** 9-11 771-83.

[2] Resch S and Karpuzcu U R Quantum Computing: An Overview Across the System Stack 2019 *Preprint* arXiv:1905.07240.

[3] Hidary J D 2019 *Quantum Computing: An Applied Approach, in Plastics* (Cham, Switzerland: Springer Nature) 47–59.

[4] Nielsen M A and Chuang I L 2016 *Quantum Computation and Quantum Information (10th Anniversary Edition)* (Cambridge, UK: Cambridge University Press) 277-82.

[5] Preparata G 1995 *QED Coherence in Matter* (Singapore, London, New York: World Scientific Publishing Co Pte Ltd) 195-217.

[6] Caligiuri L M and Musha T 2016 *The Superluminal Universe: from Quantum Vacuum to Brain Mechanism and beyond* (New York, USA: NOVA Science Publisher).

[7] Caligiuri L M and Musha T 2019 Quantum hyper-computing by means of evanescent photons *J. Phys.: Conf. Series* **1251** 012010.

[8] Caligiuri L M and Musha T 2019 Accelerated Quantum Computation by means of Evanescent Photons and its Prospects for Optical Quantum Hypercomputers and Artificial Intelligence *Proc. of the 2019 International Conference on Engineering, Science, and Industrial Applications (ICESI)*, (Tokyo) pp 199–208.

[9] Caligiuri L. M. 2020 *Frontiers in Quantum Computing* ed L.M. Caligiuri (New York, USA: NOVA Science Publisher) pp 1–37.

[10] Caligiuri L. M. 2020 *Frontiers in Quantum Computing* ed L.M. Caligiuri (New York, USA: NOVA Science Publisher) pp 37–102.

[11] Del Giudice E. and Tedeschi A. 2009 Water and Autocatalysis in Living Matter *Electromagnetic Biology and Medicine* **28** 1 46-52.

[12] Caligiuri L. M. 2018 Super-Coherent Quantum Dynamics of Zero-Point Field and Superluminal Interaction in Matter *Unified Field Mechanics II: Formulation and Empirical Tests* eds R.L. Amoroso, L.H. Kauffman, P. Rowlands and G. Albertini (Singapore, London, New York: World Scientific Publishing Co Pte Ltd) pp 331-43.

[13] Lloyd S. 2000 Ultimate physical limit to computation *Nature* **406** 6799 pp. 1047-54.

[14] Margolus N. and Levitin L. B. 1998 The maximum speed of dynamical evolution *Physica D* **120** 1-2 188-95.

[15] Del Giudice E., Spinetti P.R. and Tedeschi A. 2010 Water Dynamics at the root of Metamorphosis in Living Organisms *Water* **2** 3 566-86.
[16] Caligiuri L.M. and Musha T. 2016 Superluminal Photons Tunneling through Brain Microtubules Modelled as Metamaterials and Quantum Computation *Advanced Engineering Materials and Modeling* eds A. Tiwari, N. Arul Murugan, and R. Ahula (New York, NJ, USA: Wiley Scrivener Publishing LLC) pp 291-333.
[17] Hepp K. and Lieb E. H. 1973 Equilibrium statistical mechanics of matter interacting with the quantized radiation field *Phys. Rev. A* **8** 2517-25.
[18] Caligiuri L. M. 2020 *Frontiers in Quantum Computing* ed L.M. Caligiuri (New York, USA: NOVA Science Publisher) pp 39–56.
[19] Del Giudice E. and Preparata G. 1995 Coherent dynamics in water as possible explanation of membrane formation *J. of Biol. Phys.* **20** 105-16.
[20] Arani R., Bono I., Del Giudice E., and Preparata G. 1995 QED coherence and the thermodynamics of water *Int. J. of Mod. Phys. B* **9** 15 1813-41.
[21] Del Giudice E. and Preparata G. 1998 A new QED picture of water: understanding a few fascinating phenomena *Proc. Macroscopic Quantum Coherence* (Singapore, London, New York: World Scientific Publishing Co Pte Ltd) pp 108-29.
[22] Voeikov V. L. and Del Giudice E. 2009 Water respiration: the basis of the living state *Water* **1** 52-75.
[23] Bono I., Del Giudice E., Gamberale E., and Henry M. 2012 Emergence of the coherent structure of liquid water *Water* **4** 3 510-32.
[24] Buzzacchi M., Del Giudice E. and Preparata G. 2002 Coherence of the glassy state *Int. J. Mod. Phys. B* **16** 25 3771-86.
[25] Del Giudice E., Tedeschi A., Vitiello G., and Voeikov V. 2013 Coherent structures in liquid water close to hydrophilic surfaces *J. Phys.: Conf. Series* **442** 012028.
[26] Caligiuri L.M., 2021 Accelerated quantum computation based on quantum coherent dynamics of evanescent photons in liquid water *Intelligent Computing* ed K. Arai (Cham, Switzerland: Springer Nature) vol 1 pp 169-87.
[27] Del Giudice E., Spinetti P. R., and Tedeschi A. 2010 Water Dynamics and the Root of Metamorphosis in Living Organisms *Water* **2** 3 566-86.
[28] Franson J. D., Jacobs B.C., and Pittman T.B. 2004 Quantum computing using single photons and the Zeno effect *Phys. Rev. A* **70** 6 062302.

A "user-friendly" phantom to conduct Quality Controls on MRgFUS device

G Acri[1], F Caridi[2,a], B Testagrossa[1], S Gurgone[2], C Anfuso[3], G Paladini[2], G Vermiglio[4], C Greco[2] and V Venuti[2]

[1] Dipartimento di Scienze Biomediche, Odontoiatriche e delle Immagini Morfologiche e Funzionali, Università degli Studi di Messina, Via Consolare Valeria, 98125, Messina, Italy
[2] Dipartimento di Scienze Matematiche e Informatiche, Scienze Fisiche e Scienze Della Terra, Università degli Studi di Messina, V.le F. Stagno D'Alcontres, 31-98166 Messina, Italy
[3] IRCCS Centro Neurolesi "Bonino-Pulejo" - Messina, Italy
[4] Medical Physicist, Freelance

[a] E-mail: fcaridi@unime.it

Abstract. Magnetic Resonance guided Focused Ultrasound Surgery (MRgFUS) is a hybrid technique which uses Magnetic Resonance (MR) to obtain morphological information of the examined anatomical district and Focused Ultrasound Surgery (FUS) to ablate the body region under investigation by using MR as guidance. To ensure patients' safety is necessary to establish periodic Quality Controls (QCs) providing a complete evaluation of system status. QCs cannot be done directly on patients, for this reason, dedicated phantoms are used to guarantee the maintenance of quality over equipment lifetime. Several phantoms are available on the market, which are distributed by manufacturers and specific to the medical equipment, resulting in closed systems. In this study, we idealized, developed and created a prototype of a completely "user-friendly" phantom, to conduct QCs on FUS devices. It consists of an empty cylinder of polymethylmetacrylate (PMMA), that can be filled with different fluids and various inserts, available for the Medical Physicists, to perform test measurements on different parameters (size of FUS spot, precision of FUS spot, linearity of FUS power, linearity of electric power, evaluation of target temperature). The obtained results demonstrated that the novel phantom represents a useful and adaptable device that could be used during FUS QCs evaluations.

1. Introduction

In the last decade different medical imaging hybrid techniques were developed and used in many clinical fields, including oncology, neurology, cardiology, neuroscience. In this context, Magnetic Resonance guided Focused Ultrasound (MRgFUS) represents an important new perspective in the field of imaging/therapy, combining MR to obtain morphological information and FUS for tissue ablation. On one side, MR utilizes powerful static magnetic fields and Radiofrequency signal to obtain information from organs inside the human body and to construct the images [1,2]. On the other side, FUS uses ultrasound (US), generated by a transducer embedded in the Magnetic Resonance Imaging (MRI) device (head coil or table). The energy of the focused US beam causes significant increase in temperature of the targeted tissues, producing coagulation and necrosis [3], and, at the same time, MR images are used as guidance during the procedure and as temperature feedback [4]. This hybrid technique has been proposed for possible applications in the treatment of breast cancer [5], prostate cancer [6], uterine fibroids [7], and in mediated thalamotomy of the ventral intermediate nucleus (VIM) [8].

In order to maintain or enhance the quality of the radiological device and to ensure safe, for patients and health care staff, it is necessary to establish adequate quality-assurance (QA) procedures to identify and correct all problems that may lead to reduction of the diagnosis area in advance. [9].

In this context, Quality Controls (QCs) play a fundamental role, allowing a complete evaluation of the system status [10-12]. As can be reasonably understood, QCs cannot be done directly on patients, but they are performed by using dedicated phantoms, usually mimicking human tissues. For FUS systems, several phantoms are available, built using different materials, such as PMMA, water, Agar gel, etc., to simulate the different constituents of the human body [13-15]. The main limitation of these phantoms, commercially available, is that they are "closed" system. As a matter of fact, they are developed to analyze only few parameters, and if a complete QCs program is needed, to the use of more than one phantom is required. In addition, in many cases, the manufacturer of the device provides the phantoms together with the diagnostic equipment they are dedicated to, in order to test the functioning of the set-up according to dedicated protocols, often developed by the same companies for each machine and usually directly managed by the equipment software.

In order to overcome the aforementioned limitation, in this study we developed and built a "user-friendly" phantom to conduct a complete QCs program on MRgFUS equipment. The novel phantom was built in PMMA material, it has the shape of a cylinder, hollow inside so as to be filled with materials and inserts chosen by the medical physicist. By using this novel dedicated phantom, we are able to perform test measurements on different parameters (size of FUS spot, precision of FUS spot, linearity of FUS power, linearity of electric power, evaluation of target temperature). The reliability of this innovative phantom was evaluated during the acceptance test of a MRgFUS device. In all cases, the results obtained using the proposed new phantom were reliable and promising, confirming the flexibility of the method, its simplicity and reproducibility as an efficient tool for quick inspections, as discussed in detail in the following sections.

2. Materials and Methods

2.1. The used set-up

The QCs measurements, conducted by using the prototype of the developed phantom, were successfully performed on a 1.5 T MR scanner (Optima 450 W GE Medical Systems, Waukesha, WI) equipped with the ExAblate® 2000 (InSightec Inc., Haifa, Israel), for FUS treatments. The measurements were conducted by using the FUS transducer mounted on the MR table containing a high power phased array transducer (120 mm diameter, 211 elements, central frequency 1.1 MHz). The images were acquired setting a T2-weighted fast spin–echo in coronal, axial and sagittal planes, on MR scanner, and transferred to the ExAblate® FUS workstation.

2.2. The novel phantom

The developed phantom consists of an empty cylinder of 14.5 cm diameter and 15.5 cm height. The lateral surface, having a thickness of 0.5 mm, is realized in polymethylmetacrylate (PMMA). This choice of material and thickness is motivated by the fact that this configuration is able to simulate the cranial theca. The base surface is made with an easily interchangeable module, depending on the tests to be carried out. The material used in our case is a PMMA sheet with a thickness of 0.05 mm. This choice was made to facilitate the coupling of the ultrasonic transducer with the phantom. The cylinder can be filled with various liquids, solids or gel, chosen by the medical physicist according to the parameters to be tested. The novel phantom is coupled with movable inserts. They consist of holders, always in PMMA, designed to sustain instrumentation or small Teflon pieces (1.0 mm x 1.0 mm and 1.0 mm x 2.0 mm) in order to simulate microcalcifications, mounted on nylon wire of 0.01 mm thickness. In figure 1, the developed phantom and the movable inserts are depicted.

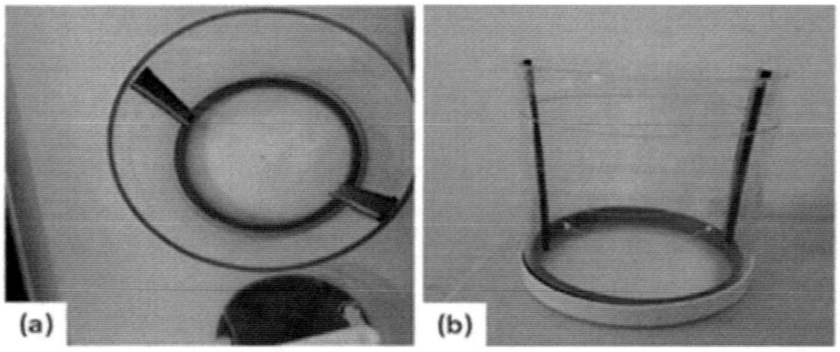

Figure 1. Upper view (a) and lateral view with movable inserts (b) of the developed phantom.

Since during treatment several sonication parameters can be manually adjusted, when performing QCs it becomes necessary to evaluate, for each parameter, the correspondence between nominal value and measured one. In particular, the novel developed phantom allows the testing of the parameters reported in table 1, and better detailed in the following subsections.

Table 1. Parameters that can be tested by using the developed phantom.

Parameter
Precision of the FUS spot
Dimension of the FUS spot
Linearity of the mass vs. electric and FUS output power
Evaluation of the target temperature

During all measurements, the phantom was placed on the MRgFUS bed, and demineralized water was placed in the interface between the bottom of the phantom and the bed in order to provide an acoustic path for the ultrasound beam.

2.3. Precision of the FUS spot

FUS precision spot was evaluated conducting two different experiments. In the first one, the phantom was filled with a piece of white meat and demineralized water. Four wood pointers were positioned in different places on the target. A series of four identical sonication were performed, and the corresponding images were acquired by MR scanner and stored for image analysis. In the second test, the micro-calcifications insert were used and the phantom was filled with demineralized water. On the nylon wires of the insert were positioned, in different position, five meat pieces (10 mm x 10 mm x 10 mm) and a series of five identical sonication were executed and centred on the meat targets. The images were acquired, stored and analysed by using MR scanner.

2.4. Dimension of the FUS spot
Dimensions of sonication spots are a function of several sonication parameters, which can influence the success of the treatment. The novel phantom was filled with a piece of white meat and demineralized water. Five identical sonication were performed at 85 W output power, 18 seconds and at 5 mm scan slice thickness. The images were acquired and stored by MR scanner for post evaluation. The dimensions of the sonication spot (length and diameter) were evaluated by using the MR software available at the console and by using ad-hoc software, previously validated by authors [16], and developed in LabView environment. The DICOM images were used as standard for acquired pictures [17]. Statistical analysis was used The one way ANOVA was used to confirm the validity of the obtained results, comparing different and independent datasets of the same variable to assume whether they are generated by the same distribution.

2.5. Linearity of the mass vs. electric and FUS output power
Linearity refers to the ability of the amplifier to produce signals that are accurate copies of the input, generally at increased power levels [19-21].

The test consisted of measurements intended to assess the pressure of the ultrasounds, which can be linked to the intensity of the same. A precision balance (PCD - KERN & SOHN GmbH, Stuttgart, Germany) with a sensitivity of 0.1 g was used, positioned at a height of 16 cm, resting on the dedicated support, inserted inside the phantom, which was previously filled with physiological solution. It was assumed that the mass, as measured by balance and due to the FUS pressure, increases with increasing the output power of the transducer. The phantom was filled with demineralized water, to increase the coupling between the transducer and the phantom itself. Seven different values of Acoustic power were set and the corresponding value of mass was measured by balance. The Pearson's correlation was used to validate the linearity of measurements.

2.6. Target Temperature
In this study, the target temperature was evaluated by using two different probes, i.e., a contact thermometer (HI83141 – Hanna Instruments) and a thermal camera (FLIR T440), and the obtained results were compared with the temperature value as given by the MR scanner. The measurements were conducted in two steps. In the first one, the phantom was filled with meat and demineralized water and placed inside the gantry, six sonication of 10 s each and at different set power input were executed, and the corresponding temperature gradients were evaluated by MR scanner. In this step, due to the high magnetic field inside the gantry, the thermal camera and the contact thermometer cannot be used. For this reason, it was necessary the second step. The phantom was always positioned on the bed coupled with the FUS transducer and the same sonication, used in the step one, were repeated with the phantom out of the gantry. The thermal camera was positioned at 50 cm from the target, while the contact thermometer was placed on the target, in correspondence of the FUS spot. The One-way ANOVA was used to assess whether a significant difference existed between data set obtained following different procedures.

3. Results
In the following, the results of the measurements conducted during the acceptance test of MRgFUS device are reported. All the parameters were acquired and measured by using the novel developed phantom.

3.1. Precision of the FUS spot

In the first test, the matching between FUS spot and wood pointers was evaluated using the MR software and the maximum spot-pointers distance turned out to be ± 0.3 mm, where the positive (negative) sign indicates a distance greater (smaller) than the specific length.

In the second one, always by using the MR software, it was verified that all small targets were centred by the FUS spot.

In all cases a good agreement was achieved.

3.2. Dimension of the FUS spot

The diameter and the length dimensions of the FUS spot were measured by using coronal and sagittal plane images respectively. In figure 2 we show the MR images acquired along the sagittal, axial and coronal scan planes.

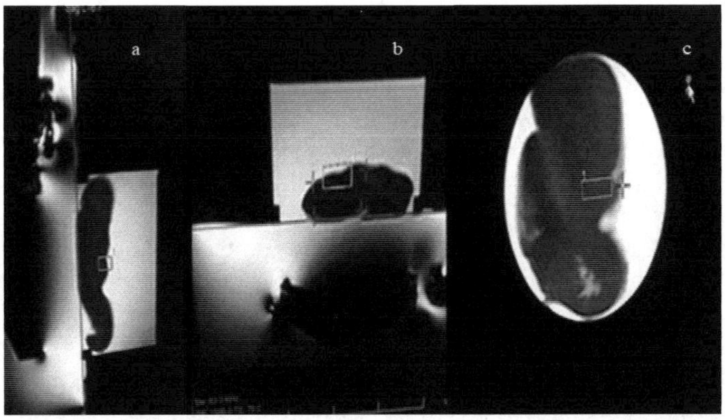

Figure 2. MR images acquired along (a) sagittal, (b) axial, and (c) coronal scan plane. The user defined Region of Interest (ROI) are indicated by yellow rectangles.

In table 2 the measurements results are reported.

Table 2. Measured diameter and length of sonication spot. The results obtained by using MR and LabView software refer to the average of the five sonication.

Dimension	Nominal value (mm)	MR software (mm)	LabView (mm)
Diameter	3.0	3.20 ± 0.30	3.14 ± 0.30
Length	6.0	6.48 ± 0.25	6.55 ± 0.30

3.3. Linearity of mass vs. Electric and FUS output power

From the control console it was possible to modify the electric power, with consequent variation of the FUS output power, and the corresponding mass was measured. Figure 3 shows the trend of the mass as a function, respectively, of electric power and FUS output power.

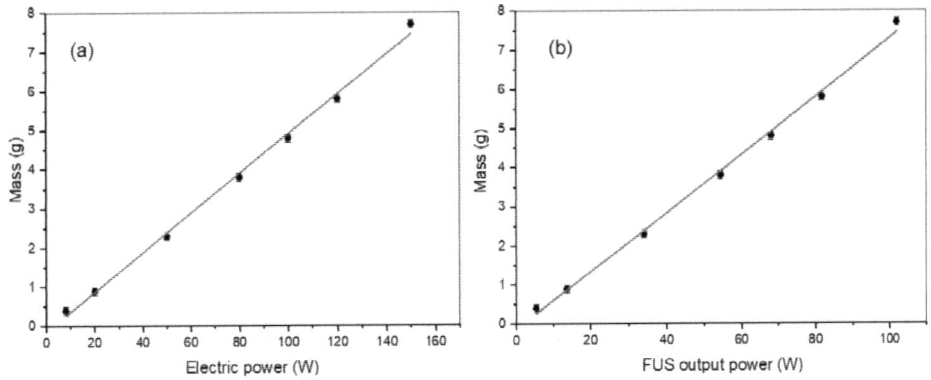

Figure 3. Plot of the mass as a function of the electric (a) and FUS output (b) power.

As can be seen, a linear fit faithfully reproduces the experimental data, being characterized by r2 ~ 0.996 in both cases.

3.4. Target temperature

The current gold standard for real-time monitoring during sonication is accomplished using the known linear dependence of proton resonance frequency as a function of temperature [21, 22], during QCs it is important to evaluate the correctness of the parameter under investigation. In figure 4 the trend of temperature gradient as a function of sonication power, as obtained by using the contact thermometer (black line), thermal camera (red line) and MR scanner (blue line), is reported.

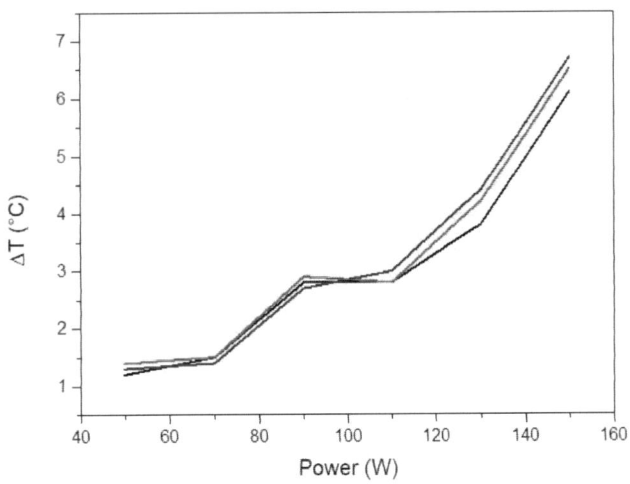

Figure 4. Temperature gradient as a function of sonication power, as obtained by contact thermometer (black line), thermal camera (red line) and MR scanner (blue line).

The one-way analysis of variance (ANOVA) was used to assess whether a significant difference existed between data set obtained following different procedures. According to it, the calculated F value is 0.02232, significantly smaller than the tabulated value ($p = 0.9779$ at $p = 0.05$ confidence level). Therefore, there is no statistically significant difference among the results obtained by the three methodologies.

4. Discussion

In this study we have described a novel phantom, completely "user-friendly", used to perform QCs on FUS devices. The phantom was designed to be used on bed and head FUS transducer, embedded on MRI scanners. We tested the novel phantom evaluating different parameters; in particular, in our measurements we have considered the size of FUS spot, the precision of FUS spot, the linearity of FUS power, the linearity of electric power and the temperature variation of the target. All measurements were made by using the FUS transducer embedded on the MRI bed, as it can emit ultrasound both inside and outside the gantry. If we had used the head transducer, it would have been able to work only inside the gantry. In fact, usually the use of head transducer makes it possible to work only inside the gantry, where, however, because of the high static magnetic field, the use of electronic instrumentation employed for QCs is forbidden.

A key challenge in designing a FUS phantom was the need of coupling the FUS transducer with the phantom. The PMMA material was chosen to simulate typical anatomical body regions for quality assurance [23] and different thicknesses are used to mimicking different body regions (greater thickness for the head, less for the abdomen). The achieved results showed the validity of this novel phantom; in fact, all the measurements related to the observed parameters reproduced the experimental data.

5. Conclusions

The use of MRgFUS for imaging/therapy treatmets represents the culmination of scientific progress and clinical research on the application of FUS to human body. Understanding of flow physics is important to execute detailed healthcare product development; theoretical (Computational Fluid Dynamics) [24] and experimental evaluation are necessary to yield quantitative results. This progress includes improved technology to provide controlled levels of ultrasonic energy that can be focused to a target noninvasively and guided by MR imaging. To ensure patients' safety is necessary to establish periodic Quality Controls (QCs) providing a complete evaluation of system status.

In the present work, a novel, completely "user-friendly" phantom was developed and used to conduct acceptance tests on MRgFUS equipment, with the final aim of providing an adaptable, reliable and robust procedure to perform a complete QCs program. The realized phantom is totally suited to the needs of the Expert in Medical Physics, responsible, in accordance with current legislation, for QCs, who can use it to evaluate the parameters representative of the state of the device.

The obtained results look very encouraging, so much so that the evaluated parameters allowed to draw up a complete QCs protocol for the used equipment, which represents the baseline for future evaluations.

The proposed FUS phantom has yielded exciting early results and serves as a prototype. Future implementations of the phantom may include new inserts mimicking other body organs.

Considering the importance of the new hybrid methodology and its prospects for applications to different targets of the human body, it is our intention to implement the developed prototype, to have always a single "user-friendly" phantom to be used for conducting QCs measurements.

References

[1] Schenck, J. F. (2005). Physical interactions of static magnetic fields with living tissues. *Progress in Biophysics and Molecular Biology*, *87*(2–3), 185–204. https://doi.org/10.1016/j.pbiomolbio.2004.08.009

[2] Bonanno, L., Marino, S., Morabito, R., Barbalace, G., Sestito, A., Testagrossa, B., & Acri, G. (2019). Evaluation of US and MRI techniques for carotid stenosis: a novel

phantom approach. *La Radiologia Medica*, *124*(5), 368–374. https://doi.org/10.1007/s11547-018-0971-7

[3] Gorny, K. R., Hangiandreou, N. J., Hesley, G. K., Gostout, B. S., McGee, K. P., & Felmlee, J. P. (2006). MR guided focused ultrasound: technical acceptance measures for a clinical system. *Physics in Medicine and Biology*, *51*(12), 3155–3173. https://doi.org/10.1088/0031-9155/51/12/011

[4] Wu, T. (2002). A quality control program for MR-guided focused ultrasound ablation therapy. *Journal of Applied Clinical Medical Physics*, *3*(2), 162. https://doi.org/10.1120/1.1459262

[5] Gianfelice, D., Khiat, A., Amara, M., Belblidia, A., & Boulanger, Y. (2003). MR Imaging–guided Focused US Ablation of Breast Cancer: Histopathologic Assessment of Effectiveness—Initial Experience. *Radiology*, *227*(3), 849–855. https://doi.org/10.1148/radiol.2281012163

[6] Gelet, A., Chapelon, J. Y., Bouvier, R., Pangaud, C., & Lasne, Y. (1999). Local control of prostate cancer by transrectal high intensity focused ultrasound therapy: preliminary results. *Journal of Urology*, *161*(1), 156–162. https://doi.org/10.1016/S0022-5347(01)62087-1

[7] Taheri, M., Galo, L., Potts, C., Sakhel, K., & Quinn, S. D. (2019). Nonresective treatments for uterine fibroids: a systematic review of uterine and fibroid volume reductions. *International Journal of Hyperthermia*, *36*(1), 294–300. https://doi.org/10.1080/02656736.2018.1564843

[8] Elias, W. J., Huss, D., Voss, T., Loomba, J., Khaled, M., Zadicario, E., Frysinger, R. C., Sperling, S. A., Wylie, S., Monteith, S. J., Druzgal, J., Shah, B. B., Harrison, M., & Wintermark, M. (2013). A Pilot Study of Focused Ultrasound Thalamotomy for Essential Tremor. *New England Journal of Medicine*, *369*(7), 640–648. https://doi.org/10.1056/NEJMoa1300962

[9] Acri, G., Tripepi, M. G., Causa, F., Testagrossa, B., Novario, R., & Vermiglio, G. (2012). Slice-thickness evaluation in CT and MRI: an alternative computerised procedure. *La Radiologia Medica*, *117*(3), 507–518. https://doi.org/10.1007/s11547-011-0775-5

[10] Chen, C.-C., Wan, Y.-L., Wai, Y.-Y., & Liu, H.-L. (2004). Quality Assurance of Clinical MRI Scanners Using ACR MRI Phantom: Preliminary Results. *Journal of Digital Imaging*, *17*(4), 279–284. https://doi.org/10.1007/s10278-004-1023-5

[11] Goodsitt, M. M., Carson, P. L., Witt, S., Hykes, D. L., & Kofler, J. M. (1998). Real-time B -mode ultrasound quality control test procedures. Report of AAPM Ultrasound Task Group No. 1. *Medical Physics*, *25*(8), 1385–1406. https://doi.org/10.1118/1.598404

[12] Acri, G; Gurgone, S; IOvane, C; Romeo, MB; Borzelli, D; Testagrossa, B. (2021). A Novel Phantom and a Dedicated Developed Software for Image Quality Controls in X-Ray Intraoral Devices. *Journal of Biomedical Physics and Engineering*, *11*(2). https://doi.org/10.31661/jbpe.v0i0.2001-1061

[13] McGarry, C. K., Grattan, L. J., Ivory, A. M., Leek, F., Liney, G. P., Liu, Y., Miloro, P., Rai, R., Robinson, A., Shih, A. J., Zeqiri, B., & Clark, C. H. (2020). Tissue mimicking materials for imaging and therapy phantoms: a review. *Physics in Medicine & Biology*. https://doi.org/10.1088/1361-6560/abbd17

[14] Hofstetter, L. W., Fausett, L., Mueller, A., Odéen, H., Payne, A., Christensen, D. A., & Parker, D. L. (2020). Development and characterization of a tissue mimicking psyllium husk gelatin phantom for ultrasound and magnetic resonance imaging. *International Journal of Hyperthermia*, *37*(1), 283–290. https://doi.org/10.1080/02656736.2020.1739345

[15] Farrer, A. I., Odéen, H., de Bever, J., Coats, B., Parker, D. L., Payne, A., &

Christensen, D. A. (2015). Characterization and evaluation of tissue-mimicking gelatin phantoms for use with MRgFUS. *Journal of Therapeutic Ultrasound*, *3*(1), 9. https://doi.org/10.1186/s40349-015-0030-y

[16] Acri, G., Testagrossa, B., Sestito, A., Bonanno, L., & Vermiglio, G. (2018). CT and MRI slice separation evaluation by LabView developed software. *Zeitschrift Für Medizinische Physik*, *28*(1), 6–13. https://doi.org/10.1016/j.zemedi.2017.09.009

[17] Rafeek Mamdouh, Hazem M. El-Bakry, Alaa Riad, Nashaat El-Khamisy, Converting 2D-Medical Image Files "DICOM" into 3D- Models, based on Image Processing, and Analysing their Results with Python Programming, WSEAS Transactions on Computers, Volume 19, 2020, pp. 10-20.

[18] Shaw, A.; Martin, K. The acoustic output of diagnostic ultrasound scanners. In The Safe Use of Ultrasound in Medical Diagnosis, 3rd ed.; ter Haar, G., Ed.; The British Institute of Radiology: London, UK, 2012; pp. 18–45.

[19] Civale, J., Rivens, I., Shaw, A., & ter Haar, G. (2018). Focused ultrasound transducer spatial peak intensity estimation: a comparison of methods. *Physics in Medicine & Biology*, *63*(5), 055015. https://doi.org/10.1088/1361-6560/aaaf01

[20] Csány, G., Gray, M., & Gyöngy, M. (2020). Estimation of Acoustic Power Output from Electrical Impedance Measurements. *Acoustics*, *2*(1), 37–50. https://doi.org/10.3390/acoustics2010004

[21] Mouratidis, P. X. E., Rivens, I., & ter Haar, G. (2015). A study of thermal dose-induced autophagy, apoptosis and necroptosis in colon cancer cells. *International Journal of Hyperthermia*, *31*(5), 476–488. https://doi.org/10.3109/02656736.2015.1029995

[22] Hindman, J. C. (1966). Proton Resonance Shift of Water in the Gas and Liquid States. *The Journal of Chemical Physics*, *44*(12), 4582–4592. https://doi.org/10.1063/1.1726676

[23] Bouwman R.W., Diaz O., Young K.C., van Engen R.E., Veldkamp W.J.H., Dance D.R. (2012) Phantoms for Quality Control Procedures of Digital Breast Tomosynthesis. In: Maidment A.D.A., Bakic P.R., Gavenonis S. (eds) Breast Imaging. IWDM 2012. Lecture Notes in Computer Science, vol 7361. Springer, Berlin, Heidelberg. https://doi.org/10.1007/978-3-642-31271-7_42

[24] Mohamed Yacin Sikkandar, Natteri M. Sudharsan, S. Sabarunisha Begum, E. Y. K. Ng, Computational Fluid Dynamics: A Technique to Solve Complex Biomedical Engineering Problems - A Review, WSEAS Transactions on Biology and Biomedicine, Volume 16, 2019, pp. 121-137.

Early-age heat flow and hydration in cementitious composites: numerical modelling based on temperature measurements

E Martinelli[1], **M Pepe**[2] and **C Lima**[3]

[1] Associate Professor, Department of Civil Engineering, University of Salerno, Italy
[2] Research Assistant, Department of Civil Engineering, University of Salerno, Italy
[3] Head of Modelling section, TESIS srl, Fisciano (SA), Italy

e-mail: mapepe@unisa.it

Abstract. The heat generated during early age cement hydration causes a semi-adiabatic temperature rise of hardening concrete, while starting to develop its physical and mechanical properties. As a matter of principle, the heat generated by the hardening mixture depends on the cement properties and its hardening conditions like for instance the type of binders, quality of aggregates, water-to-cement ratio and type of formwork. A simple 1D numerical model can be formulated for simulating the evolution in time of this reaction, in order to identify the so-called degree of hydration. The present paper proposes a detailed description of a theoretical approach which allow to assess the temperature development occurring in concrete elements at the early-age when the produced mixture is cured in semi-adiabatic boundary conditions. In this numerical procedure, the differential heat equation takes into account the heat that liberates to the environment through the formwork or concrete's surface. This is done by considering the Arrhenius Principle and assuming a pre-defined shape of the adiabatic hydration curve of the concrete mixture. Hence, an indirect identification procedure of the aforementioned adiabatic curve is ideally carried out, as the simulated temperature evolution in semi-adiabatic conditions is brought to match the temperature measurements on a hardening concrete sample. This modelling procedure, enabling various boundary conditions, ranging from semi-adiabatic to isothermal, can be used to calculate the degree of hydration of a real in-situ cast concrete. Specifically, considering the variation of concrete strength class and element size some possible applications are proposed: this represents a simplified approach for the prediction of the temperature time-evolution in the concrete elements at early-age which may be also used as a practical tool for mitigating the risk of premature cracking.

1. Introduction

The mechanisms governing the evolution of the setting and hardening processes for cementitious composites are, generally, identified through the identification of the Degree of Hydration (DoH) [1]. In addition, the DoH is directly correlated to the physical and mechanical properties' development, at the early-age, for the concrete mixture under investigation. [2]. On the other hand, it is worth to mention that the determination of the degree of hydration result to be easily evaluable only in the case in which the setting and hardening processes occurs in adiabatic condition, since the resulting DoH is directly proportional to the generated heat, leading to temperature increase within the cement-based composite [3]. Consequently, when the concrete mixture curing occurs in more general boundary conditions, such as pseudo-adiabatic and/or isothermal, it is necessary to adopt some advanced theoretical tools allowing the identification of the DoH. [4]. Beyond the complex problem characterizing the identification of the setting and hardening processes as a function of the boundary condition (i.e., adiabatic, pseudo-adiabatic and isothermal), it is worth to highlight that the early-age evolution of the cementitious composites' properties is also governed by the intrinsic characteristics of the mixtures: cement and aggregates type and amount, possible use of alternative binders, water-to-binder ratio etc. [5][6].

As reported by Fairbairn & Azenha [7], massive structures are those structures for which the effects of cement hydration in the setting and hardening phases can cause cracks in early-age concrete mixtures. In the past, such cracking risks were considered relevant only for massive concrete structures, where the dissipation of the hydration heat was normally slow and, therefore, high temperature increases were observed. Nowadays, with the diffusion of high-performance concrete, the risk of cracking in "young" concrete is no longer a peculiarity of the massive structures. The higher quantities of cementitious materials associated with lower water-to-cement ratios result, respectively, in greater heat of hydration and in microstructures with finer pores, thus generating greater amplitudes of thermal gradients and autogenous shrinkage [7]. For this reason, giving the definition of massive structure, it was done in a broad sense, considering all the concrete elements for which the effects of the hydration of the cement can involve risks of thermal cracking. A very significant case, reported by Betioli et al. [8] concerns the hydroelectric dam of Itaipù (located on the Paranà river, on the border between Paraguay and Brazil) whose construction began in 1975. During its the construction, cracking phenomena were observed. At that time, as reported by Rosso & Piasentin [9], the analyzes indicated that this crack pattern was due to the heat of hydration observed in the field, which was considerably greater than that assumed during the dam project. It was also found that the models to simulate the behavior of early-age concrete were very simplified and did not took into account the stress concentrations induced by the elements' geometry. To improve the dam behavior, various measures were taken such as changing the mix design of the concrete or introducing additional reinforcements in some areas. In addition, existing cracks were injected with epoxy and corresponding structures examined by an intensive monitoring program. Another relevant example concerns a dam built on the Blue River, in China, inaugurated in 2006. For this demanding work, as reported by Zhang et al. [10], a first report on concrete mix design was developed in 1998, in which some problems were highlighted. They mainly concerned the need to obtain, on the one hand, a concrete with high mechanical strength, so as to cope with the dynamic impacts due to the dragging of stone materials into the watercourse. On the other hand, the produced mixtures should have been able to have a structure as free as possible of cracks of thermal origin which may have been linked to an excessive dosage of cement required to achieve high mechanical strength. The results obtained in this first phase of the study were satisfactory for the mechanical strength required and for the associated resistance to erosion, but not acceptable for the high cracking state recorded on the surface of the first castings, due to the tensile stresses induced by the excessive heating in the central core of the structures and stresses induced by the hygrometric shrinkage of the concrete surface. Therefore, it was decided, in 2001, to perform new research to reduce the cracking, exploiting the possible use of different raw materials and various mixture proportioning.

In this context, this paper presents a detailed description of a numerical procedure which is capable of simulating the time evolution of a concrete's early age temperature development under semi adiabatic conditions [11][12]. In this numerical procedure the differential heat equation takes into account the heat that liberates to the environment through the formwork or concrete's surface. This is done by considering the well-known Arrhenius Principle [13] and assuming a pre-defined shape of the adiabatic hydration curve of the concrete mixture. An iterative procedure is developed with the aim to bring the simulated temperature development (which depends on both the assumed adiabatic curve and the thermal boundary conditions) to match the actual temperature measurements recorded in a sample of concrete during setting and hardening stages. This modelling/identification procedure, enabling various boundary conditions, ranging from semi-adiabatic to isothermal, can be employed to simulate the time evolution of the DoH of a concrete strucuture.

2. Theoretical formulation

2.1. Fundamental assumptions of a theoretical model for cement hydration

The hydration reaction which develops in concrete mixture can be analysed by means of a theoretical approach recently proposed in the literature by the authors of this paper [11]. As a matter of the fact, the proposed modelling approach is able to assess and simulate the time development of the *degree of hydration* (DoH, identified with the symbol $\alpha_h(t)$). The latter, in fact, is identified by the ratio among the heat **Q(t)** which is produced up to a certain time **t** and the maximum heat $\mathbf{Q_{max}}$ which can be potentially produced as a result of the whole cement reaction:

$$\alpha_h(t) = \frac{Q(t)}{Q_{max}} \tag{1}$$

As a consequence, the reaction of the cement results to occur in a heat transfer process and, consequently, in a transient temperature field **T**. which can be described by the following equation describes the **T** field when it develops in a **1D** domain (see also the schematic representation proposed in **Figure 1**):

$$\rho_c c_c \frac{\partial T}{\partial t} = \lambda_c \cdot \frac{\partial^2 T}{\partial x^2} + q_c(x,t) \ , \tag{2}$$

In the above-mentioned equation (2), the term:

- ρ_c represents the is mass per unit volume of the produced mixtures;
- c_c is the specific heat of the concrete mixtures;
- λ_c represents the thermal conductivity coefficient and;
- $q_c(x,t)$ is the rate of heat production of concrete.

Moreover, the "heat production of concrete" ($q_c(x,t)$) is defined as follows:

$$q_c(x,t) = C \cdot \frac{\partial Q}{\partial t} \ . \tag{3}$$

The analytical equations presented by equation (3) and, then, **Q(x,t)** cannot directly evaluated since they depend on the actual temperature which is developed within the concrete mixture sample.

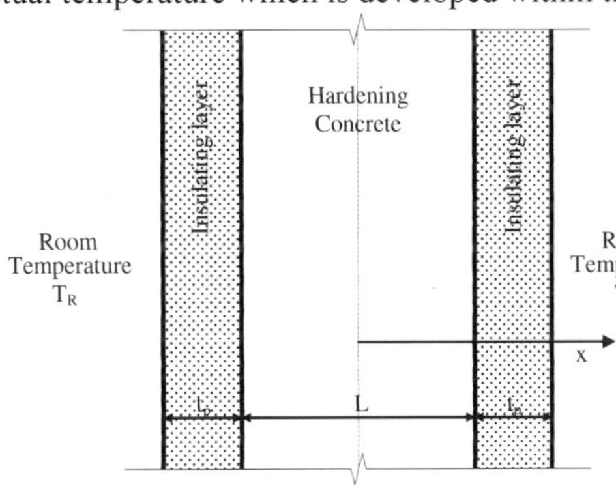

Figure 1: Schematic representation of the 1D problem (adapted from [2]).

On the other hand, the following equation (w) describes the cement heat production (labelled as $\mathbf{Q_a(t)}$) for the case in which the concrete sample is cured in adiabatic boundary conditions: [13]:

$$Q_a(t) = Q^*_{a,max} \cdot e^{-\left(\frac{\tau}{t}\right)^{\beta}} \ . \tag{4}$$

where:

- the term $\mathbf{Q^*_{a,max}}$ (which is minor than the Q_{max}) represents the actual amount of heat produced at the end of the reaction (in adiabatic conditions);
- $\boldsymbol{\tau}$ and $\boldsymbol{\beta}$ are two numerical parameters controlling the shape of the adiabatic curve.

Therefore, the rate of the produced heat $\mathbf{q_a(t)}$ is defined by introducing the expression proposed by equation (**4**) within the rate definition defined by equation (**3**). However, the heat concrete production ($\mathbf{q_c(x,t)}$) is not generally equal to the $\mathbf{q_a(t)}$ occurring in adiabatic condition. This is due to the fact the different temperatures develop inside the concrete sample cured in different boundary conditions (i.e., adiabatic and non-adiabatic). Despite this, based on the well-know Arrhenius Principle [13], it is possible to correlate the $\mathbf{q_c(x,t)}$ and $\mathbf{q_a(t)}$ where the former can be expressed in terms of the latter and substituted in eq. (**2**). As a consequence, the numerical problem turns into an integral-differential equation [11]. To solve it, initial and boundary conditions should be defined:

$$T(x, t=0) = T_R \tag{5}$$

in which the term "$\mathbf{T_R}$" represents the initial room temperature.
On the other hand, the boundary condition are defined by the characteristics of the border between the cured concrete mixtures and the surrounding external condition. For instance, when semi- (or pseudo-) adiabatic conditions occur, the following relationship can be defined in order to guarantee the continuity of heat flow throughout the insulating layer and the external environment:

$$\lambda_p \cdot \frac{T_{left}(t) - T_R}{t_p} = \lambda_c \cdot \left.\frac{\partial T}{\partial x}\right|_{x=-L/2}, \quad \lambda_p \cdot \frac{T_{right}(t) - T_R}{t_p} = -\lambda_c \cdot \left.\frac{\partial T}{\partial x}\right|_{x=L/2}. \tag{6}$$

in which:

- **L** represents the specimen's length in the **x** direction (heat flow direction);
- $\mathbf{t_p}$ is the thickness of the insulating layer;
- $\boldsymbol{\lambda_p}$ represents the thermal conductivity of the insulating layer.

It is worth to mention that, in the case in which the concrete mixture is cured under isothermal conditions and at a constant (in time) room temperature ($\mathbf{T_R}$) the following boundary conditions can be defined:

$$T(x=-L/2, t) = T_R, \quad T(x=L/2, t) = T_R. \tag{7}$$

2.2. Indirect identification of hydration reaction processes

The proposed modelling approach presented in the in the previous section, leading to the resolution of the integral-partial-differential problem (defined by equation (**2**)) once the boundary conditions are defined (see equation (**6**) and/or (**7**)) can be easily employed for simulating the time evolution of the temperature (**T**) occurring within the curing concrete sample. This simulation can be symbolically denoted by the following equation (**8**):

$$T_{th} = T_{th}(x, t; \mathbf{q}_r, \mathbf{q}_f), \tag{8}$$

where the vectors $\mathbf{q}_r$ and $\mathbf{q}_f$ include the two following sets of parameters:

$$\mathbf{q}_r = \left[T_R, Q_{max}, \alpha_{h,max}\right], \quad \mathbf{q}_f = \left[\tau, \beta, \lambda_p\right]. \tag{9}$$

where the components of the vector $\mathbf{q}_r$ can be directly measurements or defined by analytical expressions already within the literature [14]. On the other hand, the components collected in $\mathbf{q}_f$ are determined for each concrete sample by solving the following unrestrained optimisation problem:

$$\Delta(\mathbf{q}_r, \mathbf{q}_f) = \sum_{k=1}^{n} \left[T_{exp}^{(k)} - T_{th}(x=0, t_k; \mathbf{q}_r, \mathbf{q}_f)\right]^2. \tag{10}$$

$$\bar{\mathbf{q}}_f = \underset{\mathbf{q}_r}{\operatorname{argmin}}\left[\Delta(\mathbf{q}_r,\mathbf{q}_f)\right]. \tag{11}$$

The values obtained for each concrete mixtures (for both $\mathbf{q}_r$ and $\mathbf{q}_f$ vectors) are, then, use for simulating the actual hydration processed occurring within the concrete samples.

3. Model application

3.1. Definition of the case study

The formulated numerical procedure presented in the previous section, and recently validated in previous studies [11][12], remarks that the hydration processes as well as the time-evolution of the concrete's mechanical performances, are governed by the composition of the cement-based mixture (identified throught the parameters Q_{max}, $\alpha_{h,max}$, τ and β presented in section 2), the concrete element geometry (i.e., the length L) and the boundary condition both in terms of insulation characteristics (i.e., t_P and λ_P) and room/ambient temperature (i.e., T_R).

Once the key parameters are identified, it is possible to proposes some parametric analyses on representative mixtures considering the variation of the above-mentioned properties. In this context, the present section proposes the analysis of different concrete mixtures characterized by three different strength classes (low-strength C25/30, medium-strength C35/45 and high-strength C50/60 as defined by the Eurocode 2 [15]). The concrete mixtures are designed with the same the water-to-cement ratio, equal to 0.40, maximum size of the aggregates (equal to 16 mm) and the amount of the free water (equal to 215 l/m^3) which warranty the same slump class (value ranging from 100 mm and 150 mm) for all the analysed mixtures. On the other hand, in ordert to reach different strength class, three different types of Portland Cement (32.5, 42.5R and 52.5R in accordance with the EN 197-1: 2011 [16]) are considered for achieving the desired characteristic strength at 28 days (i.e., R_{ck}) [15]. The main characteristics of the mixtures under consideration are summarised in **Table 1.**

Table 1. Properties and characteristics of the concrete mixtures considered in this study.

Description		C25/30	C35/45	C50/60
	Portland Cement	32.5	42.5R	52.5R
	Mean cylindrical compressive strength - $f_{cm,28}$ *[MPa]*	34.56	45.35	57.80
	Characteristic cubic compressive strength - $R_{ck,28}$ *[MPa]*	32.00	45.00	60.00
	Q^*_{max} *[kJ/kg]*	295	380	400
Parameters defining the adiabatic temperature curve shape	τ *[h]*	15.0	12.0	10.0
	β	0.80	0.80	0.80

Based on the numerical procedure described in section 2, it is possible to evaluate the adiabatic curve for each type of employed cements, through the calibration of the parameters Q^*_{max}, τ and β which have been introduced defined in equation (**4**). It is worth to mention that these calibrations are based on the best fits of the values proposed by the Italian guidelines on structural concrete production [16]: also the results of these analysis are summarised in **Table 1**.

3.2. Temperature profiles and cracking risk assessment

As already mentioned, once the concrete mixture composition has been defined, it is possible to simulate the time evolution and the space development of the temperature within the curing concrete element through the numerical procedure summarised in section 2.. For instance, the following **Figure 2** proposes the resulting temperature curve simulation which is obtained in the central point of a 1 m concrete element composed of high-strength concrete mixtures (i.e., C50/60 in **Table 1**) which has been cured either in adiabatic and isothermal (i.e., the concrete surface is not covered by any isolating layer and the ambient temperature presents a constant value equal to 20 °C) conditions. The proposed figure highlights that, although the ambient temperature $\mathbf{T_R}$ is kept constant, the distribution of the temperature in the centre of the concrete element behaves in a "quasi-adiabatic" condition: this phenomenon is associated to the element's geometry as well as to the thermal properties of the employed cementitious mixtures. Moreover, since the temperature developments is directly associated to the time-evolution of the DoH, the latter is governed by the curing boundary conditions. This evidence is also confirmed by the curves reported in **Figure 2** (e.g., see the secondary vertical axes).

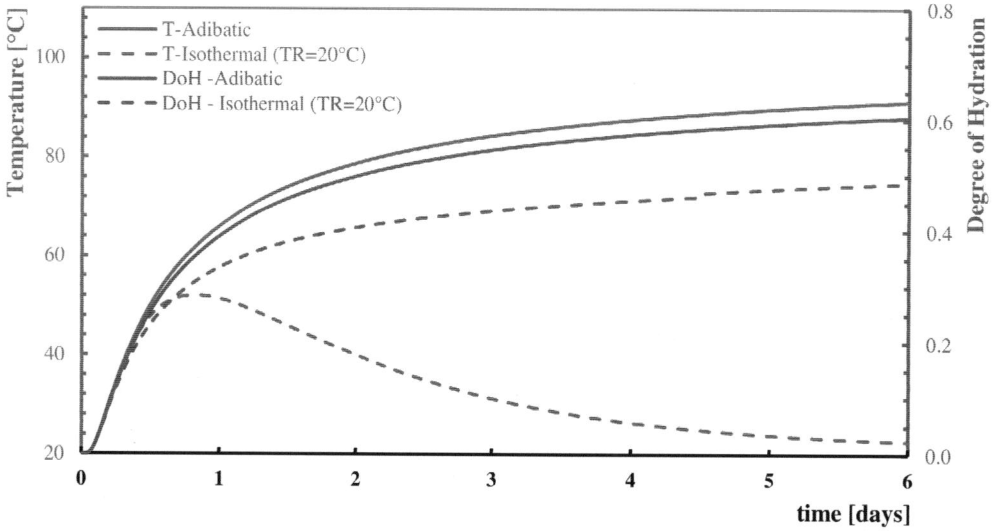

Figure 2. Temperature and DoH development: representative numerical simulation.

The simulation proposed in **Figure 2** allow to trace the temperature profiles at the early-age for concrete elements during their hydration process, as also demonstrated in **Figure 3** in which this simulation is presented for 1 m and 2 m thickness element by also considering the possible variation of the concrete strength class: low strength C25/30, C35/45 and C50/60) which were produced with different types of cement and were cured in isothermal condition (T_R=20°C). More specifically, the coloured maps proposed in **Figure 3** show the time and space (thickness) temperature variation up to 6 curing days.

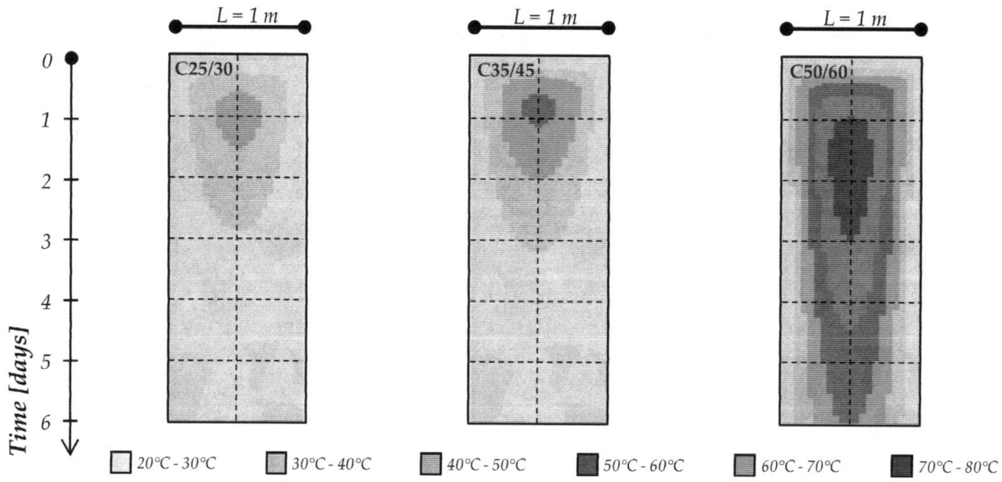

(a)

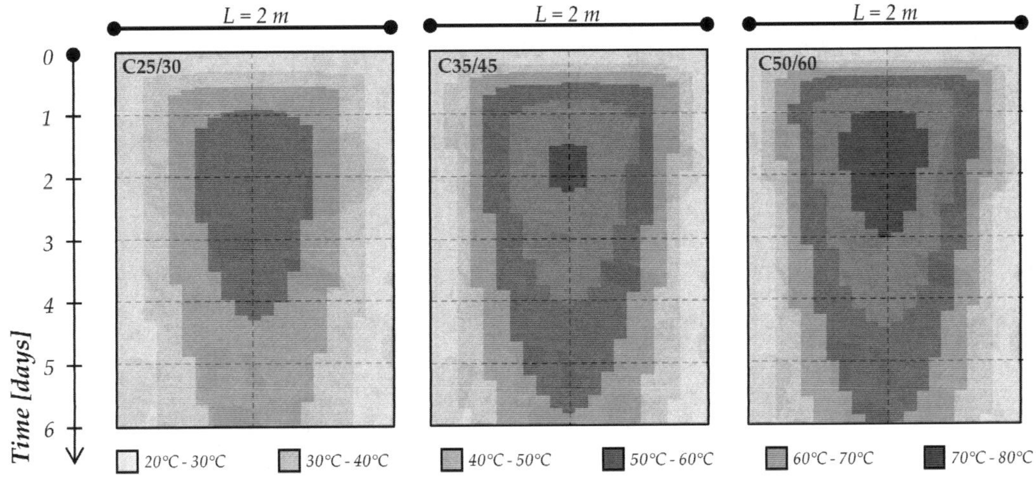

(b)

Figure 3. Temperatures development in (a) 1 m and (b) 2 m early-age concrete (adapted from [18]).

These kind of graphs (see **Figure 3**) give the possibility to perform an indirect verification of cracking risk phenomena which may occur in cementitious mixtures. In fact, in accordance with the Italian guidelines released by the Superior Council of Public Works [16], it is possible to prevent the cracks formation due to the thermal stresses, by avoiding that the maximum concrete temperature reaches above 70 °C and, moreover, the various part the concrete element present a gradient temperature below 20 °C. In fact, since the proposed procedure allow to predict the evolution of the temperature in curing concrete, it can be also applied for unveiling the key design parameters (e.g., element geometry and boundary condition) for mitigating the cracking risk occurring in structural concrete: for instance, by the possible application of thermal insulating moulds and/or the use of hot/cold water when the external ambient temperature is outside of the recommended range.

Finally, it is worth to highlight that, as also demonstrated in previous studies [12][18], the key output of the proposed model (i.e., degree of hydration in time) can be correlated with relevant mechanical properties of concrete (i.e., compressive and tensile strength and elastic modulus), by considering the various scenario in which the concrete elements is cured.

4. Conclusion

This study presented a theoretical modelling approach aimed at simulating the heat and the hydration processed occurring in hardening concrete and, the following main conclusions can be drawn out:

- the model furnishes a practical simulation of the time-evolution of either temperature and DoH in concrete elements cured in adiabatic, pseudo-adiabatic and isothermal boundary conditions;
- the 1D flow hypothesis turns out the model to be a useful tool for analysing the data obtained from early age concrete structures and, moreover, it can be easily used for the assessment of the cracking phenomena due to thermal stresses occurring in hardening concrete;
- one of the next step of the present research is the possible correlation of the key output model (i.e., DoH development) with the corresponding concrete mechanical properties. This, will allow to propose a more advanced simulation tool for the verification of the thermal cracking risk;
- consequently, the system proposed herein can be also used as a practical tool for construction site monitoring of temperature rise and forecast removal of formworks.

References

[1] Dolado JS, Van Breugel K (2011) Recent advances in modeling for cementitious materials, *Cement and concrete research*, 41(7): 711-726.

[2] Scrivener KL, Nonat A (2011) Hydration of cementitious materials, present and future, *Cement and concrete research*, 41(7): 651-665.

[3] Schindler AK, Folliard KJ (2005) Heat of hydration models for cementitious materials' *Materials Journal*, 102(1): 24-33.

[4] Ye G, Van Breugel K, Fraaij ALA (2003) Three-dimensional microstructure analysis of numerically simulated cementitious materials, *Cement and Concrete Research*, 33(2): 215-222.

[5] Jiang W, De Schutter G, Yuan Y (2014) Degree of hydration based prediction of early age basic creep and creep recovery of blended concrete, *Cement and Concrete Composites*, 48: 83-90.

[6] Badogiannis E, Makrinou E, Founta M. (2021). Durability of Normal and Lightweight Aggregate Mortars with Different Supplementary Cementitious Materials. *WSEAS Transactions on Environment and Development*, *17*, 271-281.

[7] Fairbairn EM, Azenha M (Eds) (2018). Thermal Cracking of Massive Concrete Structures: State of the Art Report of the RILEM Technical Committee 254-CMS. Springer.

[8] Betioli I, Sà SN, Fiorini AS (1997). Criteria for control and survey of the cracking of the butress dam of Itaipu. XXII Simpòsio Nacional de Grandes Barragens, CDBD (Brazilian Committee on Dams) Sao Paulo.

[9] Rosso JA, Piasentin C, (1997). Studies on the cracking of the buttress dam of the buttress dam of Itaipu. XXII Simpòsio Nacional de Grandes Barragens, CDBD (Brazilian Committee on Dams) Sao Paulo.

[10] Zhang Y, Collepardi M, Coppola L, Guan W, Zaffaroni P (2003). La Diga delle Tre Gole in Cina: la più grande del mondo. Enco Journal, Quadrimestrale, Anno VIII, Numero 22.

[11] Martinelli E, Koenders EAB, Caggiano A (2013) A numerical recipe for modelling hydration and heat flow in hardening concrete, *Cement and Concrete Composites*, 40: 48-58.

[12] Koenders EA, Pepe M, Martinelli E (2014) Compressive strength and hydration processes of concrete with recycled aggregates, *Cement and Concrete Research*, 56: 203-212.

[13] van Breugel K (1991) Simulation of hydration and formation of structure in hardening cement-based materials, Ph.D. Thesis, Delft University of Technology, Delft (NL), 171 pages.

[14] Hansen TC (1986) Physical structure of hardened cement paste. A classical approach, *Materials and Structures*, 19(6): 423-436.

[15] EN 1992-1-1 (2004) Eurocode 2: Design of concrete structures - Part 1-1: General rules and rules for buildings.

[16] EN 197-1 (2011). Cement - Part 1: Composition, specifications and conformity criteria for common cements. *European Standards Specifications*.

[17] Italian Ministry of Infrastructure and Transport (2017) Linee Guida per la messa in opera del calcestruzzo strutturale (in *Italian*). *Superior Council of Public Works*.

[18] Pepe M Lima C Martinelli E (2020). Early-age properties of concrete based on numerical hydration modelling: A parametric analysis. Materials, 13(9): 2112.

Application of the Harmonica Index for noise assessment in different spatial contexts

G Graziuso[1], A B Francavilla[1], S Mancini[2] and C Guarnaccia[1,a]

[1] Department of Civil Engineering, University of Salerno, via Giovanni Paolo II 132, 84084, Fisciano, SA, Italy
[2] Department of Information and Electric Engineering and Applied Mathematics, University of Salerno, via Giovanni Paolo II 132, 84084, Fisciano, SA, Italy

[a] E-mail: cguarnaccia@unisa.it

Abstract. Currently, noise pollution deriving from traffic, industry and leisure activities is one of the main environmental problems that affect people quality of life and health. However, it is very difficult to quantify noise effects, because they depend on people noise level tolerance, environmental noise sources, methods of definition of noise exposure and indicators. The acoustic descriptors currently used, indeed, are hardly understandable for those who do not deal with the topic, thus limiting the comprehension of the damage caused by noise. Consequently, it is essential to provide information that is easier to understand and closer to the noise pollution perceived by people. Innovative acoustic indices, like the Harmonica Index, have been defined in recent years, to overcome the aforementioned limits. Their application to different spatial contexts, with the identification of the background and the event-based components of the environmental noise, can contribute to understand the influence of human activities on the quality of an acoustic environment. In this paper, starting from data acquired both in crowdsourcing and with traditional instruments, the application of the Harmonica Index to urban, peri-urban and rural areas in south Italy will show their acoustic characteristics that could be understood also by a non-technical public.

1. Introduction

In recent decades, the demographic expansion and the increase of human activities related to traffic, industry and leisure have strongly influenced the quality of life in cities. The road traffic, in particular, is one of the main sources of acoustic pollution in urban areas. Its negative impacts on the environment and human health depend also on people noise level tolerance and systems of definition of noise exposure. In order to anticipate and adopt methods of mitigation and modelling of noise, it is necessary to study and deepen the noise impact on urban environments, because these will represent the sound environments of the future [1]. Specifically, the mitigation actions are the basis of an effective environmental policy which promote an overall environmental awareness and the use of the best practice of developed countries [2].

Generally, noise regulations are aimed at avoiding significant nuisances, disturbances and health effects. In 2002, the European Noise Directive (END) defined that all member states had to produce, every five years, noise maps and noise management action plans for urban agglomerations and major roads, railways and airports [3]. These maps and plans can be considered instruments for the management and reduction of environmental noise to protect human health and preserve the natural soundscapes [4-6].

In spite of the actions defined by local authorities with the purpose of applying the END that led to improvements in the effectiveness of policies for the management and reduction of environmental noise, it is necessary, also, to define appropriate noise indicators that can be understood by the public and can contribute to increase a general awareness in the noise pollution. The standard noise indicators, generally expressed in decibel, are hardly understandable by people, which are not able to comprehend the damage caused by noise in its entirety [7, 8]. Therefore, it is essential to define innovative indicators able to provide information that can be easier to understand and closer to the noise pollution perceived by people. As a consequence, a bottom-up approach could emerge that would lead to the definition of further strategies that could integrate the existing official ones [9, 10].

In this paper, after a preliminary brief recognition on the main noise indicators used currently, an innovative index, the Harmonica Index, will be presented and applied to different spatial contexts, using both crowdsourcing data and data acquired with sound level meters, in order to feature the different acoustic environments in different time periods.

2. Materials and methods

2.1. Main noise indicators

The noise indicators have a fundamental role for noise management because they can be considered tools for supporting authorities on the adoption of the suitable noise abatement measures.

Traditional noise indicators are defined according to different calculations for the evaluation of the environmental sound level, in relation to the input data, the weather conditions and the period over which they are calculated [11-13]. Most of the EU member states use the same noise descriptors for road, rail and industrial noise, i.e. the A-weighted long-term average sound level determined over all the day, evening and night periods of a year (L_{den}) and the night-time noise indicator (L_{night}) determined over all the night periods of a year.

Since the main aspects for the analysis of the environmental noise are energy, time and spectrum, other classifications of the noise indicators can be find in the literature. In particular, in [7, 14] the noise indicators are divided into energetic, statistical, noise variations, spectrum and emergences descriptors.

The energetic descriptors, such as the equivalent continuous level L_{AeqT} referred to a T time interval, give information about the total sound level and are, generally, used by legislations, which establish their limit values. The statistical (e.g., the percentile levels, L_{An}) and the noise variation (e.g., the standard deviation, $\square_L$) indicators of the sound levels detected in the T time interval, describe the variability of the levels (eg, L_{A10}-L_{A90}). Moreover, a statistical indicator can represent a level of the distribution of higher energy sound events (e.g., L_{A1}) or background noise (e.g., L_{A95}). The indicators representative of the spectrum of the sound environment are, for example, the sound ecology indicators, for discriminating presence of biophonic sounds and anthropogenic sounds, the Spectral Gravity Center (SGC) of the 1/3 octave band spectrum, that is very sensitive to events, the percentile levels for frequency bands and the difference L_{Ceq}-L_{Aeq}, which describes the amount of low frequencies. Furthermore, Asensio *et al.* [15] outlined a set of descriptors for emergences which better enables the application of more novel approaches to the evaluation of the effect of the new soundscape due to the measures against COVID-19 on people's subjective perception. They are based on the number of noise events and the percentage of the time, that exceed a given threshold.

In spite of the large number, no indicator can describe the complexity of sounds in different territorial contexts in its entirety. For example, the L_{Aeq}, when calculated on short periods, is highly influenced by noise peaks. Moreover, each statistical descriptor only corresponds to one point of the $L_{Aeq,1s}$ distribution and, therefore, one descriptor value can also correspond to very different sound environments [7]. Consequently, the various indicators offer different points of view of the sound levels variations and can be considered complementary to each other.

Moreover, the standard acoustic indicators are generally expressed in decibel unit, which has a logarithmic nature and is complicated to explain to people and to associate to their perception of noise [16]. As a consequence, innovative acoustic indices must be defined considering further criteria.

2.2. Innovative indices: the Harmonica project

In order to improve the noise phenomenon comprehension with the use of an appropriate descriptor, it is necessary to build an innovative index with reference to a suitable method of calculation and representation [17], which must consider the spatial and temporal specificities and highlight the background noises and noise peaks. Furthermore, the innovative index must be based on a participatory process, focusing also on the sound perception that people have of an environment. Based on these criteria, a new index, integrated with the traditional descriptors that are mandatory by legislation, could provide additional information and evaluate the quality of the acoustic environment [18].

Ribeiro *et al.* [17] described the development of four indices (P1, P2, CY, and CC), based on different approaches, which integrated both the continuous and the sporadic nature of noise. These were the precursors of an innovative Common Noise Index, based on a nuisance scale characterized by a score from 0 to 10, that were developed in the framework of the Harmonica project, supported by the European Commission (LIFE program). The HARMOnised Noise Information for Citizens and Authorities (HARMONICA) project had the main goal of creating a normalized noise index for the general public and the authorities to improve the efficiency of the EU policies on environmental noise management.

The construction of the innovative index, finally named Harmonica Index (HI), was organized in 3 phases, as shown in figure 1 [18]. First, the selection of the index parameters resulted from the principal components analysis (PCA) on 60 hourly acoustic descriptors, calculated using $L_{Aeq,1s}$ as the elementary data. All the data were acquired with a measurement campaign organized in 24 sites representative of the eight main types of noise exposure (for instance land transport noise, air traffic noise, and quiet areas) in different spatial contexts (urban, suburban, and rural). Secondly, in order to evaluate the comprehensibility and acceptability of the index, the opinion and perceptions of the public were considered by conducting both face to face and laboratorial interviews, involving different stakeholders, like residents, general public, associations and local authorities. Finally, the HI formula was developed (Figure 1), considering two components: the background noise, BGN, and the EVT, which considers the peaks linked to specific noise events.

The HI is easy to calculate since it uses measurement data usually collected by noise measurement devices. Its complete structure represents, indeed, a combination of the equivalent level, L_{Aeq}, and the L_{A95eq}, which is the level that is exceeded 95 percent of the measurement time. Specifically, L_{A95eq} is the value in [dBA] integrated over 60 minutes of the series of 3600 values of L_{A95} calculated from the single $L_{Aeq,1s}$ contained in a 10-minute time interval, shifted progressively every second [7, 19, 20].

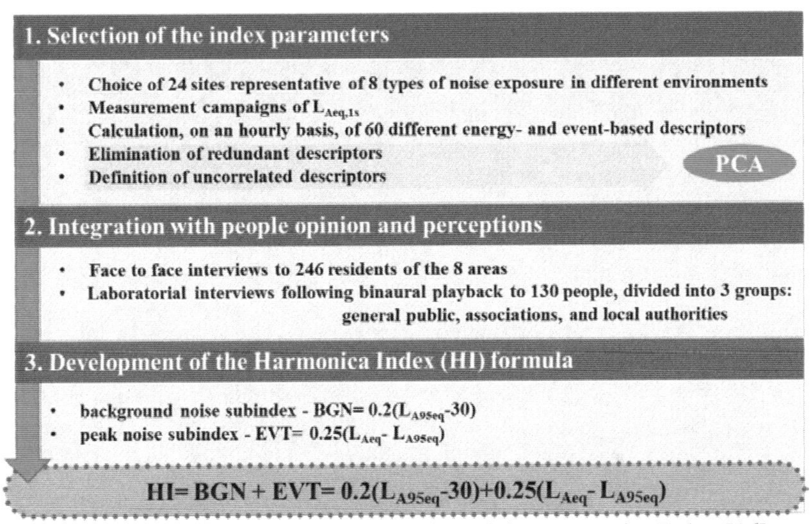

Figure 1. Methodology of construction of the Harmonica Index [16]

With the aim to understand the nature of sound, the HI is represented graphically [21] by only two basic geometric shapes, i.e. the rectangle for background noise and the triangle for the peaks of events (figure 2(a)). This representation is very effective and allows people to distinguish the two components in different spatial contexts. For example, during daytime, in an industrial area, a high value of the BGN sub-index could generally denote the presence of working technological facilities, while, in an urban centre, the increase of background noise could be mainly generated by vehicular traffic. However, the representation of the EVT could highlight specific situations which differ from the standard conditions of a place. Furthermore, the two shapes can be painted with a green, orange or red colour consistent with the value of the index. This allows the evaluation of the quality of the acoustic environment according to the critical values recognized by the World Health Organization (WHO), which varies depending on day and night time periods (figure 2(b)), because of a higher sensitivity to noise during the night.

In order to evaluate the relationship between the Harmonica Indices (BNG, EVT and HI) and the END noise indicators (L_d, L_e, L_n and L_{den}), Prascevic *et al.* [22] carried out a correlation analysis for their different combinations. They obtained high values of the Pearson correlation coefficient, thus highlighting the high strength of the association of noise indicators and Harmonica indices and the adequacy of Harmonica index for the environmental noise assessment.

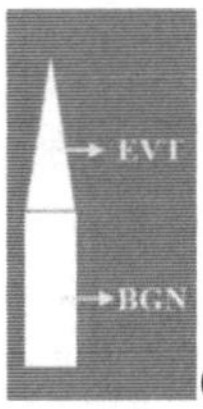

(a)

Acoustic environment	day (from 6 am to 10 pm)	night (from 10 pm to 6 am)
Very noisy	$I_H \geq 8$	$I_H \geq 7$
Noisy	$4 \leq I_H < 8$	$3 \leq I_H < 7$
Quiet	$I_H < 4$	$I_H < 3$

(b)

Figure 2. HI: (a) graphical representation and (b) acoustic environment classification [21]

2.3. Methodology

Since the aim of this study is to define the characteristics of an acoustic environment in different time periods, the HI can be applied to different spatial contexts, considering data acquired with both traditional instruments, like the sound level meters, and innovative techniques related to the crowdsourcing [1]. These applications can contribute to get information about the noise sources features of a specific spatial context, to increase people awareness in noise pollution and to define more appropriate actions by local governments.

3. Applications of the Harmonica Index to different spatial contexts

In this section, three applications of the HI in different spatial contexts are presented. The case studies (figure 3) are located in the municipalities of Palomonte, Capaccio Paestum and Nocera Inferiore in the Province of Salerno (South Italy). The three sites of investigation were defined with the aim to analyse both rural and urban contexts. Specifically, as shown in figure 4, with regards to the rural area of Palomonte, the point of measurement is about 100 m far from a local road; Capaccio Paestum point of measurement is located in one of the city centers of the municipality close to a provincial road (SP277) and the point of measurement of Nocera Inferiore is located in an agricultural residential mixed area.

The measurement campaign was organized in five days, from the end of November to the beginning of December 2020, as reported by table 1. Even though the measurements were influenced by the restrictions due to the COVID 19 pandemic and, generally, the environments are less noisy, the aim of the study was, also, to determine the habits of the residents with the calculation of the HI. Since the HI is calculated on hourly base, each campaign was organized to acquire each data per single hour. The data acquisition was based on both a traditional sound level meter, namely Fusion produced by 01dB (class 1), for the case study of Nocera Inferiore, and an innovative technique, that involve people to collect data with their mobile devices for the other two case study. The traditional measurement technique is generally more accurate than the second; anyway, the positive outputs of crowdsourced data have been already discussed in [1, 23-28].

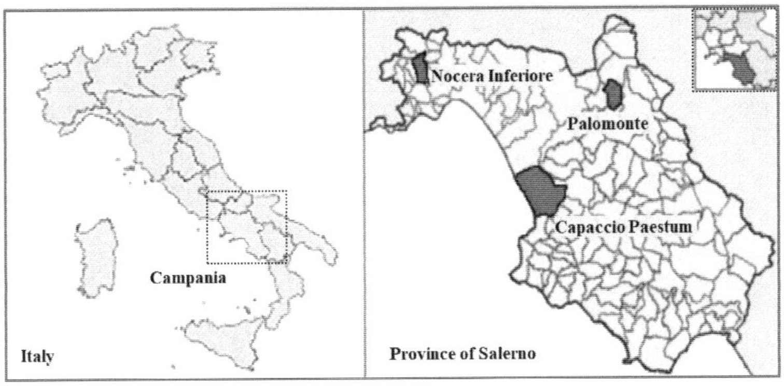

Figure 3. Territorial framework.

Table 1. Measurement campaign details

Site	Dates	Time period	Data acquisition
Palomonte	27 November 2020 29 November 2020	from 2 pm to 3 pm from 2 pm to 3 pm	crowdsourcing
Capaccio Paestum	01 December 2020 02 December 2020	from 6 am to 24 pm from 0 am to 5 am	crowdsourcing
Nocera Inferiore	28 November 2020	from 3.50 pm to 6.50 pm	sound level meter

In this study, the crowdsourcing data were acquired with the NoiseCapture application, that is a component of the Noise-Planet project [25]. This project aims to provide an integrated platform of tools, based on scientific research, dedicated to the acquisition, modelling and evaluation of environmental noise measurements. It is led by two French research teams: the Environmental Acoustics Laboratory (Eiffel University, former IFSTTAR) for environmental noise research and the DECIDE Team (Lab-STICC - CNRS UMR) for GIScience. In this project, the data is collected from the free and open-source Android NoiseCapture application and shared from the OnoM@p Spatial Data Infrastructure (SDI) [26, 27]. The Noise-Planet project is integrated also with a free GIS-based model to compute noise maps and an interactive maps viewer, to display noise data collected by the community. The NoiseCapture app allows computing each second the equivalent A-weighted sound levels and then sharing data with the community [28]. From the NoisePlanet platform, it is possible to download all the acquired data and to export them to the QGIS environment and determine the L_{Aeq} and L_{A95}, needed to calculate the HI.

3.1. Results and discussion

The application performed in the rural area of Palomonte, based on crowdsourcing data collection, had the main objective to highlight the differences of the acoustic environment in two days, a weekday and a public holiday, and to evaluate the influence of human activities on the sound levels of the study area. The data were collected using an Android smartphone and the NoiseCapture application.

Figure 4 shows the numerical values of the traditional indicators, L_{Aeq} and L_{A95} and their graphical trend over the observational period, i.e. 1 hour, with a time basis of 1 second, the numerical value of other percentile levels (10, 90 and 50) and the minimum and maximum value (L_{min} and L_{max}), and the numerical Harmonica index, with its graphic representation. In particular, results refer to measurements made starting from 2 to 3 pm on Friday 27th (figure 4 (a)) and on Sunday 29th (figure 4 (b)). The two graphs show that there is a higher level of both background noise and noise peaks on

Friday. The noises detected are mainly due to the presence of pets and chicken coops in the vicinity: the noise peaks, indeed, differ little in the two days. The difference in the peak is due to the transit of heavy vehicles and to a greater number of vehicular traffic due to a series of work activities in the weekday. Furthermore, the decrease of background noise on Sunday is due to a reduction in circulating vehicles and almost no mobility in the surroundings.

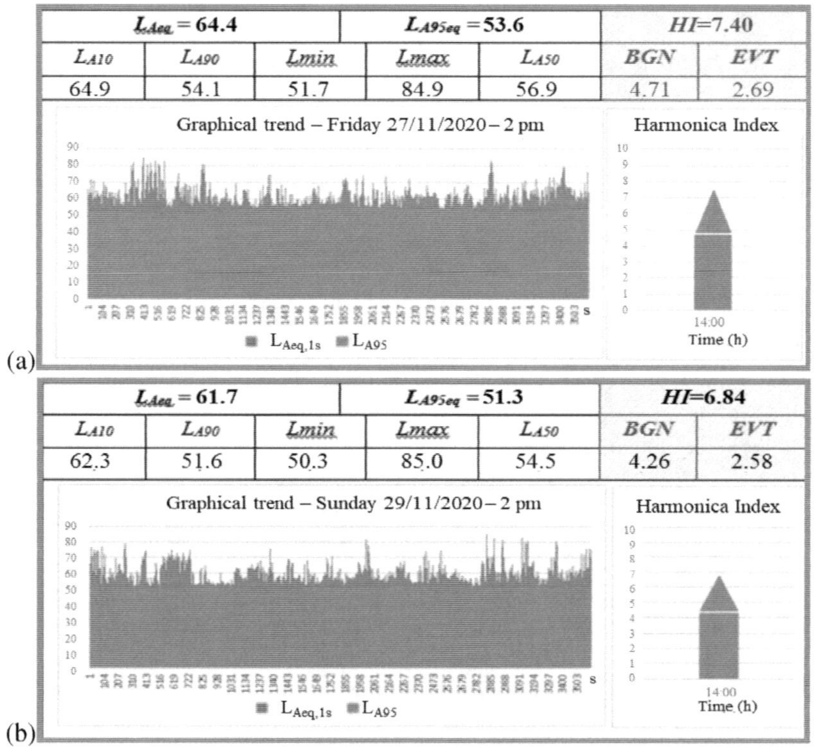

Figure 4. Rural area of Palomonte: noise indicators and HI at 2 pm on (a) Friday 27/11/2020 and (b) Sunday 29/11/2020.

In the municipality of Capaccio Paestum, 24-hour measurements were carried out, so that the representation of the HI can be extended to the entire day with the aim to have a general trend of the noise changes in the different hours of the day. Also in this case study, the data were acquired with the use of an Android smartphone. The measurements were grouped into a single 24-hour chart in which it is possible to observe how the index varies during the day and night time periods, from 6a.m. to 9p.m. and from 10p.m. to 5a.m., respectively (figure 5). Two colors, green and orange, were used according to the acoustic environment classification, reflecting the quality objectives of the WHO and the values recognized as critical for noise. In the early morning hours, a very quiet environment can be observed, with relatively low background noise and events; instead, from 11 am to 12 am a slight increase in noise is perceived, due to the increase of vehicular traffic. Then, there are no significant variations in values until the evening, where a slight increase is perceived mainly due to anthropogenic activities. During the night there is a clear increase in the sub-index EVT compared with the component BGN. This is mainly due to a reduction of the background noise in the night period and, therefore, the noise peaks due to single events will be more evident. In particular, Figure 6 (a) shows that at 1 am the BGN value is practically zero, as 30 dBA is the minimum threshold beyond which the noise can begin to be perceived as annoying in the external environment. In the time range from 1 to 2 am, as the L_{A95eq} index is less than 30 dBA, the value of the BGN sub-index is assumed to be zero. Then, in the time range from 2 to 3 am, there is a sudden increase in both background noise and peaks due to strong gusts of wind and a rainy event that began at 2:30 am and ended at 3 am, visible in the time history in figure 6 (b). Therefore, the HI calculated at 3 and 4 am reproduces an environment with a good acoustic quality.

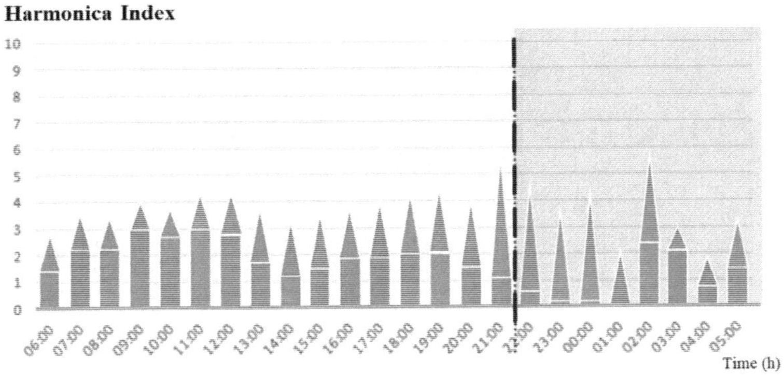

Figure 5. HI over 24 hours for the urban area of Capaccio Paestum

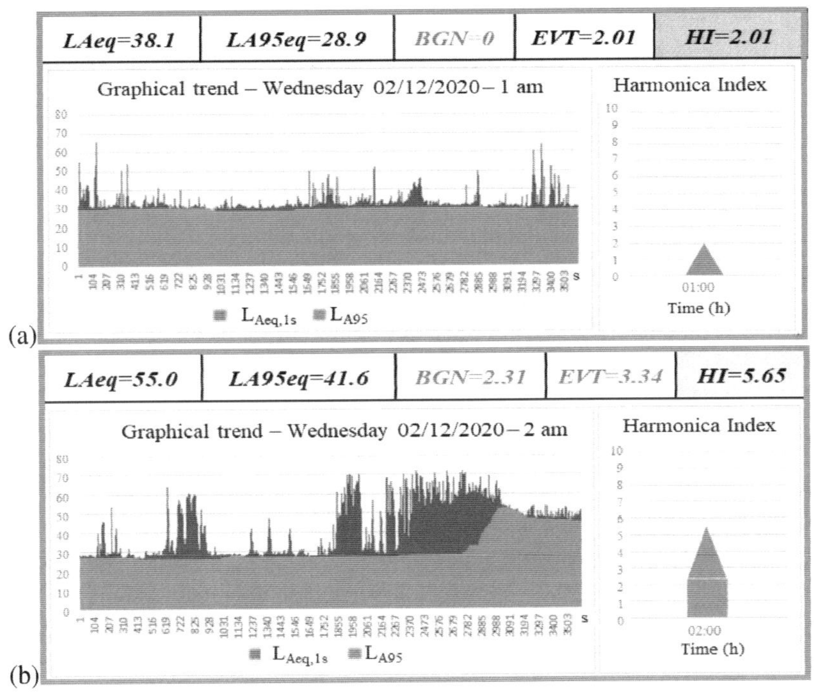

Figure 6. Capaccio Paestum: HI at 1 am (a) and 2 am (b) on Wednesday 02/12/2020

In third case study, corresponding to a mixed area (rural and residential) in Nocera Inferiore, the HI was evaluated starting from data acquired with a class 1 Fusion sound level meter. The results obtained did not highlight any particularly significant events and were coherent with the surrounding environment. Figure 7 shows the obtained results related to two different hours of measurement in the afternoon of the same day. It is possible to observe the same EVT at 3:50 and 5:50 pm, while an increase of the BGN at 5:50 pm indicates an increase of road traffic. If considering the period in which the measurements were made, this context is characterized by an overall good acoustic quality.

(a)

L_{Aeq} = 47.6		L_{A95eq} = 39.1			*HI* = 3.94	
L_{A10}	L_{A90}	*Lmin*	*Lmax*	L_{A50}	*BGN*	*EVT*
51.2	39.9	35.9	60.9	45.4	1.80	2.14

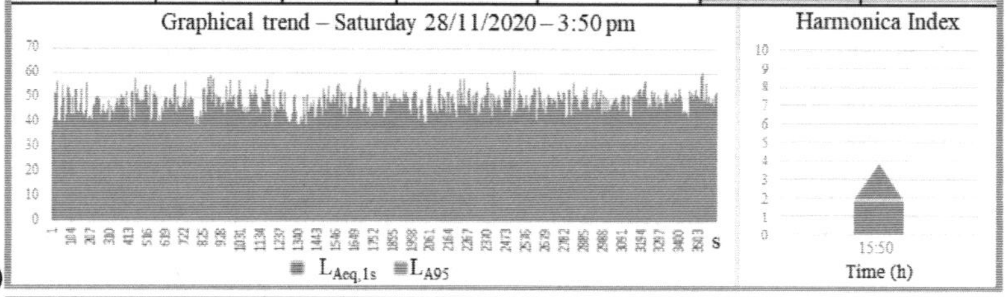

(b)

L_{Aeq} = 48.6		L_{A95eq} = 40.0			*HI* = 4.14	
L_{A10}	L_{A90}	*Lmin*	*Lmax*	L_{A50}	*BGN*	*EVT*
52.1	41.0	36.5	62.2	46.3	2.00	2.14

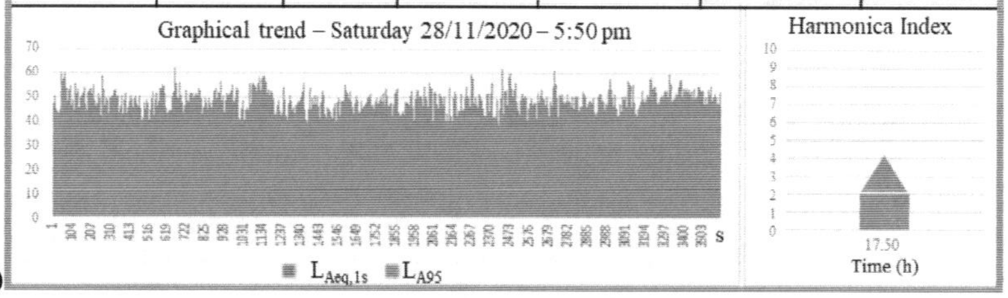

Figure 7. Mixed area of Nocera Inferiore: noise indicators and HI at 3:50 pm (a) and 5:50 pm (b) on Saturday 28/11/2020

4. Conclusions

In this paper, the Harmonica Index, developed in the framework of the Harmonica Life project, was applied to three territorial contexts, i.e. an urban, a peri-urban and a rural area, starting from data acquired both with tools that allow the acquisition of data in crowdsourcing, through applications for mobile devices, such as the NoiseCapture application, and with traditional instruments, like a class 1 sound level meter.

In the two applications based on crowdsourced data acquisition, it was possible to highlight the variation of the two components of the index in the different spatial and temporal contexts. It was also observed that, due to its structure, the HI is very sensible to noisy events over low background noise conditions. This was confirmed during night period measurements in one of the case studies, when a sudden wind and rain storm led to a strong increase of the index. This aspect of the index must certainly be studied in depth. In the third application, a class 1 Fusion sound level meter was used for correctly defining the acoustic quality of the environment, underlining, once again, the versatility of the index and the simplicity of its comprehension. Since the values of the HI were influenced by the COVID 19 restrictions, it would be interesting a comparison with the potential results under ordinary conditions in the same locations.

The choice of the Harmonica index was considered useful to give a simple reading of the acoustic characteristics of the different areas, even to a non-technical public, highlighting the influence of anthropogenic or animal activities on the natural acoustic environment of an area. Despite the positive implications of applying the index, however, two considerations can be made. Since the HI is based on an hourly integration of the two components, the impact produced by the noise in a time less than an hour could be quite inaccurate. Moreover, while the thresholds of the traditional descriptors are often connected to the subdivision of the territory into zones in relation to their destinations, the three classes of quality of the acoustic environments defined by the HI does not seem to consider this aspect yet.

Further future developments could involve the integration of the study carried out with soundscape measurements and the evaluation of psychoacoustic parameters with questionnaires administered to the visitors of the area during the measurement campaigns, in order to find the human perception and the sense of pleasantness of the area. This type of method would complete the study, giving a deepening and a broader and more precise vision of the perception of people with respect to the variation of the acoustic environment.

References

[1] Graziuso G, Mancini S, Francavilla AB, Grimaldi M and Guarnaccia C 2021 Geo-Crowdsourced Sound Level Data in Support of the Community Facilities Planning. A Methodological Proposal *Sustainability* **13**(10) 5486
[2] Bayraml G 2020 The Environmental Problems of Azerbaijan and the Search for Solutions. *WSEAS Trans Environ Dev* **16** (42)
[3] EC. Directive 2002/49/EC of the European Parliament and the Council of 25 June 2002 relating to the assessment and management of environmental noise *Off J Eur Commun* **189** pp 12–25
[4] Liguori C, Ruggiero A, Russo D and Sommella P 2017 Influence of measurement time on acoustic noise uncertainty *Proc. of I2MTC 2017 - 2017 IEEE International Instrum Meas Tech Conf. (Turin)* 7969850
[5] Liguori C, Ruggiero A, Russo D and Sommella P 2018 Accurate estimation of the environmental noise through sampling approach: Selection of the measurement time *Proc. of IEEE T Instrum Meas,* **67**(5) pp 1006-13
[6] Sebillo M, Vitiello G, Grimaldi M and De Piano A 2020 A Citizen-Centric Approach for the improvement of territorial services management. *ISPRS Int. J. Geo-Inf.* **9** 223
[7] Bellucci P, Brambilla G, Bisceglie A, Peruzzi L and Zambon G 2017 Analisi degli indicatori del rumore da traffico stradale per la comunicazione al pubblico *Proc. Of 44 Nat. Conf. Italian Acoustic Association (Pavia)*
[8] Vincent B, Gissinger V, Vallet J, Mietlicky, Champelovier P and Carra S, 2013. How to characterize environmental noise closer to people's expectations. *Proc. of Inter-Noise Conf.* (Innsbruck)
[9] González-Méndez M, Olaya C, Fasolino I, Grimaldi M and Obregón N 2021 Agent-based modeling for urban development planning based on human needs. Conceptual basis and model formulation *Land Use Policy* **101** 105110
[10] Andrea V, Tampakis S, Karanikola P, Galatsidas S, Giannakopoulos D 2020 Investigating the citizens' views and perceptions about the urban forest infrastructure and services of Komotini, Greece. *WSEAS Trans Environ Dev* **16** (61)
[11] Cvetković D and Praščević M 2000 Rating noise level as environmental noise indicator *Facta universitatis Series: Working and Living Environmental Protection* **1** (5) pp 39-50
[12] European Commission 2000 Position Paper on EU noise indicators *Office for Official Publications of the European Communities* (Luxembourg) pp 1-78
[13] Can A. and Gauvreau B. 2015 Describing and classifying urban sound environments with a relevant set of physical indicators *The Journal of the Acoustical Society of America* **137**, 208
[14] Can A, Aumond P, Michel S, De Coensel B, Ribeiro C, Botteldooren D and Lavandier C 2016 Comparison of noise indicators in an urban context *Proc. of Int. Cong. and Exp. of Noise Control Engineering (Hambourg)* pp 1-9
[15] Asensio C, Aumond P, Can A, Gascó L, Lercher P, Wunderli J M, Lavandier C, De Arcas G, Ribeiro C, Muñoz P and Licitra G 2020 A Taxonomy Proposal for the Assessment of the Changes in Soundscape Resulting from the COVID-19 Lockdown *Int. J. Environ. Res. Public Health* **17** (4205) pp 1-9
[16] Praščević M, Mihajlov D, Cvetković D and Gajicki 2017 A long-term noise measurements and harmonica index as a way of raising public awareness about traffic noise in the city of Niš *Facta universitatis Series: Working and Living Environmental Protection* **14** (1) pp 41-51
[17] Ribeiro C, Anselme C, Mietlicki F, Vincent B, Da Silva R and Gaudibert P 2013 At the heart of Harmonica project: the Common Noise Index (CNI) *Proc. of Inter-Noise Conf. 2013* (Innsbruck) pp 1-10
[18] The Harmonica website, http://www.noiseineu.eu/en/20-the_harmonica_index/subpage, last

accessed on July 28, 2021

[19] Prascevic M, Mihajlov D, Cvetkovic D, Gajicki A 2015 Assessment of Environmental Noise by Harmonica Index – Case Study: The City of Niš. *Appl Mech Mater* **801** pp 51-59

[20] Alsina-Pagès RM, Benocci R, Brambilla G, Zambon G 2021 Methods for Noise Event Detection and Assessment of the Sonic Environment by the Harmonica Index. *Appl Sci* **11**(17): 8031

[21] Mietlicki F, Mietlicki C, Ribeiro C, Gaudibert P, Vincent B and Gissinger V 2015 New tools to inform the public about environmental noise in cities and to assist decision-making *Proc. of* Euro-Noise Conf. (Maastricht)

[22] Prascevic M, Mihajlov D and Cvetkovic D 2015 The Correlation between Harmonica Indices and Noise Indicators *Analele Universităţii "Eftimie Murgu" reşiţa anul XXII* **2**

[23] Can A and Gauvreau B 2015 Spatial categorization of urban sound environments based on mobile measurement *Proc. of Euro-Noise (MAASTRICHT)* pp 5p

[24] Graziuso G, Grimaldi M, Mancini S, Quartieri J and Guarnaccia C. Crowdsourcing Data for the Elaboration of Noise Maps: a Methodological Proposal. 2020 *J. Phys.: Conf. Ser.* **1603** 012030

[25] Picaut J, Fortin N, Bocher E, Petit G, Aumond P and Guillaume G 2019 An open-science crowdsourcing approach for producing community noise maps using smartphones *Building Environ* **148** pp 20-33

[26] Noise Planet Project website, https://noise-planet.org/index.html, last accessed on July 28, 2021

[27] Bocher E, Petit G, Picaut J, Fortin N and Guillaume G 2017 Collaborative noise data collected from smartphones *Data in Brief* **14** pp 498-503

[28] Bocher E, Petit G, Fortin N, Picaut J, Guillaume G and Palominos S 2016 OnoM@p: a Spatial Data Infrastructure dedicated to noise monitoring based on volunteers measurements *PeerJ Preprints* **4**

Acknowledgments

Authors would like to acknowledge Domenico Caputo for contributing during his bachelor thesis to the elaboration of the applications to the study cases.

Predicting daily water tank level fluctuations by using ARIMA model. A case study

S Mancini[1], A B Francavilla[2], A Longobardi[2], G Viccione[2,a] and C Guarnaccia[2,b]

[1] Department of Information and Electric Engineering and Applied Mathematics, University of Salerno, via Giovanni Paolo II 132, 84084, Fisciano, SA, Italy
[2] Department of Civil Engineering, University of Salerno, via Giovanni Paolo II 132, 84084, Fisciano, SA, Italy

E-mail: [a]gviccion@unisa.it, [b]cguarnaccia@unisa.it

Abstract. The intrinsic dynamical features of water demand highlight the need of proper operational management of tanks in water distribution networks. In addition, due to the water resource scarcity, sustainable management of urban systems is essential. For this purpose, the aid of a predictive model is crucial since it allows to give short term forecasts that can be used to predict the oscillations of relevant parameters, i.e. tanks level and/or water demand. Urban water managers can use these predictions to implement actions aimed at the optimisation of the network function. Among several modelling techniques, the univariate time series analysis is instrumental since it allows forecasting the studied parameter by using the measurements of the parameter itself. In this paper, an autoregressive integrated moving average (ARIMA) model is calibrated on water levels data, measured in an urban tank in Benevento, Campania region (Italy) and then tested on a large dataset not used to tune the parameters. The validation and forecast phases show good performances of the model on a short-term forecast horizon demonstrating the excellent potentiality of this techniques. Finally, the residuals and errors analysis complete the work suggesting possible future implementations and improvements of this technique.

1. Introduction

Water supply and distribution networks belong to the ensemble of critical infrastructures, essential systems for maintaining health, safety, economic and social well-being of citizens. As a primary objective, the water distribution network (WDN) service is oriented to meet the user water demand, providing the user itself a reliable water distribution, that is providing the user with an adequate quantity of water, of good quality and with adequate pressure [1]. The management of water distribution networks (WDNs) relies on water utility operations consisting of both quick responses to either water demand or water tanks and reservoirs recharge variations or long-term management of network ageing effects [2,3]. In this context, storage tanks play a key role, actually acting as lungs [4] - that is, balancing instantaneous flow variations in the water demand pattern as well as compensating abrupt interruptions of the water recharge to the storage tank, such as in cases of drought periods or electricity shortages in pumping stations delivering water, when the tank level fluctuates within a fixed range of levels. Water tank level prediction and forecasting are crucial steps for supporting decision-making regarding operating actions as they represent the pressure heads that affect the reliable WDN operation. Water tank levels can be modelled through hydraulic models when the water demand and the management rules and operations are known. Still, in practical applications, the latter facts are not always fully known [5,6]. From a "modellistic" point of view, the AutoRegressive Integrated Moving Average (ARIMA) typology of models is well established, having been applied in the field of water demand forecasting for a long time [4,5,7-9]. This is justified by the fact that the model follows the trend at different time scales. On the other side, ARIMA models are also particularly suitable to predict and forecast groundwater levels, typically representing one of the primary sources of reservoir and tanks recharge [10]. Despite the number of applications in urban water demand, there is a gap in the literature concerning the use of ARIMA models for tank water levels. In [11], the link between the water supply, consumer demand and water level at the tank is, however, discussed, with the aim of providing a practical tool for water utilities to take prompt action based on water level variations.

In this paper, a univariate time series of water levels recorded in a urban tank is analysed. The data will be aggregated on one hour basis and an ARIMA modelling technique will be implemented, to produce short term predictions of the hourly water tank levels. The paper is organised in a "Materials and methods" section, in which the theoretical background of the ARIMA models is summarised, together with the case study and dataset presentation. Then, results of the application of the model are presented and discussed extensively, including the exploratory data analysis, the selection and calibration of the best ARIMA model, and the validation and the forecast testing phase. Since the selected model provides a one-step-ahead prediction, the forecast analysis will be performed in a 24-48 hours range. Quantitatively conclusions will be drawn analysing the residuals and the errors of the model, respectively in calibration and testing dataset. The definition of an ARIMA model applied to water levels would allow the prediction and forecasting of water tank levels for hydraulic WDN modelling as the baseline for supporting decision making regarding operating actions. Additionally, it would allow the identification of proper water tank management, improving drinking water waste reduction and drinking water saving, on one hand, and treatment costs related to chlorination or purifying techniques reduction on the other hand. This is an aspect of paramount importance as the circumstance in which the tank is not able to serve due to water scarcity is not rare, whereas there is a waste of the resource when the water inflow is not controlled [3].

2. Materials and methods

In this paper, the Box-Jenkins/ARIMA forecasting model is implemented. Autoregressive integrated moving average (ARIMA) was proposed by Box and Jenkins [12] in 1976.

The key components of the ARIMA model are the autoregressive or AR component, the differencing component and the moving average or MA component applying the Box-Jenkins method for finding the best fit coefficients. [13]

The order of an ARIMA model is usually denoted by the following notation ARIMA(p,d,q), where p, d and q are the order of the autoregressive part, the differencing and the moving-average process, respectively. The following equation expresses the general source formula:

$$\phi_p(B)(1-B)^d\, Y_t = \theta_q(B)\, e_t \tag{1}$$

Y_t : value of the series observed at the time t,

B: delay operator,

ϕ_p: autoregressive polynomials,

θ_q : moving average polynomials

e_t is the difference between the observed value Y_t and the forecasted one $\hat{Y}$ at the time t.

Any seasonal component is not supposed to be considered when using equation (1).

The general methodology implemented in this paper is summarised in Table 1. It is composed by four main steps. The preliminary phase aims to analyse the dataset to numerically and graphically describe it, in terms of data types, outliers, missing values, distributions. Then, the selection of the best ARIMA model is determined according to either Akaike Information Criterion (AIC), Akaike Information Criterion corrected (AICc) or Bayesian Information Criterion (BIC).

At least, the model validation phase to verify the forecasting abilities of the proposed model and the forecasting are performed. The entire process has been implemented in R software [14]. The

ARIMA modelling in 'R' uses a variation of the Hyndman-Khandakar algorithm [15], minimizing the standard and information criteria above mentioned to obtain the best ARIMA model. In particular, the auto.arima function provides hyperparameters and parameters of the best fitting model , using the conditional-sum-of-squares to find starting values, then the maximum likelihood for finding the best values. Then, residual tests need to be performed in order to examine the appropriateness of the model. So, the values of the autocorrelation and partial autocorrelation functions (ACF/PACF) of the residual series must all be approximately near to zero. If they do not look like white noise, another model should be tried. Once the residuals look like white noise, forecasting can be processed. [16]

Table 1. Workflow of the applied methodology.

n.	Phase	Description
0	exploratory data analysis	Dataset analysis and summary of their main characteristics
1	model calibration	Hyperparameters tuning and best ARIMA model selection, according to the information criteria
2	model validation	Test of selected model on a dataset not used in the calibration (fixed parameters but observed data used as input of the model)
3	model forecasting	Forecast with the selected model and comparison with observed data (parameters fixed and no observed data in input)

2.1. Case study and dataset presentation

This study is applied on the time series of the levels observed in one of the tanks of the water supply system of the town of Benevento, Italy, located in the area of Gesuiti. Authors already analysed data from this site in a previous paper where a preliminary study, focused on the use of a stochastic model to fill big sequences of missing data, and the calibration of a model have been described [6].

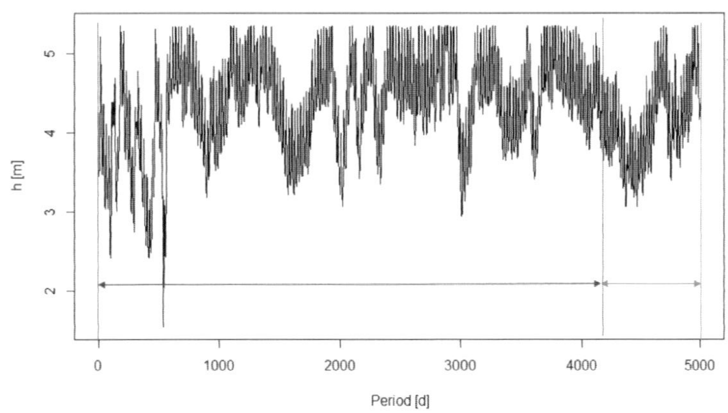

Figure 1. Time plot of the data with evidence of the subset of calibration (blue) and testing (green).

In this paper, data refer to daily average water levels, expressed in meters, registered from 2 p.m. on 6 September 2018 to 11 p.m. on 2 April 2019. The original dataset has been divided into two subsets: the first 4330 values were used to perform the calibration and, therefore, the estimation of the model parameters (calibration dataset) and the last 672 data were used for the testing (including the validation and forecasting phase). The time plot of the dataset is reported in Fig.1.

As already graphically revealed by the time plot, the dataset shows no relevant gaps. Indeed, only 13 data are randomly missing for short time ranges. Thus, the gaps were filled by using the mean value among the nearly previous and subsequent data value registered. Just in one case, when 6 hours measurements were missing, the data were imputed with the mean of the observed levels at the same hours of previous and subsequent days. The description of the final data subset is given in Table 2.

Table 2. Description of the calibration and testing dataset.

Dataset	**From**	**To**	**Sample size [n]**	**Missing values [n]**	**Imputed data [%]**
Calibration	September 6, 2019	March 5, 2020	4330	13	0.3
Testing	March 6, 2020	April 2, 2020	672	0	0

3. Results and Discussion

Following the workflow presented in Table 1, the first step of the study is an exploratory data analysis that is performed on the original dataset, as well as on the calibration and testing ones, to quantitatively and qualitatively describe the main features.

Table 3 reports a preliminary descriptive analysis of the datasets, including the main statistics of the data. It can be noticed that the calibration and the full datasets have similar distributions. The same conclusion can be drawn looking at the histograms and boxplots in Figure 2, and the correlograms and partial autocorrelation functions in Figure 3. The testing dataset differs from the calibration and the full ones because it missed the left tail (water tank levels below 3 m).

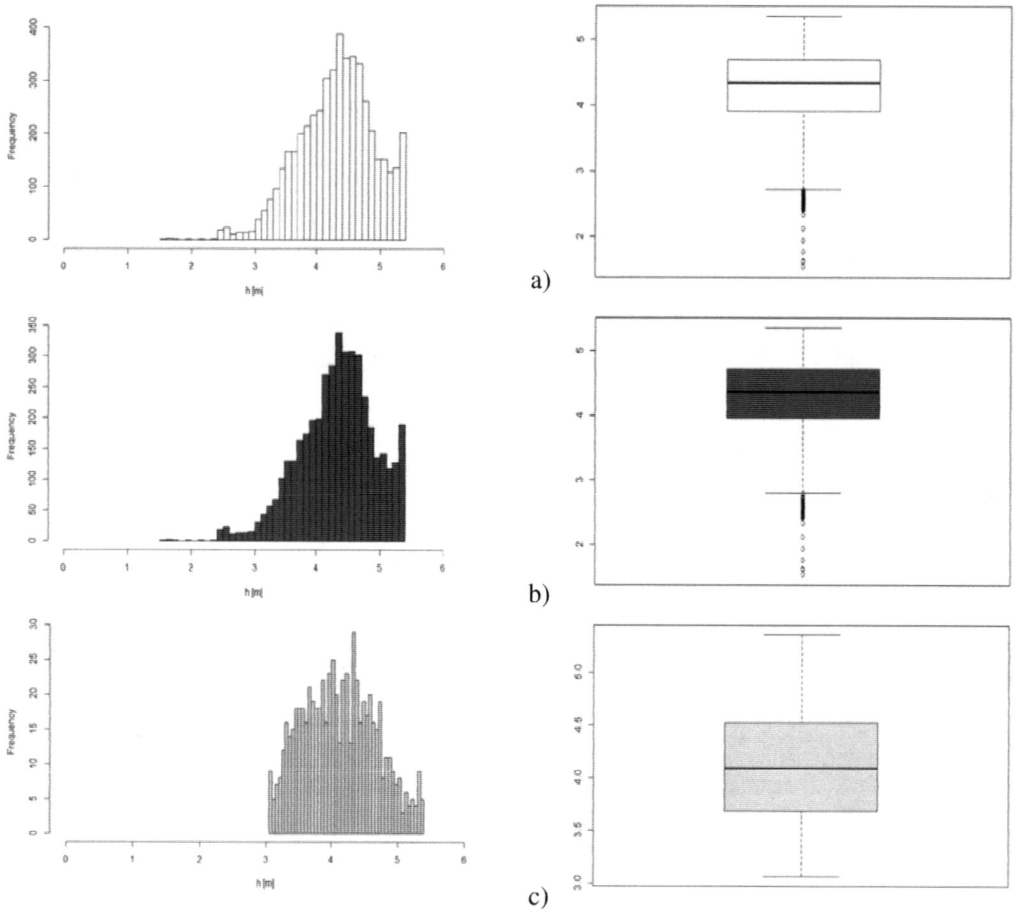

Figure 2. Histograms (left) and boxplots (right) of the data for a) full dataset, b) calibration dataset and c) testing dataset

Table 3. Summary statistics of the datasets of the water levels observed in the tank of Gesuiti.

Dataset	Mean [m]	Std. Dev. [m]	Median [m]	Min [m]	Max [m]	**Skewness**	**Kurtosis**
Calibration	4.32	0.60	4.37	1.54	5.36	-0.53	0.41
Testing	4.12	0.55	4.10	3.07	5.36	0.18	-0.74
Total	4.30	0.60	4.34	1.54	5.36	-0.43	0.18

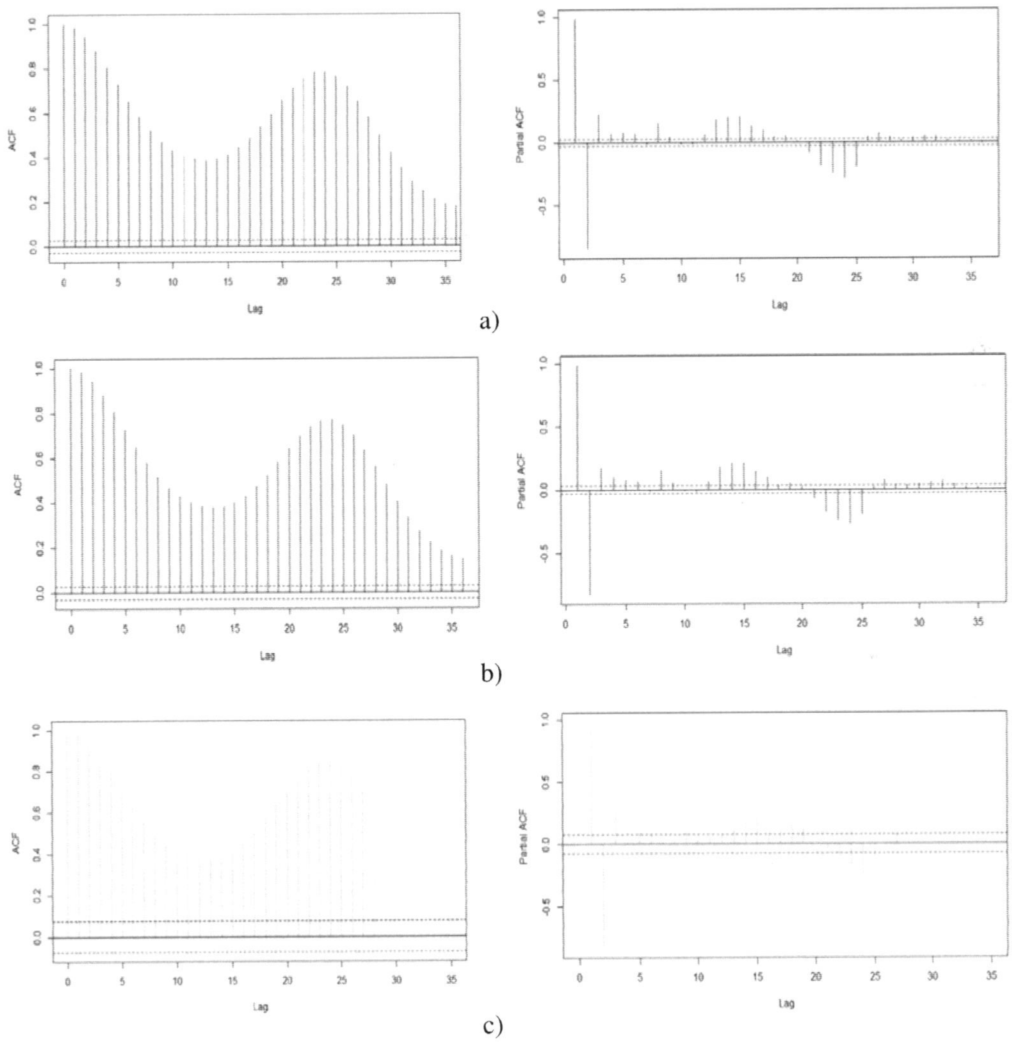

Figure 3. Autocorrelation plots (left) and Partial autocorrelation plots (right) of the data for a) full dataset, b), calibration dataset and c) testing dataset.

3.1. Model selection and calibration of hyperparameters

The best ARIMA model has been chosen according to the auto.arima function implemented in the "forecast" package of R software [14]. This function compares different models according to the common information criteria, such as Akaike Information Criterion (AIC), Corrected Akaike Information Criterion (AICc) and Bayesian Information Criterion (BIC). Running on the calibration dataset, this function suggested an ARIMA (4,1,1) model. The predicting formula of this model is:

$$\hat{Y}_t = Y_{t-1} + \varphi_1(Y_{t-1} - Y_{t-2}) + \varphi_2(Y_{t-2} - Y_{t-3}) + \varphi_3(Y_{t-3} - Y_{t-4}) + \varphi_4(Y_{t-4} - Y_{t-5}) + \vartheta_1 e_{t-1} \quad (2)$$

This model requires in input the previous observations thus it is a one-step-ahead model. The model parameters were estimated using again the statistical program R and the results are reported in Table 4.

Table 4. Estimated coefficients of the ARIMA (4,1,1) model.

	AR1, φ_1	AR2, φ_2	AR3, φ_3	AR4, φ_4	MA1, ϑ_1
Estimated Value	1.88	−1.10	0.18	-0.03	-0.91
Standard error	0.017	0.032	0.032	0.016	0.008

The plot of the simulated levels (red dots) overlapped with the observed data (black solid line) is reported in Figure 4. The single point simulated in period 536, i.e. hour 9 p.m. of September 28, 2018, completely misses the prediction. The prediction is 0.13 m and comes after a range of 6 data that were imputed because of missing observations in the raw dataset. The measurements were probably missing because of a maintenance of the tank, which caused a lowering of the levels. This oscillation was reproduced by the model, with a one-step delay, as shown in Figure 5, in periods 535 and 536.

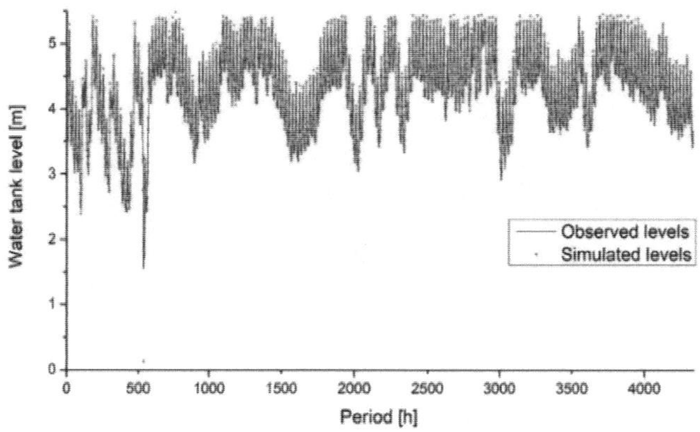

Figure 4. Comparison between observed (solid black line) and simulated (red dots) water tank levels plotted versus time in the calibration dataset.

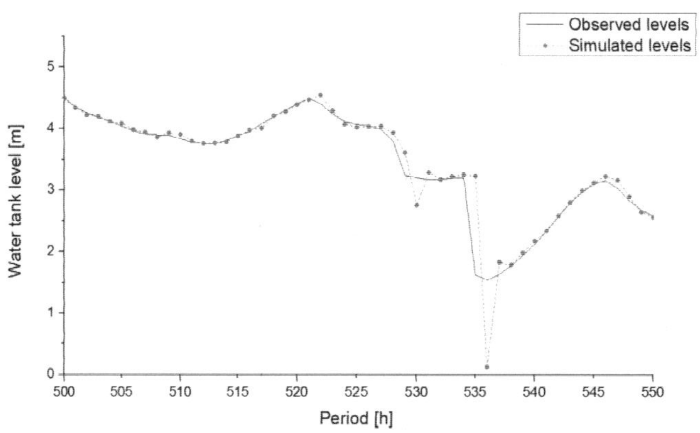

Figure 5. Comparison between observed (solid black line) and simulated (red dots) water tank levels plotted versus time in the 500-550 periods range.

3.2. Residuals analysis

In order to quantitatively evaluate the model's performances, the analysis of the difference between observed and simulated water tank levels in the calibration phase (residuals) has been executed. The results are reported in Table 5 (summary statistics) and in Fig. 6 (slope, autocorrelation and histogram).

Table 5. Summary statistics of the residuals of ARIMA (4,1,1) model in the calibration phase.

Mean [m]	Std. Dev. [m]	Median [m]	Min [m]	Max [m]	**Skewness**	**Kurtosis**
0.00	0.05	0.00	−1.60	1.41	−2.15	298.72

It can be noticed that apart from the oscillation observed in the periods 535 and 536 for the reasons mentioned above, the residuals are very close to zero. This is confirmed by the frequency distribution that is very narrow and with basically a zero mean. The autocorrelation presents a significant peak in correspondence of a 24 hours lag. This suggests that a seasonal model could be implemented in a further analysis [17].

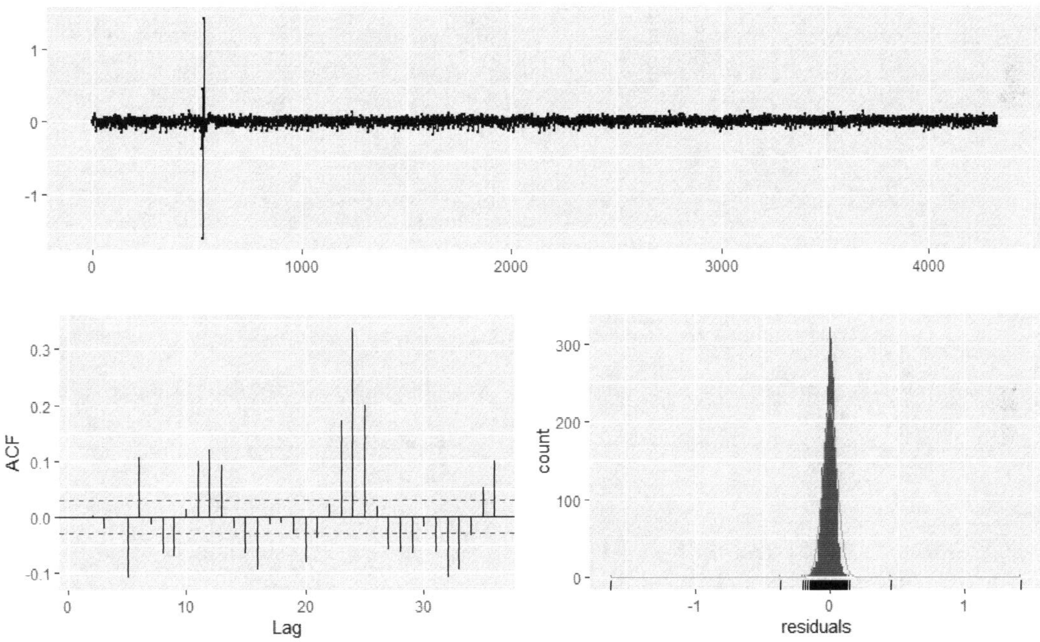

Figure 6. Time slope, autocorrelation plot and frequency distribution of the residuals in the calibration dataset.

3.3. *Model testing: validation and forecast*

Once the model is calibrated and the residuals have been analysed, the model can undergo a testing phase, i.e. a comparison between results of the models and measured data not used in the calibration phase. The key point of this phase is to check if the model has a good accuracy once the parameters have been fixed. The difference between validation and forecast subphases is in the usage of observed data in input of the prediction formula: during the validation the theoretical assumption is that the observed data are available and can be used to obtain the simulations (but not to tune the parameters). On the contrary, in the forecast, the observed data cannot be used to produce simulations because the model is predicting values in the future, but they can be compared with forecasts for performances analysis.

It is worth underlining that ARIMA models, adopting the last observed data to compute the predictions, have a short forecast horizon. The ARIMA (4,1,1) selected in this work has a one step ahead prediction and, thus, the forecast tends to a constant value, according to what suggested in TSA literature. In facts, when predicting water tank levels in the future, the model is fed by simulations and it lacks the variability of the observed data.

As already stated above, the ARIMA model was validated on the 672 periods of the testing dataset. The overlap between observed, simulated (in validation) and forecasted water tank levels is reported in Figure 7 for the entire validation dataset. It can be immediately noticed that the validation confirms the goodness of the model when working with observed data in input. On the contrary, the forecast tends to a constant value after about 48 hours. For this reason, a zoom on the first 24 hours is reported in Figure 8. Also, the error analysis presented in the following has been restricted to a two-day window.

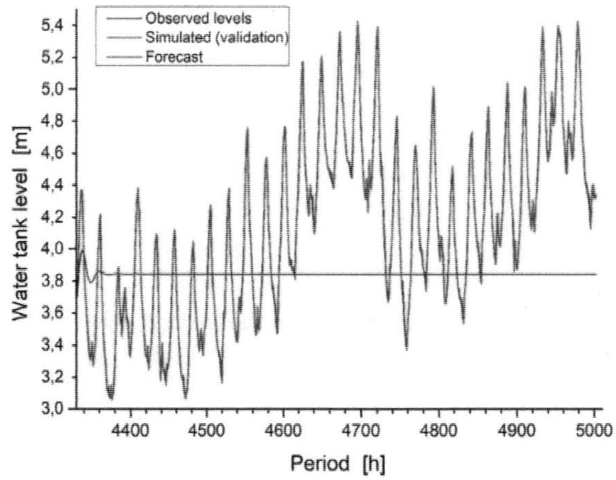

Figure 7. Comparison between observed (black), simulated in validation (red) and forecasted (blue) water tank levels plotted versus time in the testing dataset.

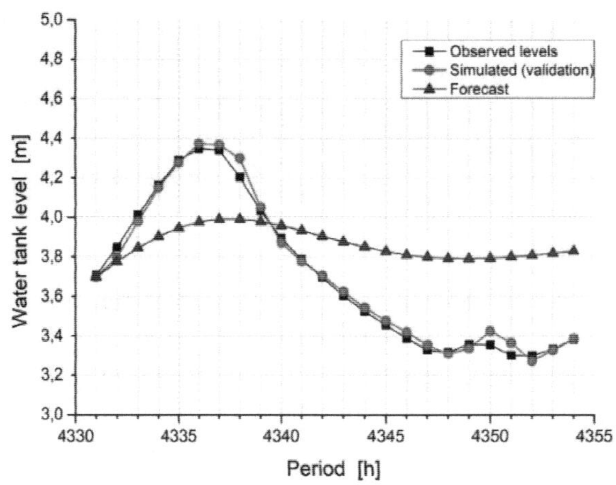

Figure 8. Comparison between observed (black), simulated in validation (red) and forecasted (blue) water tank levels plotted versus time in the first 24 hours of the testing dataset

3.4. *Error analysis in validation and forecast*

The performances of the models are tested in the validation and forecast phases by means of error analysis. The summary statistics of the validation and forecast errors are reported in Table 6. It can be noticed that the statistics of the errors are extremely good in the validation phase. In the 48 hours forecast, the mean error is reasonable, as well as the standard deviation. Figures 9 and 10 report the errors versus time, the autocorrelation and the frequency distribution, respectively for the validation and the forecast on 48 periods. The validation errors plots confirm the good performances already shown in the errors statistics, in fact the errors are almost normally distributed, with a narrow distribution and with an almost constant time slope. The autocorrelation plot again suggests that part of the variance of the data has not been explained and that a seasonal model could be an option for this dataset. The forecast errors, instead, have an oscillation over the 48 hours window, due to the convergence to a constant value of the forecast, which leads to the variation of the error.

Table 6. Summary statistics of the errors of ARIMA (4,1,1) model in the testing phase.

	Mean [m]	Std. Dev. [m]	Median [m]	Min [m]	Max [m]	**Skewness**	**Kurtosis**
Validation (672 periods)	0.00	0.04	0.00	-0.15	0.15	-0.24	0.59
Forecast (48 periods)	-0.27	0.36	-0.36	-0.77	0.37	0.27	-1.23

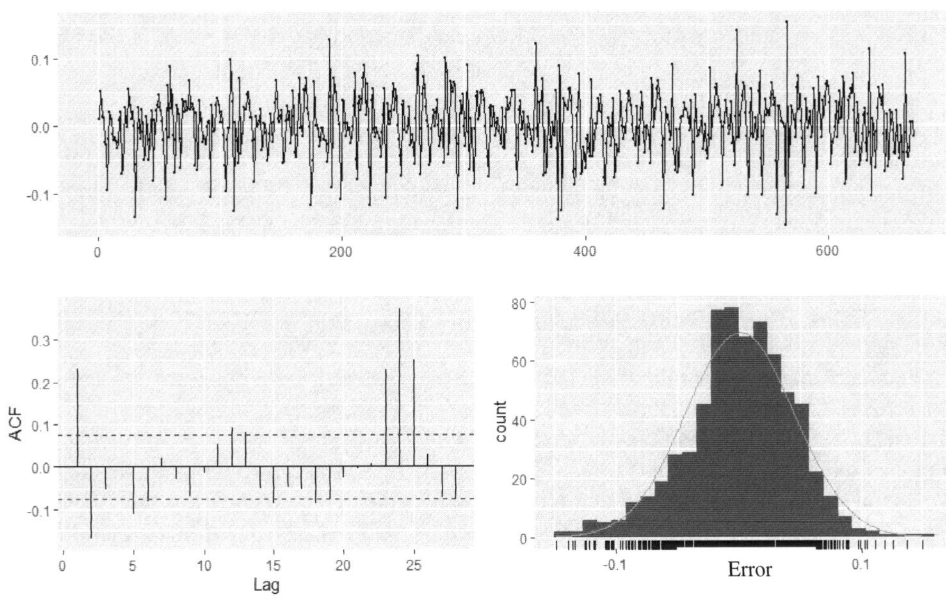

Figure 9. Time slope, autocorrelation plot and frequency distribution of the errors in validation.

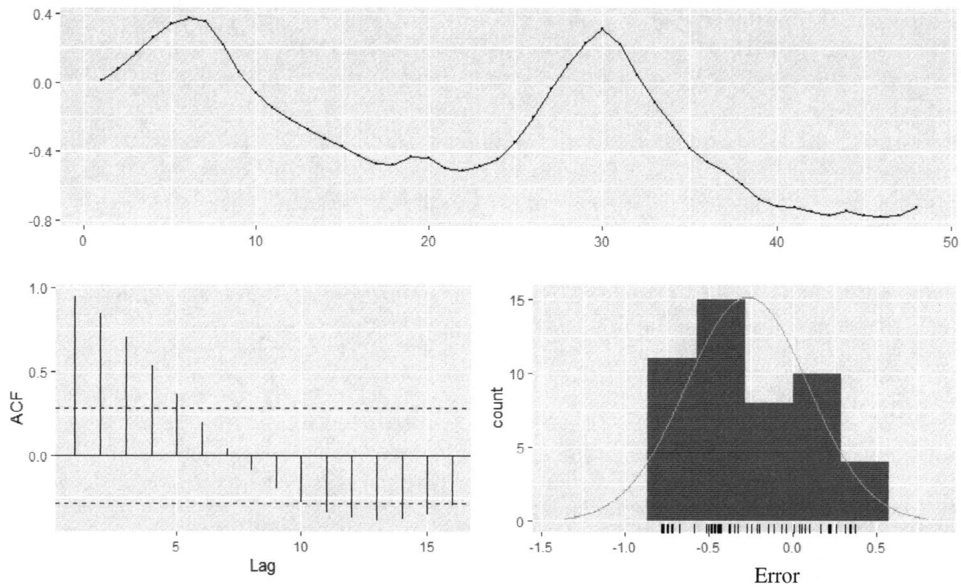

Figure 10. Time slope, autocorrelation plot and frequency distribution of the errors in the forecast.

4. Conclusions

In this paper, the modelling and forecasting of the time series of water levels in a tank connected to the supply system of a city has been discussed.

The univariate time series is related to the daily measurements of the water levels, observed from September 2018 to April 2019 (5002 hours), in the tank of Gesuiti site, connected to the urban water system of the city of Benevento (Italy). Only few measurements (13 periods) were missing, probably due to maintenance operations in the tank. These missing data have been reconstructed imputing the average value of the two temporally closest measurements, one previous and one following the missing period. In this way, a continuous time series of 5002 periods was obtained.

Once statistically characterised the time series, the best ARIMA model has been chosen, according to the main information criteria. In particular, an ARIMA (4,1,1) has been implemented.

The statistical analysis showed that the series of levels exhibits evident periodicity or seasonality, which could be used to construct a seasonal model (lag = 24 hours) as future research development.

The modelling results were very encouraging both by graphically comparing the observed level with the simulated one, and by analysing the distribution of the forecast error from a quantitative point of view. The selected model provided good predictions, even if limited to a two-day forecast horizon. The use of these techniques on water tank levels data shows, as already demonstrated in previous papers, excellent potential, presenting a possible way for the integration between the current water service monitoring systems and the forecasts obtained with this easy and simple-to-implement methodology.

References

[1] Longobardi A., G. Fattoruso, G. Guarnieri, A. Di Pietro, L. La Porta, M. Pollino. Water distribution network perspective in RAFAEL Project, a system for critical infrastructure risk analysis and forecast. In: Lecture Notes in Computer Science (including subseries Lecture Notes in Artificial Intelligence and Lecture Notes in Bioinformatics), Proceedings of the ICCSA 2021, International Conference on Computational Science and its Applications, RRS2021 Workshop, Cagliari, 13-16-Settembre 2021

[2] Bello, O. Abu-Mahfouz, A.M., Hamam, Y., Page, P.R., Adedeji, K.B., Piller, O. Solving management problems in water distribution networks: A survey of approaches and mathematical models. Water 11, 2019, 562, doi:10.3390/w11030562

[3] Viccione, G., Ingenito, L., Evangelista, S., Cuozzo, C. Restructuring a water distribution network through the reactivation of decommissioned water tanks. Water 11, 2019, 1740, doi:10.3390/w11091740

[4] Tripathi, A., Kaur, S, Sankaranarayanan, S., Narayanan, L.K., Tom, R.J. Water demand prediction for housing apartments using time series analysis. Int. J. Intell. Inf. Technol. 2019, 15, 57–75, doi:10.4018/IJIIT.2019100104

[5] Guarnaccia, C.; Tepedino, C.; Viccione, G.; Quartieri, J. Short-Term Forecasting of Tank Water Levels Serving Urban Water Distribution Networks with ARIMA Models. In Frontiers in Water-Energy-Nexus—Nature-Based Solutions, Advanced Technologies and Best Practices for Environmental Sustainability; Advances in Science, Technology & Innovation Naddeo, V., Balakrishnan, M., Choo, K.H., Eds.; Springer: Cham, Switzerland, 2018; doi:10.1007/978-3-030-13068-86

[6] Guarnaccia C., Longobardi A., Mancini S., Viccione G. Drinking Water Tank Level Analysis with ARIMA Models: A Case Study. In: V. Kanakoudis, The 4th EWaS International Conference: Valuing the Water, Carbon, Ecological Footprints of Human Activities, 2020, 1-8

[7] Muhammad, A.U., Li, X., Feng, J. Artificial Intelligence Approaches for Urban Water Demand Forecasting: A Review. In: International Conference on Machine Learning and Intelligent Communications Zhai X., Chen B., Zhu K., Eds.; Springer: Cham, Switzerland, 294, 2019, pp. 595–622, doi:10.1007/978-3-030-32388-2_51

[8] Zhao, L., Zhang, J., Chen, T. Application of product seasonal ARIMA model to the forecast of urban water supply. J. Water Resour. Water Eng. 2011, 22, 58–62

[9] Billings, R.B., Jones, C.V. Forecasting Urban Water Demand; American Water Works Association: Denver, 2008, CO, USA. ISBN 978-1-58231-537-1

[10] Boulariah O., Mikhailov P.A., Longobardi A., Elizariev A.N., Aksenov S.G. Assessment of prediction performances of stochastic models: Monthly groundwater level prediction in Southern Italy. J. Groundw. Sci. Eng., 2021, 9(2), 161-170. doi: 10.19637/j.cnki.2305-7068.2021.02.008

[11] Viccione, G., Guarnaccia, C., Mancini, S., Quartieri, J. On the use of ARIMA models for short-term water tank levels forecasting. Water Supply 20, 2020, 787–799, doi:10.2166/ws.2019.190

[12] Box, G., Jenkins, G. Time Series Analysis: Forecasting and Control. San Francisco: Holden-Day, 1976, CA,USA

[13] Khanarsa, O., Luangsodsa, A., Sinapiromsaran K. Self-Identification ResNet-ARIMA Forecasting Model. WSEAS transactions on systems and control, 2020,15, 196-211. ISSN: 2224-2856

[14] R Core Team. R: A Language and Environment for Statistical Computing; R Foundation for Statistical Computing: Vienna, Austria, 2014. Available online: http://www.R-project.org/ (last accessed on May 3, 2021)

[15] Hyndman, R. J., & Khandakar, Y. Automatic Time Series Forecasting: The forecast Package for R. Journal of Statistical Software, 2008, 27(3), 1–22.

[16] Merabet, F., Zeghdoud H. On Modelling seasonal ARIMA series: Comparison, Application and Forecast (Number of Injured in Road Accidents in Northeast Algeria), WSEAS transactions on systems and control, 2020, 15, 235-246. ISSN: 2224-2856

[17] Guarnaccia, C., Quartieri, J. Tepedino, C. Deterministic decomposition and seasonal ARIMA time series models applied to airport noise forecasting, in AIP Conference Proceedings 1836, 2017, 020079, doi: 10.1063/1.4982019

Physical Agents Quality Index (PAQI): application in a University Campus

G Graziuso[1], A B Francavilla[1], S Mancini[2], M Guida[2] and C Guarnaccia[1,a]

[1] Department of Civil Engineering, University of Salerno, via Giovanni Paolo II 132, 84084, Fisciano, SA, Italy
[2] Department of Information and Electric Engineering and Applied Mathematics, University of Salerno, via Giovanni Paolo II 132, 84084, Fisciano, SA, Italy

[a]E-mail: cguarnaccia@unisa.it

Abstract. Among the most common techniques and methodologies for the analysis of the exposure to polluting agents, the definition of environmental indexes represents a very useful tool able to describe the status of the environment quantitatively and qualitatively. In this perspective a general Physical Agents Quality Index (PAQI) describing the status of an environment in terms of presence of different physical pollutants was developed by authors in a previous work. In this paper the PAQI concept is extended from a punctual idea to a spatial analysis and a practical application in order to evaluate the potentiality of the index on a visual map is discussed. In particular, the PAQI visual map of the Campus of the University of Salerno (Italy) was redacted. The results are presented and lay down the basis for further improvements in the visual mapping and also in the calibration of the calculation of the index.

1. Introduction
The air or environmental pollution is usually quantified by means of Quality Indexes (QI). The monitoring of the pollutant concentration in a specific period allows to determine the dose of pollutant and, consequently, the computation of the QI. The challenging field of QI is to provide sufficient evidence of the problems involved in assigning values to a particular aspect of an environment such as noise, PM_{10}, CO, or more in general physical, chemical, biological pollution [1]. Indeed, QI are intended to reflect the state of any aspect or component of the environment and must respond to changes in the component it is scaling, in such a manner that it accurately reflects the magnitude of those changes, and usually are necessary to assign quantitative values to subjective aspects such as public health [2,3]. However, since the pollutants are different from each other, their effects on the human health are generally evaluated by means of specific indexes.
In order to classify the impact of the pollutants, the QI are usually identified into ranges to which a descriptor can be associated [4]. In this way, it is possible to easy inform about the environmental conditions and their variations. The estimation of the Quality Indexes is not a standard procedure since it depends on the pollutant national threshold values and quality standards adopted in the interested country. In addition, when several indexes are condensed in a single complex index, it is necessary to evaluate all the variables, the method for the calculations and the descriptor classes. Commonly, the indexes used for environmental pollutants consider all the variables separately. Nevertheless, in some studies few indexes are aggregated [5-8]. As example, mixing physical, chemical and biological pollutants, the Health Quality Index has been proposed in [5]. In [8] the development of a quality index, aggregating all the physical pollutants, as an indicator describing the overall grade of physical pollution of an environment is presented, according to the procedure proposed in [5]. This index, namely the Physical Agents Quality Index (PAQI), could be applied for the description of an outdoor and indoor environment, as well. Maps of outdoor PAQI can be used to support the strategical urban

decisions, for instance identifying healthy areas where new residential zones could be developed or addressing deteriorated areas that need improvement of activities therein.
For indoor conditions, a peculiar application is devoted to occupational health. The exposure of workers to all physical agents is usually regulated by national legislations. In Italy, for instance, the Legislative Decree 81/08 [9] defines, in section VIII, the physical agents to be controlled and fixes the thresholds values for 'danger' and for 'attention'. The Italian legislation recognizes the risk coming from exposure to noise, vibrations, ionizing and non-ionizing radiation, electromagnetic fields and, also, microclimatic conditions in the workplace. Moreover, it also defines and indicates standard techniques and methods for measuring pollutants [10]. The general idea of the PAQI index model is based on this framework but it can be easily extended to any national regulation, thanks to its modular and flexible structure.
In this paper, the PAQI concept is extended from a punctual idea to a spatial analysis. The possibility to create interpolation maps of PAQI is pursued in the case study of the Campus of the University of Salerno, located in the municipality of Fisciano, Italy. These maps will highlight potential hazardous areas and can be used to draw the quality of an outdoor environment, according to a defined classification.

2. Materials and methods

The main scope of the PAQI is to summarize in a single value the relation and the global potential impact of different physical agents, in a given site. The information packed in the PAQI can give a clear indication of the comprehensive quality of the environment, from the physical pollutant point of view.

2.1. *PAQI definition*

After selecting the relevant physical variables and the related descriptor, a sub-index is designed for each agent. The subindexes are usually designed as dimensionless ratios, with a critical value of the descriptor at the denominator. This implies that when the subindex is getting greater than 1, critical values are overcome and, consequently, the risk is increasing. A complete description of several subindexes can be found in [8], in which the physical indicators are transformed in dimensionless subindexes q_i considering the ratio between the environmental descriptors Q_i, and their reference value Q_{ref}. The evaluation of the value of Q_{ref} requires a careful analysis since it conditions the sign of the physical indicators that can be positive or negative.
When assembling the different subindexes, that have similar scales, in the single index PAQI, different quantities are put together, thus a weighting procedure is necessary. The weights p_i carry information about the relative importance between the subindexes and can be fixed, in general, by means of different approaches like the experts' involvement, epidemiological studies, statistical or other decision theory tools [5, 11].
The final formula of the PAQI is the weighted sum of the subindexes:

$$\mathrm{PAQI}=\textstyle\sum_{\mathrm{i}} p_i q_i \quad (1)$$

The first step in the determination of the PAQI is the identification of the descriptors, representative of the quality level of a certain environment. In this paper, the environmental physical agents on which the attention is focused are acoustic noise, ionizing radiation and microclimate, being considered as representative of the major hazardous physical agents for the Campus of Fisciano users, whose sub-indexes are reported in Table1. In this application, the weights have been chosen equal for each subindex, i.e. $p = 0.25$.
The other physical agents (i.e. vibrations, electromagnetic noise,...) are neglected in this application but can be easily implemented in future studies by adding more subindexes and calculating the relative weights as more detailed in [8].

Table 1. PAQI sub-indexes considered for the application in the Campus of Fisciano [8].

	Acoustic noise	Radon	Temperature	Humidity
Pollutant sub-index q_i	$\frac{L_{eq}}{L_{eq}^{ref}}$	$\frac{D}{D_{ref}}$	$\sqrt{\frac{\lvert T-T_{ref}\rvert}{T_{ref}}}$	$\frac{RH}{RH_{ref}}$

In the following subsections, the selected agents' measurements and parameters will be briefly summarized.

2.1.1. Acoustic noise

Acoustic pressure levels represent one of the central problems that must be taken into consideration in the definition of a global quality index. The selected descriptor is the equivalent A-weighted noise level L_{eq}, measured in dBA, as in [5] and in [6]. In order to develop PAQI interpolation maps of the Campus, a database of widespread measurements of the L_{eq} collected in the Campus has been used. These sound levels have been recorded using smartphones of volunteers, with the NoiseCapture free and open-source Android application [12] during two "NoiseCapture Parties" organized by the Applied Physics Laboratory (LAFIN) at the Department of Civil Engineering of the University of Salerno. The NoiseCapture approach consists of computing each second of the equivalent A-weighted sound levels along a path and then sharing data with the community. All data are aggregated in cells with the shape of a regular hexagon, to produce mean noise indicators in each one [13].

The corresponding sub-index q_i has been determined considering two different reference values depending on the characteristics of the considered environment. In fact, according to the Italian code [14], the L_{eq}^{ref} is taken equal to 50 dBA in Campus areas as study rooms, libraries and offices while the L_{eq}^{ref} is equal to 65 dBA with reference to the environments characterized by intense human activity as the canteen, the car parks and all the areas intended for the development of sports and recreational activities.

2.1.2. Natural radiation

Radon (222Rn) is an inert radioactive gas produced by the decay of uranium naturally present in rocks and soils throughout the earth's crust. As a gas, radon easily escapes from soil into the outdoor air where if diluted is of no further concern. But, in some rare case or if it reaches confined spaces radon can accumulate to harmful levels. For this reason, since the soil is generally the principal source of Radon maps of the radon potential are redacted to assess the potential risk of an area.

The assessment of the radiological risk related to the inhalation of radon and its progeny is expressed by the following expression [15]:

$$D=C_{Rn}\times F\times O\times DCF \tag{2}$$

where F is the equilibrium factor, O is the global average occupancy factor, DCF is the dose conversion factor, C_{Rn} is the arithmetic mean of radon concentration. The radon sub-index will consider the dose D, with respect to the reference value (D_{ref}) which is calculated by replacing C_{Rn} with the threshold value established by the national legislation ($C_{Rn,\ ref}$). In particular, the equilibrium factor for outdoor varies in the range 0.6-0.8. The global average occupancy factor depending by the time spent in the ambient under investigation, has been calculating considering that students spent on average about 1 hour per die in the outdoor environment when moving to and from the bus terminal, lunching and relaxing in the open air, etc. Indeed, the university campus of Fisciano is a very lively outdoor place where students enjoy studying, having break and lunch and hanging out thanks also to the typical sunny weather of the south of Italy. The DCF is a fixed values equal to 9 $nSvh^{-1}Bq^{-1}m^{-3}$. The average outdoor concentration C_{Rn} have been calculating, by assuming that generally Radon in the outdoor air, exhalating from the soil, dilute in the atmosphere until 100 times compared to the concentration measured in the soil gas. This widespread simplified assumption is very practical since measurements in the soil gas are generally easily available by means of measurements campaign or

Radon maps [16] which are mandatory developed in many countries. In this application, for example, results come from the redaction of a map of the potential radon from soil realized by researcher of the university of Salerno by means of the monitoring of the soil gas concentration in the campus area [17].

2.1.3. Microclimate

The microclimate is usually described by parameters like temperature and humidity even though wind speed could also be taken into consideration. As both low and high temperatures negatively affect the human health, the definition of a sub-index describing the temperature effects is different from the simple ratio adopted in the previous subindexes. Consequently, it is useful to define a reference range, for example from 10°C to 25 °C, rather than a single comfort value. Thus, a penalty factor should be introduced for deviations from the preferred value, both positive and negative. To overcome this problem, in [5] the authors adopted the difference between the measured temperature and a comfort reference one, under square root. The reference temperature value T_{ref} is fixed equal to 18°C since it is considered as an optimal value rather than a limit one. As for the other subindexes, the definition of the sub-index related to humidity is obtained by means the ratio of the measured value and the reference one. In the case study herein reported, the reference value of the humidity RH_{ref} is set equal to 85% since higher values create thermal discomfort in human perception [18]. Values of temperature and relative humidity have been fixed respectively at 21°Celsius and 60%, according to the values recorded during the noise level measurements.

3. Case study and results

The proposed index has been evaluated calculating the interested subindexes in 20 points widespread in the Fisciano Campus of the University of Salerno (Figure 1).

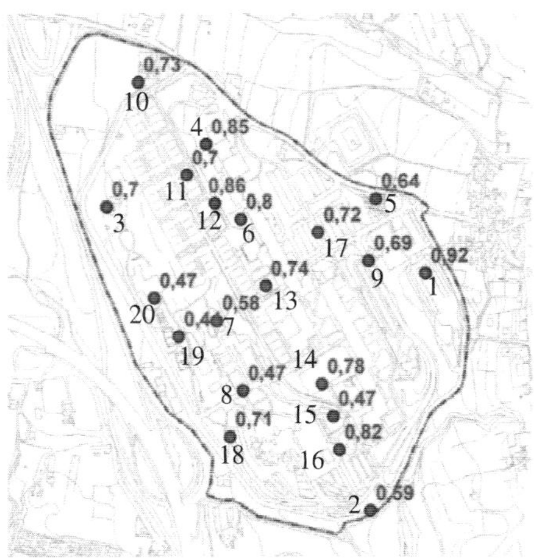

Figure 1. Case study cartography and values of the PAQI in red, estimated in each measurement point in black.

These sites have been identified on the basis of different human activities and environmental characteristics of the area (see table 2 reporting a brief description of the nearby area) as well as to mostly cover the entire area of the Campus. Data comes from different campaign of measurements performed: one campaign of three days for the measurement of natural radioactivity in the campus area and a one-day campaign for the measurement of sound pressure levels. The non-simultaneity of the measurements is, of course, not a problem since the radon potential of a soil depends on its geological features and do not vary sensibly over the time. Following the procedures and the choices proposed in the previous sections, the four subindexes and, subsequently, the PAQI values have been determined, reported in Table 2 and illustrated in Figure 1.

The spatial distribution of the index has been developed by means of the ARCMAP GIS software [19] and, interpolating the measured data by means of the Inverse Distance Weighting (IDW) method, the

PAQI unknown values has been simulated and overlapped on the cartography of the Fisciano Campus. In particular, the IDW method allows the determination of values of unknown points calculating a weighted average of the values available at the known scattered set of points.

Table 2. Measured values of the investigated physical agents and determined PAQI

# Point	Description	L_{eq} [dBA]	$C_{rn,soil}$ [Bq/m^3]	T [°C]	H [%]	PAQI
1	student residence	55	619879	21	60	0.92
2	multilevel car park	63	104068	21	60	0.59
3	masseria	66	266085	21	60	0.70
4	transit area buildings E and F	58	451359	21	60	0.85
5	bus terminal	63	193749	21	60	0.64
6	square	66	294664	21	60	0.80
7	botanical garden	55	44938	21	60	0.58
8	engineering laboratories	59	256230	21	60	0.82
9	canteen	58	294664	21	60	0.69
10	environmental laboratories	65.7	166549	21	60	0.73
11	library parking	60	294664	21	60	0.70
12	square to access to the library	58	463185	21	60	0.86
13	square in front of the E2 building	54.3	294664	21	60	0.74
14	printing centre	63	403069	21	60	0.78
15	theatre	53	294368	21	60	0.47
16	rectorate	52	449388	21	60	0.82
17	parking	65	294664	21	60	0.72
18	M2 building	62	186259	21	60	0.71
19	tennis courts	66	12161	21	60	0.44
20	swimming pool	59.2	226665	21	60	0.47

In figure 2a has been reported a PAQI map of the Campus where 9 subclasses of the index have been introduced automatically, showing how the index degrades over the distance. Then, results of the interpolation phase have been elaborated and the classes rearranged in order to convert the numerical information in terms of level of risk.

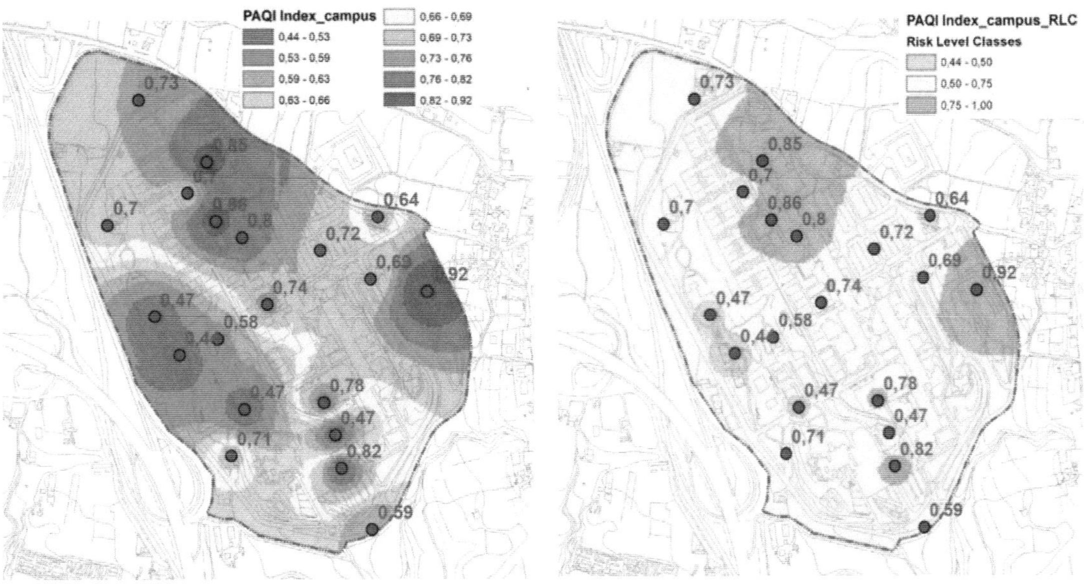

a) **b)**

Figure 2. Campus of Fisciano, Italy: a) Interpolation map of the PAQI; b) map of the environmental quality according to PAQI

In particular, according to the classification reported in Table 3, where P_{tot} is the sum of the weights, in our case equal to 1, a map of the overall quality of the environment related to the presence of physical agents has been drawn and reported in Figure 2b.

Table 3. Example of environmental classification according to PAQI values [5]

Class	Colour	Status	Range
VI		Very highly unsafe	$PAQI \geq 1.5\ P_{tot}$
V		Highly unsafe	$1.25\ P_{tot} \leq PAQI < 1.5\ P_{tot}$
IV		Unsafe	$P_{tot} \leq$ PAQI $< 1.25\ P_{tot}$
III		Alert	$0.75\ P_{tot} \leq$ PAQI $< P_{tot}$
II		Moderate safe	$0.5\ P_{tot} \leq PAQI < 0.75\ P_{tot}$
I		Safe	PAQI $< 0.5\ P_{tot}$

This map gives a quick and qualitative overview of the safety and health quality of an environment, clearly and rapidly highlighting the points with greater presence of physical agents affecting the environment. In this application, the areas in orange belong to the alert level of risk which means that some pollutant overcome reference levels. They correspond to the canteen (point 1, PAQI = 0.92), the area below big conditioning machines (points 4, 6 and 12, PAQI ≥ 0.8) and the rectorate square (point 16, PAQI = 0.82) where, in fact, acoustic noise is the pollutant affecting widely the area. On the contrary, points 19 and 20, that are located close to the park and to some sport facilities, report the lowest values of PAQI and influence the surrounding area, consequently classified as safe. According to the measured and simulated data, and according to the choice of equal weights in this preliminary application carried out in order to assess the potentiality of a map, no dangerous zones (PAQI ≥ 1), are highlighted in the map. Thus, microclimate, natural radioactivity and acoustic noise do not affect the campus area up to harmful levels for human health, but some area in orange should be taken under control, of course, in order to avoid the overcoming of reference levels.

4. Conclusions

In this paper an application of the PAQI index developed in [8] has been presented in order to evaluate the potentiality of a map describing the overall environmental quality of an area related to the presence of physical agents.

Of course, this kind of map could be a useful tool not only in urban planning. If based on real time measurements it could give additional information to authorities for the developing, adopting, implementing and evaluating environmental policy, and also easily evidence to the general public the quality of the environment they live.

Further development regarding the implementation of the map will focus on the adding of information on the number of subindexes greater than 1, with indication of the specific physical agent overcoming the threshold. Moreover, a map of the impact could be easily redacted by integrating the number of people exposed, for instance with circles whose diameter is the density of population and the filling colour is the PAQI classification. Other further applications will focus on the optimization of the general formula by defining a general method for the definition of the values of the weights coherently with the severity of the effects on human health. Indeed, since it was not the aim of this study to calibrate the methodology but to test the applicability on map it has been adopted the same weight to all the physical hazards. So, for example, since noise and radon are well recognized hazard for human health, they could be weighted differently to microclimate parameters which impact mostly on the human wellbeing without fatal consequences.

5. References

[1] Ouarda Layachi, 2020 International Liability for Pollution Damage in Outer Space Environment, *WSEAS Trans Env & Dev* **16**, 149-157

[2] Alattar N., Yousif J., Jaffer M., Aljunid S. A., 2019, Neural and Mathematical Predicting Models for Particulate Matter Impact on Human Health in Oman, *WSEAS Trans Env & Dev* **15**, 578-585

[3] Thomas W.A. Indicators of Environmental Quality: An Overview. 1972. In: *Thomas W.A. (eds) Indicators of Environmental Quality*. Springer, Boston, MA

[4] Thom G, Ott W, *Air pollution indices*, 1975, *U.S. Govern.Print. Off.* **1** 164

[5] Quartieri J, Troisi A, Guarnaccia C and D'Agostino P 2009 Development of an Environmental Quality Index Related to Polluting Agents, *Proc. Int. Conf. on Environment, Ecosystem and Development*, December 14-16, 2009, Tenerife, Spain, 153-161

[6] Quartieri J, Troisi A, Guarnaccia C, D'Agostino P, D'Ambrosio S and Iannone G 2010 An Environmental Quality Index Related to Polluting Agents and its application in the framework of a GIS platform. *Proc. 2nd Int. Conf. on Engineering Optimization*, Sept 6-9, 2010, Portugal, 1-10

[7] Thach TQ, Tsang H, Cao P, Ho LM. A novel method to construct an air quality index based on air pollution profiles. *Int J Hyg Environ Health*. 2018 Jan, **221**(1), 17-26.

[8] Mancini S, Guida M, Graziuso G, Guarnaccia C. 2020 Development of an Environmental Quality Index to Assess Physical Agents Pollution *J. Phys.: Conf. Ser.*, Applied Physics, Simulation and Computing (APSAC 2020) 23-25 May 2020, Rome, Italy, 1603, 012005

[9] Legislative Decree 9 April 2008, no. 81 Implementation of Article 1 of Law no. 123, concerning the protection of health and safety in the workplace

[10] Liguori, C., Ruggiero, A., Russo, D., Sommella, P. Task-based measurements for estimating the workers' exposure to acoustic noise. I2MTC 2018 - 2018 IEEE International Instrumentation and Measurement Technology Conference: Discovering New Horizons in Instrumentation and Measurement, Proceedings, 2018, 1–6

[11] Hongrui Y, Feier W, Jie Y, Kuan H, Huichun Z, Zhihui F. An improved weighted index for the assessment of heavy metal pollution in soils in Zhejiang, China, *Env Research*, **192**, 2021, 110246, ISSN 0013-9351.

[12] Picaut J., Fortin N., Bocher E., Petit G., Aumond P., Guillaume G. 2019.An open-science crowdsourcing approach for producing community noise maps using smartphones. B*uil Environ*. 148, 20-33
[13] Graziuso, G.; Mancini, S., Francavilla, A.B., Grimaldi, M., Guarnaccia, C. Geo-Crowdsourced Sound Level Data in Support of the Community Facilities Planning. A Methodological Proposal. S*ustainability* 2021, **13**, 5486
[14] Italian Ministry Decree, DPCM 14.10.1997 on noise pollution. 1997.
[15] UNSCEAR. Sources, effects and risks of ionizing radiation. 1993. *United Nations Scientific Committee on the Effects of Atomic Radiation*, United Nations New York
[16] Bossew P, Tollefsen T, Gruber V, De Cort M. The European Radon Mapping Project. In Conference Proceedings: Sociedada Brasileira de Proteção Radiológica, editor. IX Latin American IRPA Regional Congress on Radiation Protection and Safety - IRPA 2013. Sociedada Brasileira de Proteção Radiológica; 2013. JRC84757
[17] Guida D., Guida M., Cuomo A., Siervo V. Assessment and Mapping of Radon-prone Areas on a regional scale as application of a Hierarchical Adaptive and Multi-scale Approach for the Environmental Planning. Case Study of Campania Region, Southern Italy. WSEAS TRANSACTIONS ON SYSTEMS. **12**. 105-120, ISSN:1109-2777.
[18] Zhang L., Wei D, Hou Y., Du J., Liu Z., Zhang G., Shi,L. 2020. Outdoor Thermal Comfort of Urban Park—A Case Study, S*ustainability*, MDPI, Open Access Journal, **12**(5), 1-23.
[19] ArcGIS [GIS software]. Version 10.8.1 Redlands, CA: Environmental Systems Research Institute, Inc., 2020.

Acknowledgments
Authors would like to thank Barbara Ida Cannavale for participating in this research, during her bachelor thesis development.

An Application of ARIMA modelling to air pollution concentrations during covid pandemic in Italy

S Mancini[1,a], A Francavilla[2], G Graziuso[2] and C Guarnaccia[2,b]

[1] Department of Information and Electric Engineering and Applied Mathematics, University of Salerno, via Giovanni Paolo II 132, 84084, Fisciano, SA, Italy
[2] Department of Civil Engineering, University of Salerno, via Giovanni Paolo II 132, 84084, Fisciano, SA, Italy

E-mail: [a]smancini@unisa.it , [b]cguarnaccia@unisa.it

Abstract. Since the COVID-19 pandemic began, space and ground-based observations have shown how Earth's atmosphere has observed significant reductions in some air pollutants. Many studies, all over the world, demonstrated how the governmental restrictions imposed because of the spreading of the virus had positive and negative effects on the environment. In this paper, authors discuss how the levels of concentrations of some pollutants varied, in two case studies in Italy, because of the imposed lockdown during the coronavirus pandemic. The extent of the variations CO and PM_{10} has been evaluated by comparing data registered by local monitoring stations, related to the baseline February-May, of three different years, 2018, 2019 and 2020. In order to better assess the variation of the temporal trend of pollutants before (2018, 2019) and during COVID-19 lockdown (2020) proper physic-mathematical models have been applied to the datasets. The calibration and validation of AutoRegressive Integrated Moving Average (ARIMA) models on interesting series of CO and PM_{10} data complete the work.

1. Introduction

Air pollution is one of the most hazardous environmental problem. People health is affected by chemical and physical agents, especially in areas in which high concentrations of pollutants are observed. Thus, it is crucial to develop large and effective monitoring networks, to control the time evolution of the pollutants' concentrations. The data recorded can be used to calibrate and validate predicting models based on various techniques [1-3]. These models can be used to track sudden changes with respect to the seasonal slope and to implement mitigation actions. In this scenario, an unusual behaviour has been observed during the COVID pandemic of 2020. Several studies highlighted a decrease of some pollutants concentration in urban areas [4]. The most relevant decrease has been observed for NOx. On the contrary, in some regions CO and PM_{10} didn't show a decrease, due to the relevant increase of heating systems usage and other pollution sources [5].

In this paper, the "Time Series Analysis" (TSA) techniques are applied to a dataset of pollutant concentrations recorded in an Italian city during the first lockdown (spring 2020). These techniques are largely used in several research areas. In [6-10], deterministic decomposition is applied to noise recorded in different zones, produced by various sources, such as traffic noise, airport, etc.. Poisson processes combined with TSA models are also applied to noise in [11]. In [12] respectively the deterministic decomposition and the "Seasonal AutoRegressive Integrated Moving Average" (SARIMA) model of a CO dataset collected in the Metropolitan Area of Monterrey (Mexico) are presented, showing that for a short-term precise prediction SARIMA should be preferred.

After a preliminary statistical description of the dataset, a comparison with average concentrations of previous years will be showed, to discuss possible decrease or increase related to the pandemics. Then, a predictive model will be calibrated and validated on data not used in the hyperparameters and parameters tuning phase.

2. Material and methods

2.1. Time Series Analysis techniques

The Time Series Analysis is the mathematical study of the slope over time of a selected physical variable, measured with a given time basis. When the database includes a single variable measured over time, it can be considered a univariate time series.
The most popular techniques adopted are based on a deterministic decomposition or on Auto-Regressive Integrated Moving-Average (ARIMA) procedures [13].
Generally, when non-stationarity is an issue, the stochastic autoregressive integrated moving average (ARIMA) models are used. They are part of the family of non-stationary linear stochastic processes and, in their most complete version, can be divided into two categories: the ARIMA and the seasonal ARIMA, called SARIMA, used when there is a periodicity in the data series instead. ARIMA predictions evolve over time, using as input recent data, close to the period to be predicted, following the changes of the process. Therefore, the ARIMA models adapt quickly to possible variations of the series, but they pay this quality in terms of short forecast horizon.
The key components of the ARIMA model are the autoregressive or AR component, the differencing component and the moving average or MA component applying the Box-Jenkins method for finding the best fit coefficients. The order of an ARIMA model is usually denoted by the following notation ARIMA(p,d,q), where p, d and q are the order of the autoregressive part, the differencing and the moving-average process, respectively.
The general formula of these models, not including seasonality, can be expressed as following:

$$\phi_p(B)(1-B)^d\, Y_t = \theta_q(B)\, e_t \tag{1}$$

in which:
Y_t is the value of the series observed at the time t,
B is the delay operator,
ϕ_p is the autoregressive polynomials (polynomial AR),
θ_q is the moving average polynomials (polynomial MA),
e_t is the residual at the time t, i.e. the difference between the observed and forecasted values at the time t.

Generally, the elaboration work involves four main steps: the dataset analysis and statistical summary, the model calibration, i.e. the tuning of hyperparameters and best ARIMA model selection, the model validation, i.e. the test of the selected model on a dataset not used in the calibration and the model forecasting.

In this paper, the authors will present the calibration of two ARIMA models applied to air pollutants concentrations recorded on a large time range, in two sites described in subsection 2.2. The estimation of the hyperparameters and of the coefficients of the models is done in "*R*" software [14], in which also the validation on data not used in the calibration and the forecasting can be run.

2.2. Case studies presentation

The datasets used in this paper have been collected in two fixed monitoring stations installed by the regional agency for environmental protection (Agenzia Regionale per la Protezione dell'Ambiente Campania, ARPAC). The two stations are located in the cities of Nocera Inferiore and Solofra (Fig. 1).

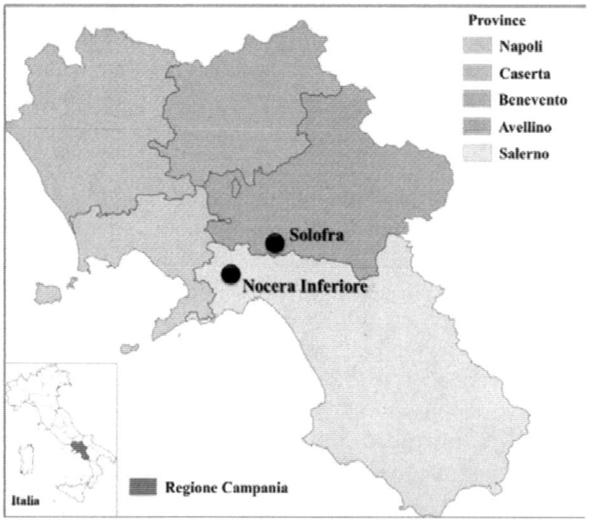

Figure 1. Sites location highlighting the Campania region (in red) and the provinces.

The area of the Solofra station is industrial, with a large presence of tanneries. The tanning, in fact, is a specific expertise of the area, with a large production of leather goods. The production process has several polluting sources related to the refining of leather, with emission of Volatile Organic Compounds (VOCs), Particulate Matter (PM), Hydrogen sulphide (H2S) that is responsible of the typical bad smell of the area. CO and NO_x are produced by the thermal plants that continuously provide hot water and steam for the processes. Besides the industrial sources, road traffic, agriculture and other anthropic sources should be considered as well.

The monitoring station of Nocera Inferiore is settled in a strategic urban position to control the changes in emissions mainly related to road traffic, since it is placed in proximity of a highway and close to local roads with large volumes of vehicles. The position of the station is in the residential downtown, with many houses and buildings in the surroundings. Nocera Inferiore is one of the most polluted cities of the Campania Region, being the most polluted city in the province of Salerno. In 2020, for instance, it had 67 thresholds exceedances for PM_{10} concentrations (50 $\mu g/m^3$), with respect to the 35 exceedances allowed by the Italian regulation [15].

The two stations record, continuously, Benzene, NO_2, SO_2, PM_{10}, $PM_{2.5}$ and CO, together with meteorological data such as temperature and humidity. For the analysis presented in this paper, the authors selected PM_{10} and CO daily concentrations, recorded from February to May 2020. These datasets have been firstly compared with the same time range in 2018 and 2019, aggregated in monthly levels (Figure 2) and then graphically reported in a time plot (Figure 3) with evidence of the governmental restriction phases.

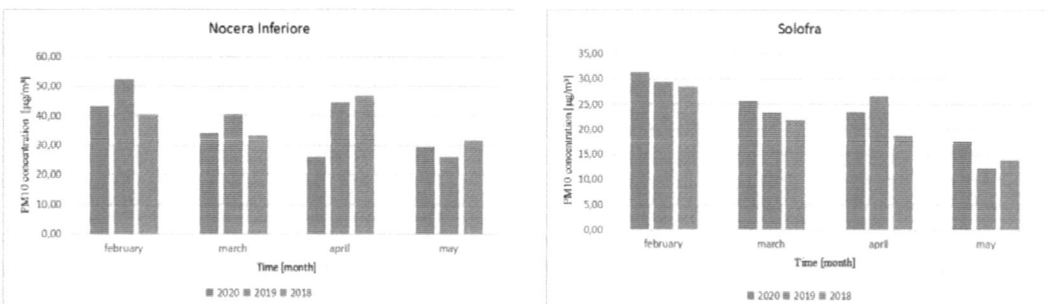

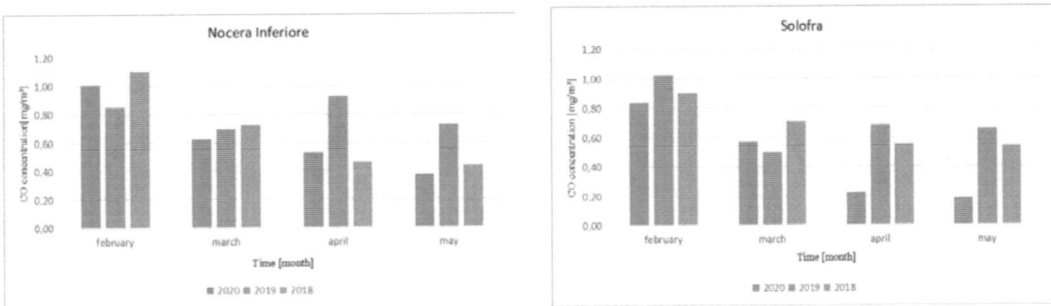

Figure 2. Average monthly levels of PM_{10} and CO registered in Nocera Inferiore and Solofra from February to May 2018, 2019 and 2020

In particular, the first green vertical line is placed in correspondence of the first publication of national restrictions on the 4th of March 2020 defining the overall closure of schools and the beginning of the first phase of lockdown. The green area ends at the start of phase 2, when a general relaxation of restrictions was implemented.

In particular, the Italian governmental restrictions from the beginning of the pandemics (February 2020), specified by the publication of ministerial decree, can be summarized as in Table 1.

Despite of the decrease of road traffic and production plants operation during the lockdown, it can be noticed that, except for CO concentrations in Solofra and PM_{10} concentrations in Nocera Inferiore, the other curves are not drastically different from previous years observations. This is probably due to the compensation of home heating systems increase, since people stayed at home longer than during the same months in 2018 and 2019. In standard conditions, a large part of the population during working days would have been in working places or at schools; these buildings are generally more effective than private houses from the energetic and pollution point of view.

Table 1. Timeline of the Italian governmental restriction from February to May 2020.

Phase	DPCM[1] date	Adopted containment Measures
1	4th March 2020	suspension of all educational activities (all levels schools and universities)
1bis	8th March 2020	total lockdown of all the cities
1ter	23rd March 2020	closure of all activities excepted essential industrial and commercial ones
2	17th May 2020	end of restriction on displacement among cities and regions

[1] Decree of the President of the Ministry Council

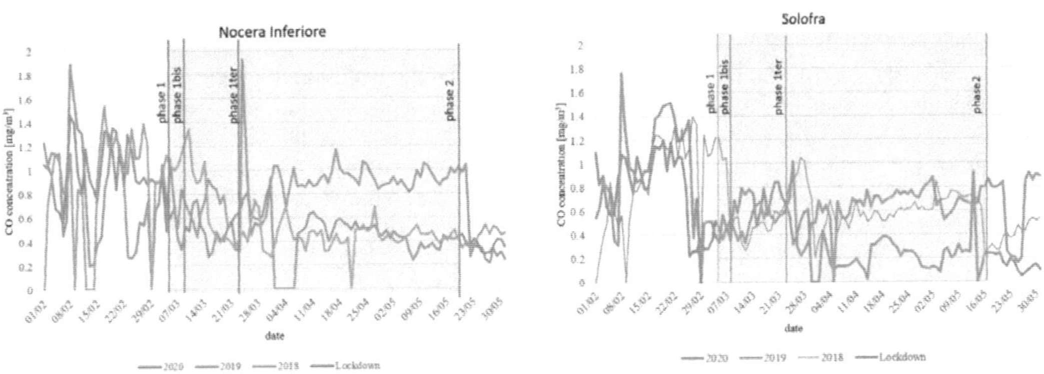

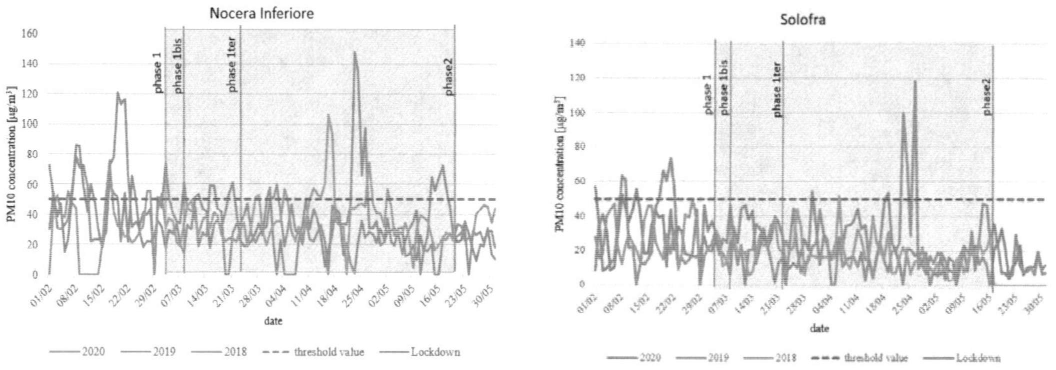

Figure 3. Average daily levels of PM_{10} and CO registered in Nocera Inferiore and Solofra from February to May 2018, 2019 and 2020 with evidence of the different restriction phases. The red dashed line in PM_{10} plots is the threshold for daily concentration that the Italian regulation allows to overcome 35 days per year.

University of Salerno, for example, currently self-produces about 30% of its energy needs, thanks to photovoltaic roofs and plants, solar thermal, cogeneration and solar cooling [16] providing necessities for more than one thousand persons. Indeed, the University employs about 1000 professors and researchers, about 500 technicians and administrative clerks, plus a large number of PhD students and postdoc fellows, usually working in the labs and in the departments' rooms, and about 35000 students. All these persons stayed at home, increasing private heating boilers usage, of course. Very often these boilers are based on old functioning systems, with large emissions and gas consumption. Basically, during the lockdown, the pollution sources increased in number, with a worsening in quality.

For the reasons mentioned above, the following analyses will be pursed on CO for the city of Solofra and on PM_{10} datasets for Nocera Inferiore.

2.3. Datasets presentation

As mentioned in the previous subsection, the datasets used in this paper are 'CO-Solofra' and 'PM_{10}-Nocera'. The two datasets have been used for the calibration phase.

Each dataset is composed of three dataset of 121 daily concentrations, both for CO and PM_{10} measured in 2018, 2019 and 2020 from the 1st of February to the 31st May. The 'CO-Solofra' dataset presented 3 days of missing data in 2020, 2 in 2019 and 3 in 2018. The 'PM_{10}-Nocera' presented 5 days of missing data in 2020, 8 in 2019 and 9 in 2018. Since time series analysis needs to be performed on continuous dataset missing data has to be imputed. More in details, the missing data (4% of the total calibration data for both pollutants) has been imputed with the mean of the previous and successive measurements, and, for PM_{10}, in some cases with the "cold deck" technique, i.e. imputing with the concentrations observed in the same period in the past years. This method of imputation process, did not altered the distribution of the data, preserving the mean and the standard deviation, as showed in Tables 2 and 3.

Table 2. Summary statistics of the CO concentrations measured in Solofra.

Dataset	Mean [mg/m³]	Std. Dev. [mg/m³]	Median [mg/m³]	Skew	Kurt
Observed	0.45	0.32	0.33	0.71	-0.77
Reconstructed	0.45	0.32	0.35	0.72	-0.79

Table 3. Summary statistics of the PM_{10} concentrations measured in Nocera Inferiore.

Dataset	Mean [μg/m³]	Std. Dev. [μg/m³]	Median [μg/m³]	Skew	Kurt
Observed	33.44	17.13	29.75	0.81	0.23
Reconstructed	33.53	17.04	29.50	0.80	0.19

3. Results and discussion

The TSA models, based on ARIMA techniques, calibrated and adopted in this analysis have been selected by using the "auto.arima" function implemented in "R" project software environment minimizing the AIC and BIC criteria, as well as according to the parsimony principle. In particular, the above mentioned automatic function provides hyperparameters and parameters of the best fitting model, using the conditional-sum-of-squares to find starting values, then the maximum likelihood for finding the best values. Then, residual tests need to be performed in order to examine the appropriateness of the model. Thus, the values of the autocorrelation and partial autocorrelation functions (ACF/PACF) of the residual series must all be approximately nil.

For the PM_{10} concentrations in Nocera Inferiore, the best model resulted a simple AR(1), with only order 1 autoregressive component. This choice is suggested by the slope of the autocorrelation and partial autocorrelation functions showed in Figure 4, as well as by the routine "*auto.arima*" implemented in the forecast package of "*R*" software.

The graphical comparison between the results of the ARIMA model and the observed data is shown in Figure 5. The overlap between observed PM_{10} concentration and simulation of the model and a one-day delay in the predictions can be easily observed.

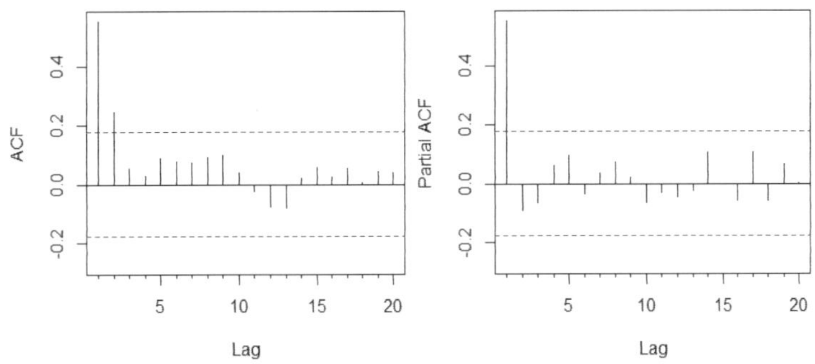

Figure 4. Autocorrelation and Partial autocorrelation for PM_{10} observed in Nocera Inferiore.

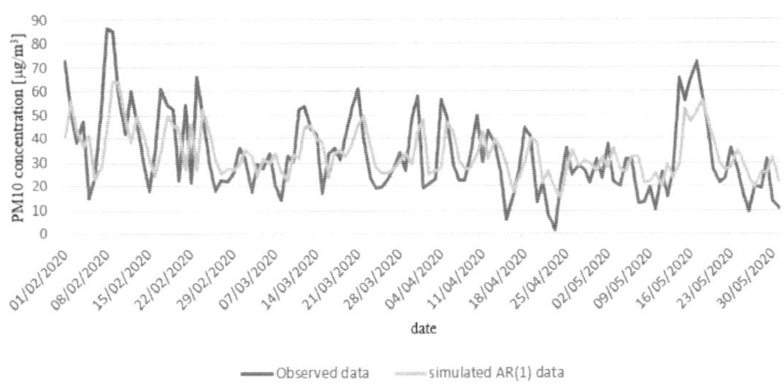

Figure 5. Plot of observed and simulated PM_{10} concentrations in Nocera Inferiore.

The good performance of the model is highlighted also in Figure 6, reporting the bi-sector scatter plot between observations and simulations, and showing that the 80.2% of simulations fall in an interval of one standard deviation from the observations, with some overestimations in the low concentration range and underestimations in the high concentration range.

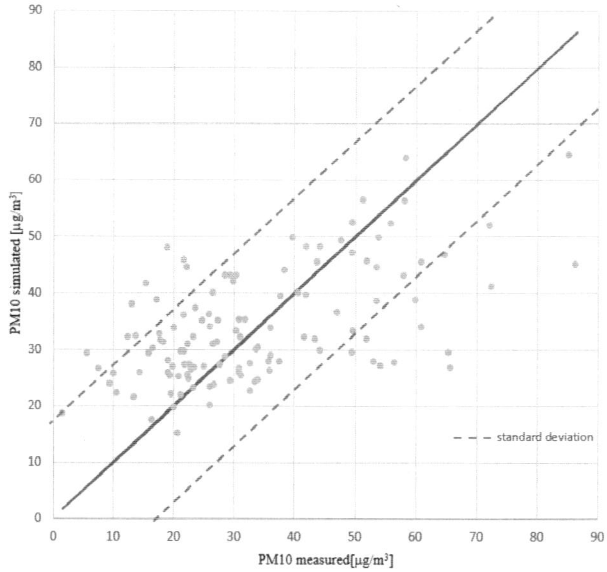

Figure 6. Scatterplot of observed and simulated PM_{10} concentrations in Nocera Inferiore

Comparison between observed and simulated data in order to assess the goodness of the selected models has been done, by using similar plots reported in Figures 7, 8 and 9, for CO concentrations in Solofra.

In this case, the series is nonstationary, thus a differentiation has been performed on the data, to work with a smoother time series. The autocorrelation and the partial autocorrelation plots of the differenced series (Figure 7b) suggest a ARIMA(14,1,14) model, that is tested together with the ARIMA(0,1,1) simple model suggested by the BIC criterion (a ranking has been obtained with the "*arimaId*" function of the "*ast*" package in the "*R*" software).

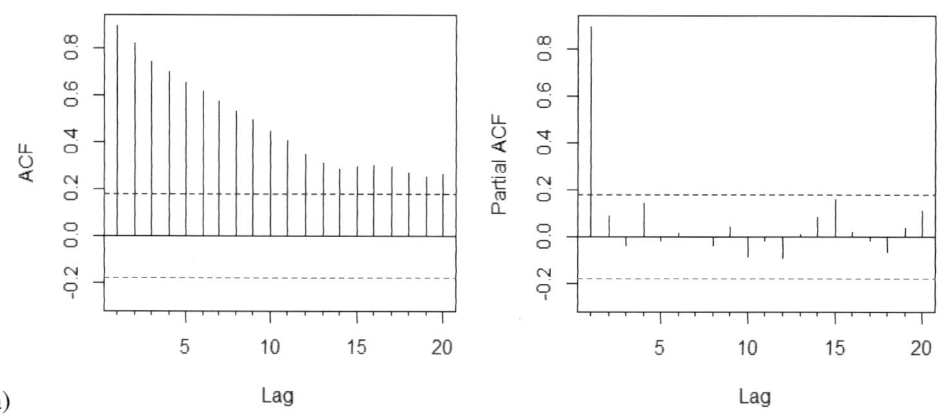

a)

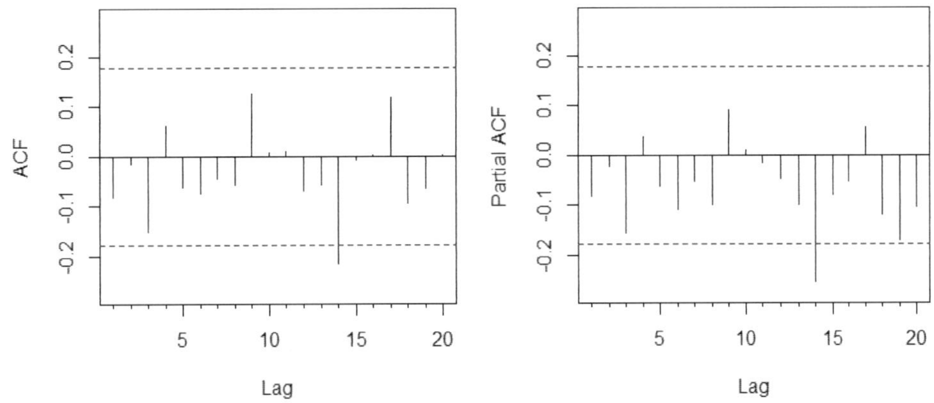

b)

Figure 7. Autocorrelation and Partial autocorrelation for CO a) observed series and b) differenced series in Solofra.

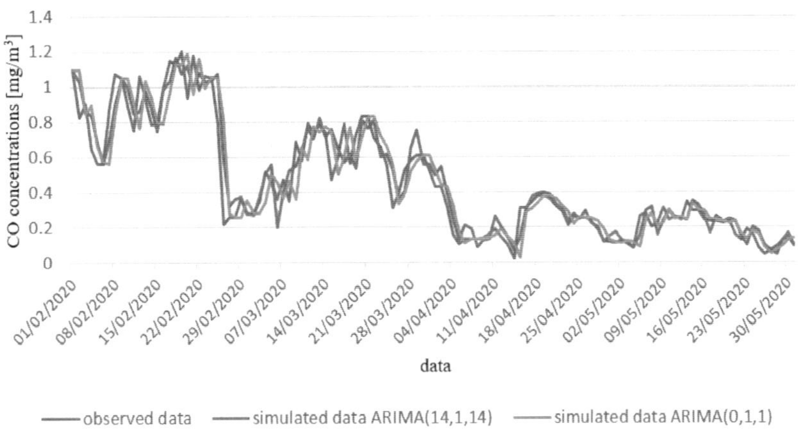

Figure 8. Plot of observed and simulated CO concentrations in Solofra, with ARIMA(0,1,1) and ARIMA(14,1,14).

The bisector plot shows that basically all the simulations obtained both with ARIMA(0,1,1) and ARIMA(14,1,14) fall within a one standard deviation interval from the observations, demonstrating that both model have very good performance.

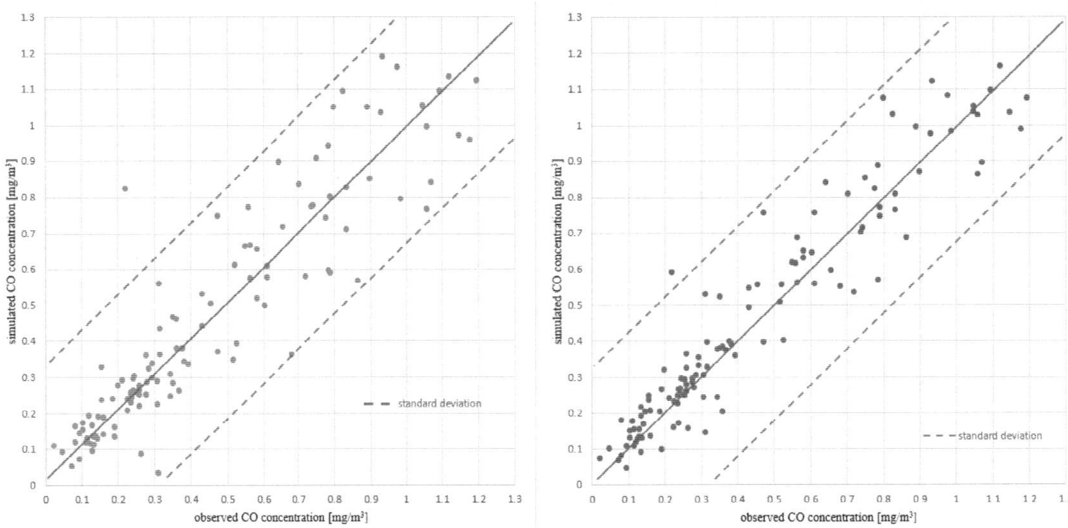

Figure 9. Scatter plot of observed and simulated CO concentrations in Solofra, with ARIMA(0,1,1) on the left, and ARIMA(14,1,14) on the right.

4. Conclusions

Italy has been one of the first European nations to be affected by COVID-19 in 2020 [17]. A wide range of governmental restrictions were adopted from February 2020 to mitigate the pandemic. In this regard, the present research set out to evaluate two aspects: the impact of lockdowns on the concentrations of particulate matter PM_{10} and CO in the Campania region, and how physics-mathematical models, as ARIMA, can perform good simulation of data. Thus, in order to compare short term effects on air quality before and after the coronavirus pandemic two monitoring stations in the Campania region, south Italy, were selected. They are located respectively in the industrial district of Solofra, a famous leather tanning area in the province of Avellino, and in the town of Nocera Inferiore, one of the most densely populated area in the province of Salerno with almost 2200 inhabitants per km^2. The aim of this choice was to analyse the effects in two different urban areas: one mainly industrial and another one, Nocera Inferiore, very mixed, characterized by the close presence of residential areas, schools, shops, highways and small industries.

The extent of the variations of the above mentioned pollutants has been evaluated by comparing data registered by local monitoring stations, related to the baseline February-May, of three different years, 2018, 2019 and 2020. As result is evident the limited or insufficient effectiveness of lockdowns in hardly reducing PM_{10} and CO concentrations. Then, by means of ARIMA models, the variation of the temporal trend of pollutants before (2018, 2019) and during COVID-19 lockdown (2020) has been assessed through the calibration and validation of models on interesting selected series: CO for Solofra and PM_{10} for Nocera Inferiore. In particular, the applied ARIMA models showed good performance in the simulation of data.

For further developments, authors will implement the selected models in order to assess the performance in the forecasting phase. Moreover, short and long term effects could be assessed, including also other pollutants, in order to evaluate if and how anti-COVID-19 restrictions influenced Italian citizens' habits.

References

[1] Cabaneros S, Calautit J K, Hughes B R. 2019.A review of artificial neural network models for ambient air pollution prediction.*Env Modelling & Software*, **119**, pp.285-304,ISSN 1364-8152

[2] Liao K ,Huang, X, Dang H, Ren Y, Zuo S, Duan C. 2021. Statistical Approaches for Forecasting Primary Air Pollutants: A Review. *Atmosphere*, **12**, 686.

[3] Guarnaccia C, Lenza TLL, Mastorakis NE and Quartieri J 2011 A comparison between traffic noise experimental data and predictive models results *Int J of Mech* **5**(4), pp. 379-386

[4] Bray CD, Nahas A, Battye WH, Aneja VP. Impact of lockdown during the COVID-19 outbreak on multi-scale air quality. 2021. *Atmos Environ (1994),* **254**, 118386.

[5] Thieriot H., Myllyvirta L. Air pollution returns to European capitals: Paris faces largest rebound.2020. Centre for Research on Energy and Clean Air (CREA)

[6] Guarnaccia C, Bandeira J, Coelho MC, Fernandes P, Teixeira J, Ioannidis G and Quartieri J 2018 Statistical and semi-dynamical road traffic noise models comparison with field measurements *AIP Conf. Proc.* **1982** pp1-6

[7] Guarnaccia C, Quartieri J, Tepedino C. 2017. Deterministic decomposition and seasonal ARIMA time series models applied to airport noise forecasting. Proc. of the Int. Conf. on Applied Mathematics and Computer Science, AIP Conference Proceedings 1836, pp. 1-7.

[8] Zhang X, Zhao M, Dong R 2020. Time-Series Prediction of Environmental Noise for Urban IoT Based on Long Short-Term Memory Recurrent Neural Network *Appl. Sci.* **10**, 1144, 1-18

[9] Guarnaccia C, Quartieri J, Rodrigues ER and Tepedino C 2014 Acoustical noise analysis and prediction by means of multiple seasonality time series model, *Int. J. Math. Mod. Meth. Appl. Sci.* **8**, 384-393

[10] Steele C 2001 A critical review of some traffic noise prediction models *Appl. Acoust.* **62** pp 271

[11] Guarnaccia C, Quartieri J, Tepedino C, Rodrigues E R. 2016. A time series analysis and a non-homogeneous Poisson model with multiple change-points applied to acoustic data. *Appl. Acou,* 114, pp. 203–212.

[12] Guarnaccia C, Cerón Bretón J G, Quartieri J, Tepedino C, Cerón Bretón C R. 2014. An Application of Time Series Analysis for Forecasting and Control of Carbon Monoxide Concentrations, *Int. J. of Math. Models and Methods in Applied Sciences*, **8**, pp 505-515.

[13] Cryer P D, Chan K. Time Series Analysis, with applications in R. 2008.Second Edition, Springer.

[14] R Core Team (2018). R: A language and environment for statistical computing. R Foundation for Statistical Computing, Vienna, Austria

[15] Gazzetta Ufficiale della Repubblica Italiana (Italian Republic) 2010. Ministerial Decree August 13, 2010 n.155. Available online at: https://www.camera.it/parlam/leggi/deleghe/testi/10155dl.html (last access July 20, 2021)

[16] https://web.unisa.it/en/campus-life/campus/photovoltaic-park (last access July 20, 2021)

[17] Rovetta A. 2021. The Impact of COVID-19 Lockdowns on Particulate Matter Emissions in Lombardy and Italian Citizens' Consumption Habits. Front. Sustain. 2:649715.

Acknowledgments
Authors would like to acknowledge Marialuisa Ciccone for contributing to this research during her bachelor thesis.

Morphological and chemical characterization of atmospheric particles PM10 in an urban site in Leon, Guanajuato, Mexico

J G Cerón Bretón[1], R M Cerón Bretón[1], A A Espinosa Guzmán[2], R C Lara Severino[1], M L Espinosa Fuentes[3], S E Carranco Lozada[4], E Ramírez Lara[5], C Guarnaccia[6] and V G Bautista Nolasco[1]

[1] Facultad de Química, Universidad Autónoma del Carmen (UNACAR), Calle 56 Num. 4, Esq. Ave. Concordia, Col. Benito Juárez, CP 24180, Ciudad del Carmen, Campeche, México

[2] Conacyt-Universidad Autónoma de Campeche, Av. Héroe de Nacozari No. 480, Campus 6 de Investigaciones, C.P. 24070 San Francisco de Campeche, Campeche, Mexico

[3] Centro de Ciencias de la Atmósfera (CCA), Universidad Nacional Autónoma de México (UNAM), Investigación Científica s/n, C.U., Coyoacán, 04510 Ciudad de México, CDMX

[4] Instituto Politécnico Nacional, IPN- CECyT 15, Doctor Gastón Melo 41, Tenantitla, Milpa Alta, 12100, San Antonio Tecómitl, CDMX, Mexico

[5] Facultad de Ciencias Químicas, Universidad Autónoma de Nuevo Léon, Ave Universidad S/N Cd. Universitaria, C.P. 66455 San Nicolás de los Garza, Nuevo León, Mexico

[6] Civil Engineering Department, University of Salerno, Via Giovanni Paolo II, 132 84084 Fisciano (SA), Italy

Abstract. Start PM10 concentrations and their trace metals content (Cd, Co, Cu, Fe and Zn) were determined in an urban site of Leon, Guanajuato during the rainy season 2018. PM10 exceeded the maximum permissible levels established by the Mexican Standards and WHO, constituting a potential risk to population health. Trace metals levels in PM10 were analyzed by Atomic Absorption Spectrometry, and in addition, morphology and elemental content were studied for some selected particles by Scanning Electronic Microscopy and Energy Dispersive Spectrometry. Chemical and morphological characterization revealed that Fe was the more abundant metal (0.986 $\mu g\ m^{-3}$), followed in order of importance by Zn (0.243 $\mu g\ m^{-3}$), due to theses metals are related to the crustal. Lower concentrations were found for Cu (0.140 $\mu g\ m^{-3}$), Cd (0.119 $\mu g\ m^{-3}$), Mn (0.070 $\mu g\ m^{-3}$) and Co (0.043 $\mu g\ m^{-3}$). Meteorological analysis showed that sources located at the SW and NE of the sampling site (industrial parks and facilities related to metals and mineral extraction) influenced the measured concentrations. Enrichment factors showed that all the measured metals were highly influenced by anthropogenic activity. Cancer Risk (CR) and Non-Cancer Risk Coefficients (cardiovascular and respiratory diseases) did not exceed the maximum permissible level established by EPA.The first section in your paper.

1. Introduction

The particulate matter present in the atmosphere is a set of substances in solid and liquid state, which is suspended in the air or, which are deposited on the surface very slowly when they reach a larger diameter. Some examples of particulate matter are soot, aerosols, sand, dust, etc. [1]. The increase in the concentration of atmospheric particles in large urban areas is related to negative effects on human health [2], adding complexity to the fact that these particles contain a variable and complex mixture of different chemical elements and compounds that come from different sources [3]. Thus, PM10 atmospheric particles refer to particulate matter in size less than or equal to 10 micrometers, which may contain various elements, among which are trace metals, which play a very important role in human health problems because they are highly bio reactive and some of them are toxic for humans [4], and vegetation [5]. The metals present in the particles are found in the fine and ultrafine fractions (in the order of micrometers in diameter) [6], [7], [8], that facilitates their entry into the alveolar region of the lungs when they are breathed [9]. Despite its importance from a health point of view, the content of metals in PM10 particles is not regulated in Mexico, only the mass concentration of particles expressed in $\mu g\ m^{-3}$ is regulated (Mexican standard: NOM-025-SSA1-2014) [10].

Atmospheric monitoring is one of the main indicators of air quality, in addition to representing a management tool to implement prevention and control actions for pollutants present in the atmosphere. In Mexico, automatic monitoring is carried out in the main cities of the country, usually measuring CO, O_3, NO_2, SO_2, PM10 and PM2.5; however, only the mass concentration of the particles is reported and no chemical speciation of their content is made. Some studies have been reported in Mexican cities on the levels of heavy metals in PM10, but most of these studies have focused on large metropolitan areas such as Monterrey and Mexico City [11], [12], [13], [14] and in the south of the Mexican Gulf [15]. In the center of the country there are some commercial and industrial areas with high demographic density, such is the case of Bajio region, where Leon is the most important city in this area. The city of León, has experienced an accelerated economic and industrial development that has resulted in increased levels of pollution and deterioration in air quality, reaching concentrations that may exceed the maximum permissible limits. Leon city is famous worldwide thanks to its shoe industry, and it is among the 5 cities with the highest number of inhabitants in the entire Mexican Republic (1 million 578 thousand 626 inhabitants). In this city, diverse economic vocations are consolidated, such as highly specialized health care and diagnosis institutions, higher education and research centers, as well as tourism and business activities. Therefore, the relative potential effects of the degradation of air quality on the health of population constitute one of the main concerns in this region. For this reason, this research project focuses on determining the concentrations in the atmosphere of particulate matter (PM10) of an urban site in the city of Leon, its content of heavy metals (Cd, Co, Cu, Fe and Zn) and morphology, to infer its possible sources and evaluate the risk to the health of the population in the study site. This study represents the first project focused on the speciation of some trace metals contained in PM10 particles in the city of Leon, Guanajuato in order to know in a preliminary way the health risk associated with the concentration levels found. This information will be useful to know not only the gravimetric concentration of the particles but also their metal content, since currently, none of the air quality monitoring stations in the city of Leon determine the content of metals in PM10. In addition, this information will serve as a basis for the future expansion of this monitoring network of trace metals in PM10 to other cities of the state of Guanajuato such as Celaya and Salamanca, since in the latter there are important sources of pollutant emissions to the atmosphere: an oil refinery and a thermoelectric plant. This will allow a comprehensive evaluation of health risks, since currently only the mass concentration of the particles is considered to evaluate the risk that this pollutant represents to human health.

2. Materials and Methods

2.1. Study Area

The municipality of Leon, is located in the state of Guanajuato. The territorial area of this municipality is 1,219.67 km2, with an average height of 1,798 masl. This city limits to the north with the municipality of San Felipe, to the east with Guanajuato and Silao, to the south with the municipalities of Silao, Romita and San Francisco del Rincon and to the west with the municipalities of Purisima, Lagos de Moreno and the Unión de San Antonio (Municipal Government Program of Leon,

Guanajuato, 2018) [16]. The study site is located in an urban area in the center of the city of Leon, within the facilities of one of the 3 air quality monitoring stations belonging to the State Air Quality Information System (SEICA) at 21.133° N and 101.680° W (See Figure 1). In the surroundings near the study site is located the General Hospital of Leon, leather tanning factories and streets that connect with the main avenues of the city.

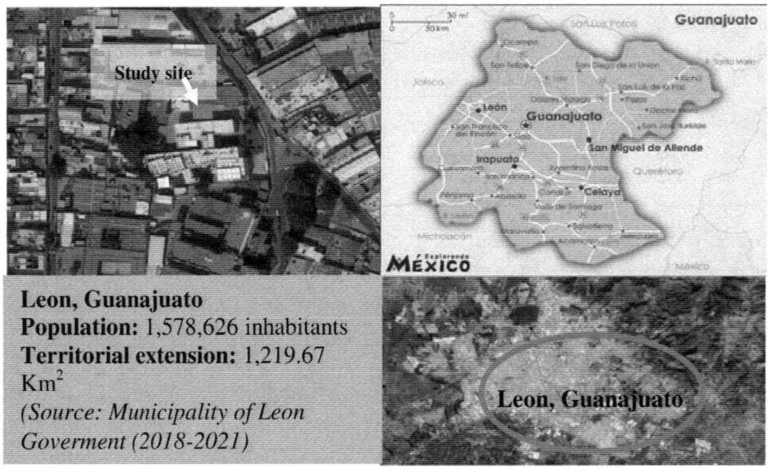

Figure 1. Sampling site location.

2.2. *Sampling Methodology*

The sample collection process was carried out using an air sampler, MiniVol TAS from Airmetrics, which is a portable sampler for particulate matter and non-reactive gases. The sampling technique used by the MiniVol is a modification of the PM10 reference method described in the United States Code of Federal Regulations (40 CFR part 50, Appendix J) [17]. The sampling was carried out during the rainy season 2018. The sampler was placed in the place of interest, then a Whatman® brand 47 mm quartz fiber filter was placed into the filter holder and the equipment was operated at a controlled flow of 5 L / min of ambient air, which passed through a particle size separator and then through the filter during a sampling period of 24 h. This process was made every day during the 8-sampling days from August 6 to August 13, 2018, obtaining a total of 8 samples during the study period.

2.3. *Analysis Methodology*

The filters were subjected to acid digestion with 10 mL of aqua regia (25mL HNO_3 + 75 mL HCl) and 1.065 mL of $HClO_4$ and heating at 60 °C for 70 minutes, until dryness. Then 20 mL of hot water were added, filtered and made up to 50 mL using deionized water, storing the extracts in polypropylene containers until later analysis [18]. Standard solutions of Cd, Co, Cu, Mn, Fe and Zn were prepared from 1000 ppm HYCEL brand standard solutions for atomic absorption. For each metal, from each stock solution, 7 dilutions were prepared at different concentrations and calibration curves were constructed. The calibration range adopted was 5–500 μg L^{-1} (seven point calibration) for each metal; the correlation coefficients (R^2) for the calibration curves were all greater than 0.998. A Thermo Scientific ™ iCE 3000 ™ Series AAS Atomic Absorption Spectrophotometer was used for samples analysis. The measurements were carried out according to the standard conditions recommended at the specific wavelengths for each metal according to standard analytical US EPA Methods IO.3.1 [17] and I.O.3.2 [20]. A deuterium lamp was used as a background corrector in all measurements [19]. Limits of detection (LODs) of trace metals were in the range of 0.003–0.06 μg L^{-1} and recovery percentages of trace metals from the spike method (n = 3) ranged from 97.4% to 104.56%, and expanded uncertainties ranged from 0.01262 to 0.05234. In order to verify reproducibility and low background metal concentrations of reagents and filters, 5% of the total number of samples were taken as blanks and analyzed for the presence of trace metals.

2.4. *Morphological characterization of PM10 using Scanning Electron Microscopy*

The morphology of the particles and their elemental composition with respect to metal content was evaluated using a Hitachi FLEXSEM-SU1000 (Scanning Electron Microscope) scanning electron microscope equipped with an Energy Dispersive (EDS) X-ray detection system TM 40000 Quantax 75/80 of Bruker that works at 20 kV, located in the Corrosion Research Center of the Autonomous University of Campeche. The low vacuum scanning electron microscope was calibrated with a copper (Cu) grating, a filament current of 300 mA, and a working distance of 5 cm. A 1 mm x 0.5 mm rectangle was cut in each filter for analysis and these sections of the filters were analyzed by SEM-EDS. Ten particles from each sample were randomly selected and images were taken at magnifications of 500X, 1500X, 2500X, 5000X 8000X, 10,000X, and 25,000X. Spectra for individual particles were obtained by EDS, with a measurement time of 100 s for each particle.

2.5. *Morphological characterization of PM10 using Scanning Electron Microscopy*

Pearson's correlation was used to identify bivariate relationships between measured metals, criteria pollutants, and meteorological variables. A variance analysis was applied to the data set using non-parametric Friedman test ($\alpha = 0.05$) in order to determine if there were significant differences among the trace metals contained in PM10. Principal component analysis (PCA) was performed to explain multivariate variation and discover the structure of the data set. This method has been widely used in environmental studies to identify patterns in data [22]. ACP results generally are represented in bi-plots or factor loading tables, which reveal correlations between observations. The information disclosed by ACP is useful to identify whether a pollutant is secondary or primary, or to identify the specific source of the pollutants.

2.6. *Meteorological Analysis*

Wind roses were built for each of the sampling days, in order to know the distribution of the wind direction and speed at the study site, using the WRPLOT software (Wind Rose Plots for Meteorological Data developed by Lakes Environmental, available at: https://www.weblakes.com/products/wrplot/index.html) [23]. This analysis let know the dominant direction of wind, which was related to the levels of concentrations of measured air pollutants in order to infer the probable location of sources of this pollutants.

2.7. *Health Risk Assessment*

Exposure has been expressed in terms of lifetime average daily dose (LADD). This value makes it possible to calculate the corresponding level of cancer risk (CR) for each metal, considering two representative groups: adults and children. LADD is used to determine the amount in which a pollutant has adverse health effects when it is absorbed by the body over a long period of time, and is calculated with the following equations [24]:

$$LADD = E \times C \tag{1}$$

$$E = \frac{IR}{BW} \times \frac{ET \times EF \times ED}{AT \times 365} \tag{2}$$

Where for equation (1), C (mg m^{-3}) is the concentration of the metal in question in PM10, which is assumed to be the same at the point of exposure, while E (m^3 Kg^{-1} day^{-1}) is obtained from equation (2), where IR (m^3 h^{-1}) is the air inhalation rate, ET (24 h day^{-1}) is the exposure time, EF (350 days $year^{-1}$) is the exposure frequency, ED (years) is the duration of exposure, BW (Kg) is the body weight and finally AT (days) is the average time, used as ATc for carcinogenic risk and as ATn for non-carcinogenic risk. The parameters used for the estimation of cancer risk (CR) and non-cancer risk coefficients (HQ: hazard quotient) are shown in Table 1.

Table 1. Parameters used for the estimation of cancer risk (CR) and non-cancer risk coefficients (HQ: hazard quotient)

Parameters	Symbol	Unit of measurement	Values	
			Adult	Children
Inhalation rate	IR	m^3/h	0.9	0.7
Body weight	BW	Kg	70	15
Exposure time	ET	h/day	24	24
Exposure Frequency	EF	Day/year	350	350
Exposure duration	ED	years	24	6
Average Time	*ATc	days	25550*	25550*
Average time	*ATn	days	210240**	52560***

* 25550 days corresponds to the age according to the typical life expectancy (70 years x 365 days / year); ** ED (24 years) multiplied by 365 days / year x 24 h / day; *** ED (6 years) multiplied by 365 days / year x 24 h / day

CR represents the increased probability of occurrence of diseases caused by tumors above the general average due to the impact of compounds of those that produce carcinogenic effects. The CR for carcinogenic substances is considered 10^{-6} (lifetime risk of developing cancer is 1 in 1,000,000). Values below 10^{-6} for individuals are considered negligible. For carcinogenic substances the Cancer Risk (CR) is determined with the following equation:

$$CR = LADD \times CSF \tag{3}$$

where CR is the probability of cancer occurrence in the exposed population during a life time of 70 years; It is determined by multiplying the LADD (mg / Kg day) and CSF (cancer slope factor) (mg / Kg day). Carcinogenic risk is defined as the increased probability of a person experiencing cancer during a lifetime as a result of exposure to a specific carcinogenic potential. The CSF is calculated with the following equation:

$$CSF = IUR \times \frac{BW}{(IR \times ET)} \times 1000 \tag{4}$$

IUR (Inhalation Hazard Unit) is a reference value reported in the database of the Environmental Protection Agency [25]. Table 2 shows the values for IUR and RfC. Regarding the inhalation risk units (IUR) and the reference concentrations (RfC), only values for Cd, Co and Mn are reported. Regarding the non-carcinogenic risk, the HQ (risk coefficient) is calculated as reported in equation 5:

$$HQ = \frac{ADI}{RfDi} \tag{5}$$

For HQ it is considered that there is an exposure level (RfDi) below which it is unlikely even for a sensitive population to experience adverse health effects. When the exposure level (ADI) exceeds the value indicated as 1, there may be concern about possible non-carcinogenic health effects; higher THQ values (> 1) suggest higher levels of concern. RfDi (mg / Kg day) represents the inhalation dose at which there are no adverse effects and is defined by equation 6:

$$RfDi = RfC \times \frac{20\, m^3}{day} \times \frac{1}{70\, Kg} \; (Using\ 7.6\, m^3\ and\ 15\, Kg\ for\ children) \tag{6}$$

Using 7.6 m^3 and 15 Kg for children. ADI (mg/ kg day) is the estimated dose that the recipient receives from exposure to polluted air [24], and is calculated with the same variables used for cancer risk:

$$ADI = E \times C \tag{7}$$

Table 2. Inhalation Risk Unit (IUR) and Reference Concentration (RfC) (EPA) [22]

Metal	**CAS**	**IUR(μg m^{-}□□ ^{1}**	**RfC (mg m^{-}□**
Cd	7440-43-9	1.80×10^{-3}	1.00×10^{-5}
Co	7440-48-4	9.00×10^{-3}	6.00×10^{-6}
Mn	7439-96-5	…………..	5.00×10^{-5}

* CAS unique number of the chemical element

3. Results and Discussion

3.1. Gravimetric concentrations of PM10

The morphology of the particles and their elemental composition with respect to metal content was evaluated using a Hitachi PM10 concentrations recorded during the study period were within the range of 17.48 (μg m^{-3}) to 45.32 (μg m^{-3}), with an average of 28.04 (μg m^{-3}). The maximum permissible limit for PM10 established in the official Mexican standard (NOM-025-SSA1-2014) [10] is 75 (μg m^{-3}), whereas, the guidelines of the World Health Organization (WHO, 2005) [24] establish a limit of 50 (μg m^{-3}). During the study period these limits were not exceeded; the maximum value recorded was 45.32 (μg m^{-3}).

3.2. Trace metals concentrations in PM10

Six metals were determined in this study: Cd, Cu, Co, Mn, Zn and Fe. The results can be observed in Figure 2. Fe and Zn were the metals with the highest concentrations 0.986 μg m^{-3} and 0.243 μg m^{-3}, respectively. Major elements such as iron and zinc can come from the re-suspension of dust when passing vehicles and from its combustion in the case of diesel vehicles. Its higher proportion compared to other metals is explained by its geological origin, being part of the dust re-suspended in the ground by the wind and transport [27]. This agrees with the results observed in other studies as shown in Table 3, where these metals were also reported with higher concentrations. According to Table 3, the results observed for Copper (Cu) are higher than those reported in other studies carried out in Mexico (Hidalgo, Colima, Tamaulipas and Puebla), however, they reach similar concentrations with respect to studies carried out in Bogotá, Colombia and Agra, India. Cu is a metal that occurs naturally throughout the environment, in rocks, soil, water, and air. It is an essential element in plants and animals (including humans). Copper compounds are commonly used in agriculture to treat plant diseases such as mold, to treat water, and as preservatives for wood, leather, and fabrics.

Sources of exposure are smoke, copper smelting and related metallurgical operations, welding, and metal powders and copper salts [28]. In particular, copper emissions into the atmosphere can be related to the burning of fossil fuels (vehicle sources), confirming results with other studies [29]. The high concentrations of copper can be attributed to the presence of metallic copper particles, which is emitted by nearby bronze smelting processes [30]. The average concentration of copper present in the air is 0.05 μg m^{-3} but in polluted areas it can reach up to a thousand times that value. The cadmium (Cd) concentrations obtained in this study are higher than those observed in other cities in Mexico (Colima, Tamaulipas and Puebla) and in other cities around the world (Venice, Seoul and Bogotá). On the other hand, the results obtained are similar to those found in a city in China (Nanchang) and Iran (Agra) (Table 3). Cadmium is naturally present in the environment in the air, soils, sediments and even in uncontaminated seawater [31]. The main sources of pollution are non-ferrous metal mining metallurgy, iron and steel metallurgy, the manufacture of phosphate fertilizers, the incineration of waste wood, coal or plastics, the combustion of oil and gasoline and the applications industrial cadmium. Cd concentration in air in industrial areas ranges from 9.1 to 26.7 μg m^{-3}, from 0.050 to 0.060 μg m^{-3} in urban areas and from 0.0001 to 0.006 μg m^{-3} in air in rural areas. Cd does not degrade and, once released into the environment, it remains in circulation [32]. The average value of Cd found in this study exceeds the maximum permissible levels established by the US EPA (11.8 times more), WHO (23.8 times more) and the European Community (23.8 times more) (See Table 4). Manganese (Mn) concentrations were higher than those found in other states of Mexico such as Hidalgo, Colima and Puebla and cities around the world such as Seoul and Venice, but lower than those found in Tampico, Bogotá and Agra (Table 3). Mn is used in the production of steel, carbon steel, stainless steel and high-temperature steel, cast iron, superalloys, dry cell batteries, matches, and fireworks [28]. It is a natural element that is found in low levels in water, air, soil and food. Manganese can be released into the air by industry or by burning fossil fuels. Elevated levels of manganese can be present in the air near iron and steel production plants, power plants and coke ovens and dust generated by uncontrolled mining operations. Manganese released from burning a gasoline additive also represents a source of manganese in the air. The most common route of exposure to high doses of manganese is through inhalation of contaminated air [33]. According to the Toxicological Profile of the ATSDR (Agency for toxic substances and disease registry) the average levels of manganese in the air are approximately 0.02 μg m^{-3}. On the other hand, Chen and Lippmann (2009) [34] determined that the average concentration of manganese in fine particles (PM 2.5) in 13 cities of the USA (AIRS sites) was of 0.004 μg m^{-3}. The average manganese concentrations found at the study site are 4.6 times higher than the maximum allowable level established by the World Health Organization (WHO) (Table 4). The Cobalt (Co) concentrations obtained in the present study are higher than those found in Puebla, the Metropolitan area of Mexico City and Venice. Lesser results are observed with respect to those found in Agra, India (Table 3). Cobalt is produced primarily as a by-product of mining and processing minerals such as copper and nickel. Various cobalt compounds are used as pigments in the glass and ceramic industries, as drying agents for paints, varnishes, printing inks, catalysts in the petroleum and chemical industries, and as trace metal additives for agricultural and medical uses. Traces of cobalt can be present in cement and various household products. Cobalt concentrations in ambient air are usually below 0.01 μg m^{-3} [35]. Higher cobalt levels can occur in air and water near industrial areas, particularly near industrial hard metal sites (0.61 μg m^{-3} has been reported in an industrial area) [28].

Table 3. Comparison of PM10 concentrations and trace metal content with other studies.

	Location	PM10	Cd	Cu	Zn	Fe	Mn	Co
		μg m^{-3}						
This study	Leon, Guanajuato (urban)	28.04	0.12	0.14	0.24	0.99	0.07	0.04
Studies in Mexico	Tampico, Tamaulipas [11] (urban)	23.44 - 60.67	0.0003	0.03	0.04	0.24	0.26	-
	Metropolitan Area Of Mexico City [36] (urban)	50 - 56	0.0010	0.0750	0.10	1.00	0.02	0.0005
	Puebla,	55.9	0.005	0.09	-	1.20	0.03	0.002

	Mexico [12] (industrial)							
	Tlaxcoapan, Hidalgo [13] (industrial)	45	-	0.012	0.02	0.51	0.015	-
	Colima, México [37] (rural)	48	0.0002	0.026	0.03	0.80	0.018	-
Studies in other regions around the world	Bogotá, Colombia [27] (industrial)	65 - 100	0.026	0.14	0.93	3.15	0.41	-
	Agra, India [38] (urban)	154.2	-	0.1000	0.50	2.90	0.90	-
	Seoul, Korea [39] (residential)	66.4	0.0041	-	0.12	-	0.02	-
	Venice, Italy [40] (urban)	46	0.003	0.04	0.10	0.91	0.02	0.0007
	Nanchang, China [41] (industrial)	-	0.11	0.05	0.24	-	-	-
	Ahvaz City, Iran [42] (urban)	116.5	0.19	-	2.81	-	-	0.29

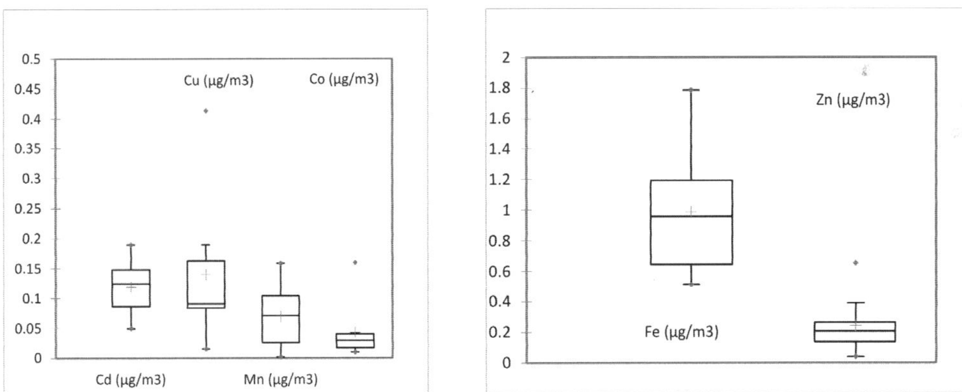

(a)

(µg/m^3)	Cd	Cu	Mn	Co
Minimum	0.049	0.016	0.002	0.010
Maximum	0.190	0.413	0.159	0.160
Mean	0.119	0.140	0.070	0.043
Standard deviation (n-1)	0.051	0.122	0.053	0.049

(b)

(µg/m^3)	Fe	Zn
Minimum	0.512	0.040
Maximum	1.785	0.653
Mean	0.986	0.243
Standard deviation (n-1)	0.423	0.197

◆ are the maximum and minimun values; + are the mean values, the center line in the box are the medians, the edges of the box represent the first and third quartiles. The box is only represented for better visualization.

Figure 2. Boxplot and descriptive statistics for (a) Cd, Cu, Co and Mn concentrations; (b) Fe and

Zn concentrations; present in PM10 in the study site

Table 4. Maximum permissible limits in ambient air for some of the trace metals present in PM10 established in international standards

Trace metal	Maximum permissible level (μg m^{-3})	Average Time	Regulation Organism
	0.0100	Annual	USEPA [43]
Cd	0.0050	Annual	WHO [44]
	0.0050	Annual	European Community [45]
Mn	0.0150	Annual	WHO [46]

3.3. Influence of wind conditions on the trace metals concentrations

From the wind roses analysis it was determined that the dominant wind during the study period blew from the SW. Figure 3 shows a representative wind-rose for the study site, as it can be observed the dominant wind came from SW. To the SW of the study site there are three tanneries (facilities focused on process that converts the skin of animals into leather), this process is associated with the release of copper into the atmosphere due to the additives that are used to preserve the leather [28]. In this direction is also Adolfo Lopez Mateos Boulevard, one of the main avenues of the city with high vehicular traffic. The burning of fossil fuels such as gasoline contributes significantly to the emissions of heavy metals into the atmosphere (Co, Cu, Mn, Zn and Cd). To the NW of the study area there is another important avenue: Paseo de los Insurgentes, which connects to the Lagos de Moreno highway where around thirteen thousand five hundred cars circulate daily. In this direction there are also foundry facilities for metal smelting such as iron, bronze and aluminum.

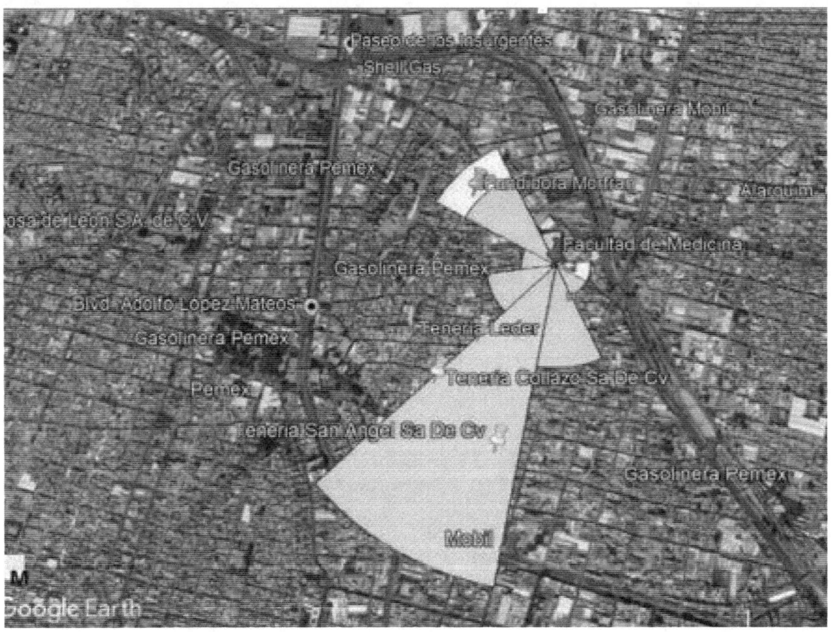

Figure 3. Representative wind rose for the study site during the sampling period.

Figure 4 shows the behavior of the concentrations of the metals measured in PM10 with respect to the prevailing winds during the study period. Fe showed higher concentration values when the wind blew from the West. Cd and Zn, showed higher levels when the winds blew from the South. Cu and Co concentrations were higher when the winds blew from the Southwest. Mn concentrations were higher when the wind came from the East. Sources located in these directions could contribute to the trace metals measured in PM10 in the study site.

It is important to mention that the study site is located within the facilities of the Faculty of Medicine and next to the General Hospital of the city, which is why it is considered that vehicular activity is important in that area. Being an urban area, many gasoline service stations are located near the sampling site. All these sources could contribute to the levels of trace metals in PM10 particles during the study period.

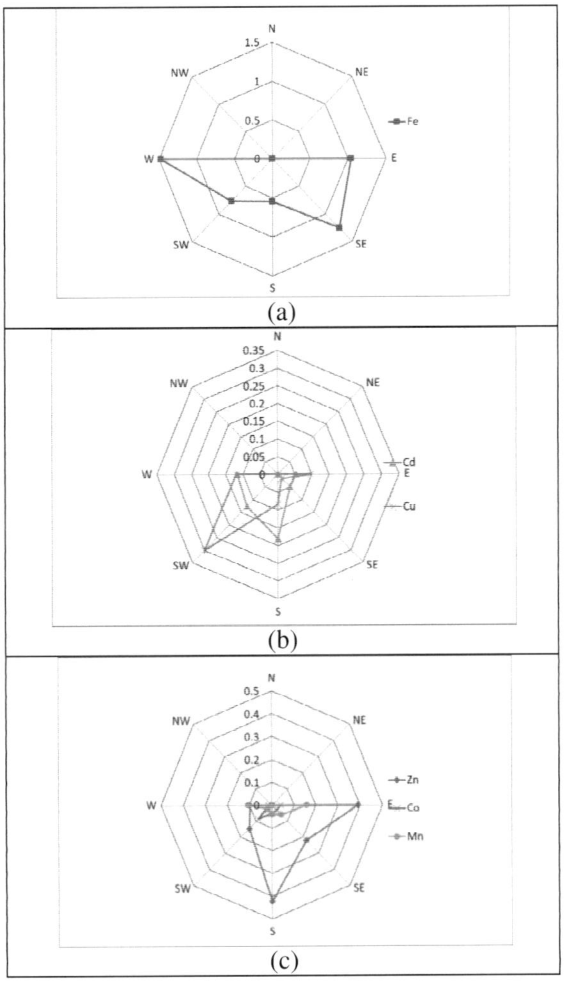

Figure 4. Relation between the prevailing winds and the trace metals concentrations for (a) Fe, (b) Cd and Cu and (c) Zn, Co and Mn

3.4 Statistical Analysis

The results of the non-parametric Friedman test at a level of significance of $\alpha = 0.05$ showed that there were significant differences between Fe with Cu (p-value = 0.047) and Fe with Mn (p-value= 0.009) and between Co with Fe (p-value <0.0001) and Co with Zn (p-value=0.039), indicating that these metals had their origin in different sources. Pearson's analysis represented in Table 5 shows that there is a positive correlation between Cd, O_3, and relative humidity (RH), as well as negative correlations between Cd-Mn, and temperature (TEMP) -wind speed (WS). A negative correlation with wind speed may indicate that, at higher speeds, the concentrations of pollutants decrease due to the phenomenon of dispersion by the action of the wind. With respect to Cu, it can be observed that it had an important positive correlation with Co, which indicates that they could have sources in common. Cu was also significantly correlated with PM10, indicating that these pollutants could have originated from the same source. Table 5 shows that Mn positively correlates with Fe, suggesting that they were originated in common sources, since both are tracers of the cortex. Mn negatively correlated with RH indicating that manganese is a soluble element, therefore it can be inferred that a washout occurs in the column of the atmosphere, which causes Mn to decrease with increasing relative humidity. Fe positively correlated with SO_2 and TEMP, which indicates that Fe may also have had anthropogenic sources of enrichment. Zn negatively correlated with SO_2, CO, NO_2 and PM10, indicating that these pollutants could be precursors of the Zn present in PM10 particles. SO_2 correlated positively with NO_2, PM10 and TEMP, indicating that these pollutants could have a strong influence from sources of combustion at high temperatures and sources of combustion that use highly sulfurized fuels such as diesel and fuel oil. CO significantly correlated positively with NO_2, HR and WD, indicating that these pollutants had their origin in vehicular sources and could be transported from nearby regions. NO_2 showed significant positive correlations with PM10 and WD, indicating that they had sources in common at the study site.From the analysis of principal components, it can be observed that two factors F1 and F2 were necessary to explain the 65.49% of the total variability of the data. Figure 5 shows the biplot including the two principal components F1 and F2. Factor F1 includes Cd, Mn, CO, HR, WS and WD (Table 6), which correlated with each other, indicating that they could have originated from common sources. CO is a tracer of vehicular combustion sources, for which Cd and Mn could be originated from vehicular emissions. The correlation with WS, WD, HR indicates that these pollutants were influenced by the transport of air masses from other areas and by the high moisture content. Factor F2 includes Fe, Zn, SO_2, NO_2, PM10 and TEMP (Table 6), which correlated with each other, indicating that they could have originated from common sources. NO_2 is a tracer of vehicular combustion sources and sources of combustion at high temperatures, which indicates that Zn, Fe and PM10 were enriched due to the vehicular activity that exists around the area. SO_2 is a tracer of industrial emissions from combustion sources that use highly sulphur fuels, so Zn and Fe could also have their origin in industrial sources. Factor F3 includes Co and O_3, showing a significant correlation with each other, indicating that these pollutants were influenced by photochemical activity. Factor F4 (Table 6) included only Cu, indicating that this metal had its origin in a different source, possibly from the metal-mechanical industry and the assembly of metal parts.

Table 5. Pearson Correlation Matrix for Cd, Cu, Mn, Zn, Fe, Co, criteria air pollutants (O_3, SO_2, CO, NO_2 and PM10) and meteorological parameters (Temperature, Relative Humidity, Wind speed and Wind direction).

	Cd	**Cu**	**Mn**	**Co**	**Fe**	**Zn**	**O_3**	**SO_2**	**CO**	**NO_2**	**PM10**	**TEMP**	**HR**	**WS**	**WD**
Cd	**1**	0.11	**-0.53**	-0.08	-0.48	0.12	**0.82**	-0.34	0.08	-0.03	-0.09	**-0.59**	**0.58**	**-0.55**	0.419
Cu	0.11	**1**	-0.37	**0.82**	-0.09	-0.43	-0.14	0.42	0.25	0.42	**0.53**	0.07	-0.01	-0.17	0.436
Mn	**-0.53**	-0.37	**1**	-0.30	**0.54**	0.28	-0.06	0.23	**-0.60**	-0.49	-0.07	**0.72**	**-0.75**	0.46	-0.283
Co	-0.08	**0.82**	-0.30	**1**	-0.16	-0.18	-0.32	0.02	0.05	0.12	0.21	-0.03	-0.10	-0.24	0.085
Fe	-0.48	-0.09	**0.54**	-0.16	**1**	-0.49	-0.12	**0.58**	-0.17	0.07	0.14	**0.59**	-0.19	0.43	0.219
Zn	0.12	-0.43	0.28	-0.18	-0.49	**1**	0.32	**-0.70**	**-0.51**	**-0.79**	**-0.54**	-0.29	-0.18	-0.39	-0.490
O_3	**0.82**	-0.14	-0.06	-0.32	-0.12	0.32	**1**	-0.27	-0.17	-0.32	-0.15	-0.43	0.45	**-0.58**	0.459
SO_2	-0.34	0.42	0.23	0.02	**0.58**	**-0.70**	-0.27	**1**	0.34	**0.63**	**0.78**	**0.66**	-0.24	0.48	0.491
CO	0.08	0.25	**-0.60**	0.05	-0.17	**-0.51**	-0.17	0.34	**1**	**0.90**	0.70	-0.33	**0.56**	-0.17	**0.572**
NO_2	-0.03	0.42	-0.49	0.12	0.07	**-0.79**	-0.32	**0.63**	**0.90**	**1**	**0.78**	0.004	0.36	0.13	**0.583**

PM10	-0.09	**0.53**	-0.07	0.21	0.14	**-0.54**	-0.15	**0.78**	0.70	**0.78**	**1**	0.24	0.02	0.06	0.707
TEMP	**-0.59**	0.07	**0.72**	-0.03	**0.59**	-0.29	-0.43	**0.66**	-0.33	0.004	0.24	**1**	**-0.83**	**0.84**	-0.187
HR	**0.58**	-0.01	**-0.75**	-0.10	-0.19	-0.18	0.45	-0.24	**0.56**	0.36	0.02	**-0.83**	**1**	**-0.65**	**0.545**
WS	**-0.55**	-0.17	0.46	-0.24	0.43	-0.39	**-0.58**	0.48	-0.17	0.13	0.06	**0.84**	**-0.65**	**1**	-0.397
WD	0.41	0.43	-0.28	0.08	0.21	-0.49	0.45	0.49	**0.57**	**0.58**	0.70	-0.18	**0.54**	-0.39	**1**

Values in bold are different from zero at a significant level of alfa=0.05

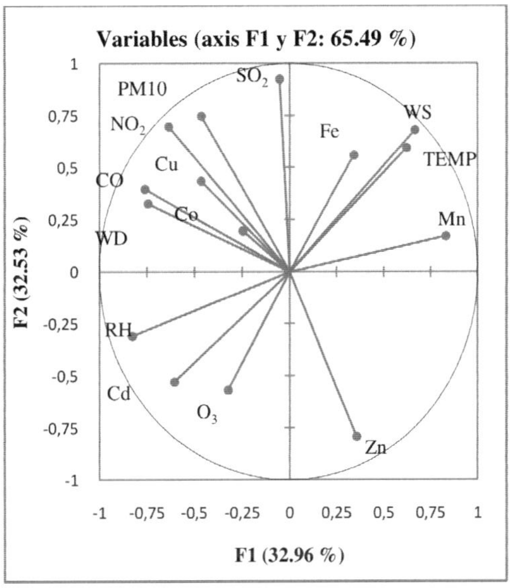

Figura 5. Biplot resulting of the multivariate statistical analysis applied to the measured variables, showing the two principal components: F1 and F2.

Table 6. Principal Component Analysis for Cd, Cu, Mn, Co, Fe, Zn, criteria pollutants and meteorological parameters.

	F1	F2	F3	F4
Cd	**0.364**	0.280	0.056	0.070
Cu	0.216	0.190	0.265	**0.294**
Mn	**0.692**	0.029	0.106	0.077
Co	0.061	0.038	**0.632**	0.191
Fe	0.116	**0.313**	0.243	0.024
Zn	0.126	**0.625**	0.008	0.014
O^3	0.105	0.322	**0.345**	0.210
SO_2	0.003	**0.857**	0.059	0.029
CO	**0.580**	0.157	0.000	0.186
NO_2	0.402	**0.487**	0.000	0.102
PM10	0.214	**0.559**	0.005	0.022
TEMP	0.443	**0.464**	0.006	0.043
HR	**0.683**	0.096	0.067	0.059
WS	**0.386**	0.354	0.003	0.084
WD	**0.555**	0.107	0.185	0.130

Values in bold correspond for each variable to the factor for which the cosine squared is the highest

3.5. Morphological characterization of PM10 and SEM-EDS Analysis

The properties of the particles such as size, density and surface can have an important influence on their behavior in the air and their effects on the environment and health [47]. On the other hand, the properties of the particles (shape, size and texture) are also indicative to determine the origin of the particle, that is, if it is the result of natural or anthropogenic emissions. The particles that come from a natural source tend to have a different structure and composition than those emitted by anthropogenic sources [48]. The individual particles analyzed in this work present different morphologies, which include irregular, amorphous, spherical and groups of small particles that form larger particles (clusters / agglomerates). For this study, around 75 particles were analyzed, for practical purposes only the 3 most representative ones are shown. Figure 6 shows the images of the SEM-EDS analysis of said particles and Table 7 shows the elemental content of these selected particles. Figure 6a shows a point particle with an irregular shape with a shiny appearance identified as 1705 which has a high content of Pb, Si and O, in a lesser proportion it has C, Na, Cd, Al, Ca and Mg. The mass percentages found are shown in Table 7. The particle with a regular spherical shape with a shiny appearance identified as 1736 shown in figure 6b, is a particle rich in Pb and O, to a lesser extent it is composed of C, Si, Na, Al and Ca. In figure 6c, two particles identified as 1745 and 1746 are observed, both with a regular spherical shape with a shiny appearance, which appear to be adhered one on top of the other. Both particles are rich in Cu and O, and to a lesser extent are composed of Si, C, S and Al. According to Aragon Piña (2011) [47] anthropogenic particles are those that contain heavy elements in their chemical composition, such as Pb, Cu, Cd, Mn, Zn, Ni and As; among others. They commonly present irregular shapes (Figure 6a), spherical (Figure 6b), and also in complexes of agglomerates of fine particles that can be globular or acicular (Figure 6c) and above all, they generally present a highly variable chemical composition. This type of particles is associated as a result of activities such as the burning of fuels, industrial processes, agriculture, mining, etc. It should be noted that most of the material that constitutes the earth's crust are silicates, and contain elements such as Si, Al, Ca, Mg, Fe and O, so it is not surprising that these elements are part of the composition of most of the particles presented in this study. Several particles containing lead were found from the SEM-EDS analysis, as shown by the morphology of particles 1745 and 1746 shown in Figure 6. The presence in the air of these lead-rich particles is related to emissions generated from the metallurgical industry (non-ferrous materials) and dense vehicular traffic [47]. Short-term exposure to lead can cause brain damage, kidney damage, and gastrointestinal problems [28]. In Mexico, the air quality standard NOM-026-SSA1-1993 [48] indicates that lead levels in ambient air must not exceed the exposure limit of 1.5 μg m^{-3} within an arithmetic average of three months.

Table 7. Elemental analysis results: mass percentage of metals contained in the selected PM10 particles

		PM10 Particle identification Code			
		1705	**1736**	**1745**	**1746**
Metals (mass percentage)	**Al**	0.64	0.29	0.11	0.13
	C	1.24	6.52	9.75	8.64
	Ca	0.27	0.25		
	Cd	0.66			
	Cu			30.3	31.8
	Mg	0.22			
	Na	0.74	0.48		
	O	22.52	25.7	33.3	33.2
	Pb	47.65	60.9		

Journal of Physics: Conference Series **2162** (2022) 012010 doi:10.1088/1742-6596/2162/1/012010

	S			8.68	8.3
	Si	26.07	5.8	17.7	17.8
Total		100	100	100	100

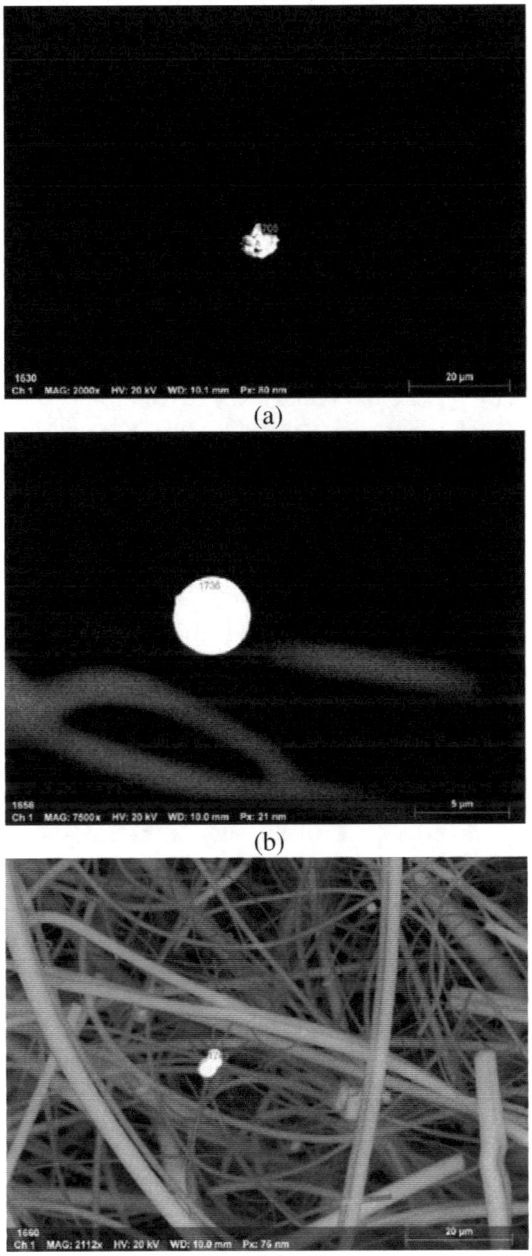

(a)

(b)

(c)

Figure 6. SEM-EDS Analysis of PM10 selected particles identified as 1705 (a), 1736 (b), 1745 and 1746 (c)

3.6. Health Risk Analysis

For the health risk analysis, two sectors of the exposed population were considered: adults and children. The carcinogenic risk (CR) was determined for the metals Cadmium and Cobalt, which have been reported to have carcinogenic effects. Table 8 shows the results of carcinogenic risk obtained in this study and their comparison with that reported by other authors, it can be observed that the limit value established by the EPA of 10^{-6} was not exceeded, indicating that the risks of cancer by inhalation of Cd and Co in the study is low. Cancer risk coefficients (CR) in adults are higher than those obtained in children, this because the extramural exposure of adults is greater than in the child population, due to mobility related to their work and occupations. According to Table 8, the CR value calculated for Cd in adults and children in the city of Torino, Italy [49] exceeds the established permissible limits. Comparing the CR values found in this study with other studies around the world (Table 8), it can be observed that CR values for Cd were lower than those reported for Torino, Italy [49], Gijón, Spain [50] and Acerra, Italy [52]. The CR values found for Co in the study site were lower than those reported for Torino [49] and Acerra [52] cities in Italy. Table 9 shows the results obtained for non-carcinogenic risk (HQ) for Cd, Co and Mn. These values must be <1 in order not to represent a health risk. During the present study, no value above that established was observed, which suggests that the risk of contracting diseases other than cancer (respiratory and cardiovascular diseases) in the exposed population is low. According to Table 9, the values observed in the city of Torino, Italy for the case of Cd in exposure to the child population exceed 6 times the legal limit and 1.69 times in exposure for the adult population. Comparing these results with those obtained in the present study, the value calculated for the city in Italy is 12.93 times higher than the value registered in children in the present study and 3.56 times higher in adults. It is worth mentioning that the results obtained for exposure to children are higher than those obtained for exposure to adults, this because the child population may be more sensitive to non-carcinogenic substances than adults [49].

HQ values for Cd in the study site were higher than those reported in Acerra, Italy [52], and lower than those reported for Torino, Italy [49], Gijón, Spain [50], Baoutou, China [51] and Kuopio, Finland [53]. Co showed HQ values in this study higher than those reported in Acerra, Italy [52], Athens, Greece [53] and Kuopio, Finland [53] and lower than those reported for Torino, Italy [49], Gijon, Spain [50], Baoutou, China [51] and Lisbon, Portugal [53]. The HQ values for manganese in this study were similar than those reported for Torino, Italy [49] and Kuopio, Finland [53], but lower than those reported in Acerra, Italy [52], Gijón, Spain [50], Baoutou, China [51] and Athens, Greece [53]. The integrated HQ values (the sum of the three metals considered) in the present study were less than 1 for adults and children. This suggests that exposure to these metals means a low risk of suffering or aggravating respiratory and cardiovascular diseases in the study site.

Table 8. Carcinogenic risk evaluated in the study area and its comparison with two cities in Europe.

Cancer Risk Coefficients (CR)			
City	**Trace Metal**	**Adult**	**Children**
Gijon, Spain (Urban) [50]	Cd	5.44E-07	1.36E-07
	Co	5.03E-07	1.26E-07
Torino, Italy (Urban) [49]	Cd	1.00E 05	9.46E 06
	Co	8.00E-08	7.00E-08
Acerra, Italy (Urban) [52]	Cd	2.54E-06	6.35E-07
	Co	7.4E-06	1.85E-06
Leon, Guanajuato, Mexico (This Study)	Cd	1.92E-07	4.81E-08
	Co	3.47E-07	8.67E-08

Table 9. Non-cancer risk coefficients evaluated in the study area, and its comparison with two cities in Europe and one city in Asia.

Non-cancer Risk Coefficients (HQ: Hazard Quotient)			
City	**Metal**	**Adult**	**Children**
Baoutou, China (Urban) [51]	Cd	0.2900	0.7800
	Co	0.0500	0.1800
	Mn	0.9000	0.9600
Gijon, Spain (Urban) [50]	Cd	0.0881	0.0881
	Co	0.0272	0.0272
	Mn	0.5060	0.5060
Torino, Italy (Urban) [49]	Cd	1.6900	6.1300
	Co	0.0050	0.0100
	Mn	0.0300	0.1000
Acerra, Italy (Urban) [52]	Cd	0.1400	0.4100
	Co	0.1400	0.400
	Mn	0.2900	0.8400
Athens, Greece (Urban) [53]	Cd	-	-
	Co	0.4510	0.4510
	Mn	0.2300	0.2300
Kuopio, Finland (Urban) [53]	Cd	0.0050	0.0050
	Co	0.0118	0.0118
	Mn	0.0643	0.0643
Lisbon, Portugal (Urban) [53]	Cd	-	-
	Co	0.0855	0.0855
	Mn	-	-
Leon, Guanajuato, Mexico (This study) (Urban)	Cd	0.4740	0.4740
	Co	0.2850	0.2840
	Mn	0.0560	0.0550

4. Conclusion

The concentrations of PM10 particles did not exceed the maximum permissible limits established by the Mexican standard (75 µg m^{-3}) or those established by the WHO (50 µg m^{-3}). The studies carried out on the collected samples showed that of the six metals that were analyzed, Fe and Zn were the ones with the highest concentrations, which was to be expected due to their higher proportion compared to the other metals due to their geological origin. It should be noted that the concentrations found for Cu in the present study were higher than in other studies carried out in Mexico, although the values were similar to those found in cities such as Bogotá, Colombia and Agra, India. The results obtained for Cd showed that the concentrations were higher in the study area than in other studies carried out in Mexico and similar in studies carried out abroad. Regarding Mn, it was found that the results were lower than those reported in other studies carried out in Tampico, Tamaulipas and in the same city of León, higher than those found in Hidalgo, Colima and Puebla and cities outside the country such as Seoul, Korea and Venice Italy. The concentrations of Co that were obtained were higher than those registered in other studies carried out in Mexico such as Puebla and the Metropolitan Area of Mexico City, and Venice, Italy, however, they were lower than those obtained in Agra, India and in the simultaneous study carried out in the same city (León, Guanajuato). The enrichment of the particles with heavy metals could be attributed to emission sources located to the Southwest of the study site, in this direction there are two main avenues, on which a significant percentage of vehicles travel daily, in addition there are leather tanneries and gas stations, which may have acted as sources contributing to the levels of the metals measured in PM10. It is worth mentioning that around the city of Leon there are important mining and industrial parks, therefore its contribution to the enrichment of the particles with heavy metals is not ruled out. In the statistical analyses carried out, significant correlations were found with the speed and direction of the wind, indicating the influence of the transport of pollutants from nearby local areas. Cu and Co showed a significant positive correlation, which indicates that both could have sources in common, this is corroborated with the wind roses and pollutants carried out for the present study, which shows that both metals had a higher concentration when the wind blew from the SW. Mn negatively correlated with RH, this could be due to the fact that manganese is a soluble element, therefore it can be inferred that a washing of the column of the atmosphere occurs at higher humidity values. In the analysis of principal components, the correlation of CO with Mn and Cd stands out, since CO is a tracer of vehicular combustion sources, it is inferred that Cd and Mn could be originated in vehicular emissions and combustion sources at high temperatures. In the morphological analyses, the presence of metals such as Al, Cd, Pb, Cu, among others, could be observed, finding particles with irregular shapes and regular shiny spherical associated with anthropogenic sources. Regarding the estimation of the risks to the health of the population, the cancer risk index in the life time (CR) did not exceed the value of 10^{-6} established by the US-EPA nor the value of 10^{-5} established by the WHO, so the concentrations of heavy metals in PM10 measured at the study site do not represent a carcinogenic risk to the health of the exposed population. Regarding the non-carcinogenic risk during the present study, no value above that established was observed, indicating that the risk of contracting diseases other than cancer (respiratory and cardiovascular diseases) by inhalation of PM10, containing metals, at the study site is low.

As future work, it is recommended to consider the concept of bio accessibility of metals to carry out a comprehensive assessment of health risks, since the total values of CR and HQ may overestimate the dangers they pose to humans. It is also necessary to design a bio monitoring plan in the future in the city of Leon, also considering other nearby cities such as Celaya and Salamanca (in the latter there is an oil refinery and a thermoelectric plant). In this study, only some metals in PM10 particles and the associated risk due to the exposure of the population to the concentration levels found were considered, however, it is necessary to include in the future in the evaluation of health risk, other toxic metals such as Pb, Cr, Al, Si, as well as organic pollutants (dioxins, PCBs, among others) to evaluate a broader matrix of pollutants that may represent an inhalation risk in the study population.

References

[1] Alejandrina G, Anselmo G, Benjamin S, Pilar, F, Maria G, Carlos B, Ángel, P, Reyes, N and Senén, D 2012 *Contaminación atmosférica* (Ed. UNED, Madrid, Universidad Nacional de Educación a Distancia).

[2] Moreno T, Querol X, Alastuey A, Viana M, and Gibbons W 2005 *Atmospheric Environment* **39** 6109.

[3] Adamson I, Prieditis H, Hedgecock C, and Vincent R 2000 *Toxicology and Applied Pharmacology* **166** 111.

[4] Gorelova SV, Gorbunov AV, Frontasyeva MV and Sylina AK 2020 *WSEAS Transactions and Development* **16** 608.

[5] Tzortzi N, Georgi J, and Sophocleous J 2018 *WSEAS Transactions and Development* **14** 270.

[6] Birmili W, Allen A, Bary F, and Harrison R 2006 *Environmental Science and Technology* **40** 1144.

[7] Milford J and Davidson C 1985 *Journal of the Air Pollution Control Association* **35** 1249.

[8] Utsunomiya S, Jensen K, Keeler G, and Ewing R 2004 *Environmental Science and Technology* **38** 2289.

[9] Schaumann F, Borm P, Herbrich A, Knoch J, Pitz M, Schins R, Luettig B, Hohlfeld J, Heinrich J, and Krug N 2004 *American Journal of Respiratory and Critical Care Medicine* **1708** 898.

[10] Diario Oficial de la Federación Norma Oficial Mexicana NOM-025-SSA1 2014 *Valores Límites Permisibles Para la Concentración de Partículas Suspendidas PM10 y PM2.5 en el Aire Ambiente y Criterios Para su Evaluación* (Secretaría de Salud: Ciudad de México, México).
[11] Flores Rangel R, Rodríguez-Espinosa P, de Oca-Valero J, Mugica-Álvarez V, Ortiz-Romero-Vargas M, Navarrete-López M, Dorantes-Rosales H and Morales-García S 2015 *Air Quality, Atmosphere and Health* **8** 367.
[12] Morales-García S, Rodríguez-Espinosa P, Jonathan M, Navarrete-López M, Herrera-García M, and Muñoz-Sevilla N 2014 *Environmental Monitoring and Assessment* **186** 55.
[13] Martínez Carrillo M, Solís C, Andrade E, Isaac-Olivé K, Beltrán-Hernández R, Medina Moreno G, Martínez-Resendiz G, Ramírez Reyes A, Lucho-Constantino C, and Razo L 2010 *Aerosol Revista Mexicana de Física* **56** 62.
[14] González-Santiago O, Badillo-Castañeda C, Kahl J, Ramírez-Lara E, and Balderas-Renteria I 2011 *Journal of the Air and Waste Management Association* **61** 573.
[15] Espinosa A, Miranda J, Hernández E, Reyes J, Alarcón A, Torres M and Sosa R 2019 *Air Quality, Atmosphere & Health* **12** 1267.
[16] Gobierno del municipio de León *Programa de Gobierno Municipal de León, Guanajuato* 2018 (2028-2021).
[17] Code of Federal Regulations AE 2.106/3:40/ 40 CFR 2012 *Protection of Environment Chapter I Subchapter C Air Programs, Part 50- National Primary and Secondary Ambient Air Quality Standards Appendix J to Part 50 Reference Method for the Determination of Particulate Matter as PM10 in the Atmosphere* (US Environmental Protection Agency).
[18] Machado A, Velazquez H, García N, García C, Acosta L, Córdova A, and Linares M 2007 *Interciencia* **32** 312.
[19] United States Environmental Protection Agency (US EPA) 1999 *Compendium of Methods for the Determination of Inorganic Compounds in Ambient Air-Compendium Method IO-3.1* In Selection, Preparation and Extraction of Filter Material (US EPA: Cincinnati, OH, USA).
[20] United States Environmental Protection Agency (US EPA) 1999 *Compendium of Methods for the Determination of Inorganic Compounds in Ambient Air-Compendium Method IO-3.2* In Determination of Metals in Ambient Particulate Matter Using Atomic Absorption (AA) Spectroscopy (US EPA: Cincinnati, OH, USA).
[21] Machado A, García N, García C, Acosta L, Córdova A, Linares M, Giraldoth D, and Velásquez H 2008 *Revista Internacional de Contaminación Ambiental* **24** 171.
[22] Abdul-Wahab S, Bakheit C, and Al-Alawi S 2005 *Environmental Modelling and Software* **20** 1263.
[23] Lakes Environmental 2011 *WRPLOT View Version 7.0 Wind Rose Plots for Meteorological Data* http://www.weblakes.com/products/wrplot/index.html
[24] Di Vaio P, Magli E, Caliendo G, Corvino A, Fiorino F, Frecentese F, Saccone I, Santagada V, Severino B, Onorati G, et al 2018 *Atmosphere* **9** 58.
[25] United States Environmental Protection Agency 2009 *Part F Supplemental Guidance for Inhalation Risk Assessment* In Risk Assessment Guidance for Superfund—Volume I: Human Health Evaluation Manual (US EPA Washington DC USA).
[26] World Health Organization (WHO) 2000 *Air Quality Guidelines for Europe 2nd ed* Regional Publication European Series No. 91 (WHO Regional Office for Europe, World Health Organization Copenhagen, Denmark).
[27] Pachón J and Sarmiento H 2008 *Revista Facultad de Ingenieria* **43** 120.
[28] Geiger, A. and Cooper, J 2010 *Overview of Airborne Metals Regulations, Exposure Limits, Health Effects, and Contemporary Research* (USEnvironmental Protection Agency Portland, USA) https://www3.epa.gov/ttnemc01/prelim/otm31appC.pdf.
[29] Wang Y, Huang L, Li C, Mi H, Luo J and Tsai P 2003 *Atmospheric Environment* **37** 4637.

[30] Aragón-Piña A, Campos-Ramos A, Leyva-Ramos R, Hernández-Orta M, Miranda-Ortiz N, and Luszczewski-Kudra A 2006 *Revista Internacional de Contaminacion Ambiental* **22** 5.

[31] Morais S, Garcia e Costa F and Pereira M, 2012 *Health, Environmental Health - Emerging Issues and Practice* ed Prof. Jacques Oosthuizen ISBN: 978-953-307-854-0 (InTech) Available from: http://www.intechopen.com/books/environmental-health-emerging-issuesand-practice/heavy-metals-and-human-health

[32] Ramírez A 2013 *Anales de La Facultad de Medicina* **63** 51.

[33] ATSDR 2016 *Resumen de Salud Pública: Cadmio (Cadmium)* | PHS | (ATSDR Agencia Para Sustancias Toxicas y El Registro de Enfermedades) https://www.atsdr.cdc.gov/es/phs/es_phs5.htmlX

[34] Chen L and Lippmann M 2008 *Inhalation Toxicology* **21** 1.

[35] Lauwerys R and Lison D 1994 *The Science of the Total Environment* **150** 1

[36] Querol X, Pey J, Minguillón M, Pérez N, Alastuey A, Viana M, Moreno T, Bernabé R, Blanco S, Cárdenas B, Vega E, Sosa G, Escalona S, Ruiz H and Artíñano B 2007 *Atmos. Chem. Phys. Discussions* **7** 10589.

[37] Campos-Ramos A, Aragon-Piña A, Alastuey A, Galindo-Estrada I, and Querol X 2011 *Atmospheric Pollution Research* **2** 409.

[38] Kulshrestha A, Satsangi P, Masih J and Taneja A 2009 *Science of the Total Environment* **407** 6196.

[39] Park E, Kim D and Park K 2008 *Environmental Monitoring and Assessment* **137** 441.

[40] Rampazzo G, Masiol M, Visin F and Pavoni B 2008 *Water Air, and Soil Pollution* **195** 161.

[41] Chen R, Wang B, Wang Z and Yao S 2015 *Biomedical and Environmental Sciences* **28** 44.

[42] Heidari-Farsani M, Shirmardi M and Goudarzi G 2013 *Journal of Advances in Environmental Health Research* **1** 1.

[43] Cook M and Morrow H 1995 *Anthropogenic Sources of Cadmiun Transport into Plants* (Canadian Network of Toxicology Centres Ottawa, ON, Canada).

[44] Chandler A 1996 *Characterizing Cadmium in Municipal Solid Waste, Sources of Cadmium in the Environment* In: Inter-Organization Programme for the Sound Management of Chemicals (IOMC) Organization for Economic Co-operation and Development (OECD Paris, France). 1996.

[45] World Health Organization (WHO) 2000 *Air quality guidelines for Europe Second edition* ((WHO regional publications European series No. 91).

[46] World Health Organization (WHO) 1981 Environmental Health Criteria No. 17 Manganese (World Health Organization Geneva Technical Report Series).

[47] Aragón Piña A 2011 *¿Cómo son las partículas atmosféricas antropogénicas y cuál es su relación con los diversos tipos de fuentes de contaminantes? (*Universidad Autónoma de San Luis Potosí Instituto de Metalurgia Facultad de Ingeniería San Luis Potosí S.L.P., México) 1–177.

[48] Norma Oficial Mexicana NOM-026-SSA1 1993 *Salud Ambiental Criterio para evaluar la para evaluar la calidad del aire ambiente con respecto al Plomo (Pb) Valor normado para la concentración de Plomo (Pb) en el aire ambiente, como medida de protección a la salud de la población* (Secretaría de Salud: Ciudad de México, México).

[49] Romanazzi V, Casazza M, Malandrino M. Maurino V, Piano A, Schilirò T and Gilli G 2014 *Chemosphere* **112** 210.

[50] Megido L, Suarez-Pena B, Negral L, Castrillon L, and Fernandez-Nava Y 2017 *Chemosphere* **177** 284.

[51] Li K, Liang T and Wang L 2016 *Environ Geochem Health* **38** 843.

[52] Di Vaio P, Elisa Magli e, Caliendo G, Corvino A, Fiorino F, F Frecentese F, Saccone I, Santagada V, Severino B, Onorati G, D'Onofrio G, Manzo, C, and Perissutti, E 2018 *Atmosphere* **9** 1.

[53] Chalvatzaki E, Chatoutsidou SE, Lehtomäki H, Almeida SM, Eleftheriadis K, Hänninen O and Lazaridis M 2019 *Atmosphere* **10** 1.

Machine Learning techniques applied to Road Health Status Recognition through Tyre Cavity Noise Analysis

Gloria Schiaffino[1,a], Lara Ginevra Del Pizzo[2,b], Stefano Silvestri[1,c], Francesco Bianco[2,d], Gaetano Licitra[3,4,e] and Filippo Giammaria Praticò[1,f]

[1] Università Mediterranea di Reggio Calabria, Reggio Calabria (RC), Italy
[2] IPOOL Srl, Pisa (PI) Italy
[3] CNR - IPCF, Via Moruzzi, 1, 56124 Pisa, Italy
[4] ARPAT, Via Vittorio Veneto, 27, 56127 Pisa, Italy

E-mail: [a]gloria.schiaffino@unirc.it, [b]lara.delpizzo@i-pool.it, [c]silvestri.sfn@gmail.com, [d]francesco.bianco@i-pool.it, [e]g.licitra@arpat.toscana.it, [f]filippo.pratico@unirc.it

Abstract. This paper proposes a system based on Neural Networks (NN), designed for providing an efficient, non-invasive and automated method for monitoring the health status of road pavements by using features derived from Tyre Cavity Noise (TCN) analysis. Indeed, visual inspection remains to date the most common choice for evaluating the condition of road pavements; however, this method is both labor-intensive and time consuming. The system presented in this work uses a microphone placed inside the vehicle tyre that measures TCN while travelling normally, and an embedded data acquisition system based on a Raspberry Pi which feeds the NN tools to recognize and classify road deterioration. We also present a preliminary analysis of features based on temporal and spectral characteristics of TCN signals generated by tyre/road interaction and acquired on three different kind of road distresses. The results show good classification capability and, moreover, the sound pressure measured inside the tyre was correlated accelerometric data measured on-board.

1. Introduction

Regular maintenance of road pavements is of paramount importance to ensure both traffic safety and passenger comfort. Swift actions can be crucial not only to prevent further damages to road pavements, caused by the constant and intensive use of road infrastructures and other environmental factors, but also because with proper scheduling, road maintenance costs can be reduced up to three times [1]. Indeed, the main cause of pavement distresses is the repeated passage of heavy traffic loads, temperature variations, adverse weather conditions and frequent temperature oscillations. The quality of asphalt components and of the laying process is also fundamental to ensure the durability of the road surface [2, 3].

Traditional monitoring methods can be divided in destructive and non-destructive methods. Among the destructive ones, pavement coring is the most popular since it has the advantage to provide an in-depth analysis of the pavement status; however, it is expensive and requires closing the road to traffic. Non-destructive techniques rely mainly on visual inspection, which can require an on-site "walk and look" approach or measurements made with mechanical devices. This techniques can provide both an accurate estimation of the road health status and a proper identification of the distresses, but have the disadvantage of being time-consuming and lack repeatability, since they depend on the operator's judgment [1]. On the other hand, aside visual inspection, other mechanical devices are used, such as laser profilometry [4] or Ground Penetration Radar (GPR), which exploits the propagation and reflection of electromagnetic waves in underground media [1, 5]. However, both destructive and non-destructive traditional methods have the disadvantage of being expensive, due to the costs related to the operator and to the measurement equipment.

In the last decades, new sensors which rely on accelerometric [6, 7], photographic [8, 9] and acoustic [10] measurements combined to Artificial Intelligence (AI) technologies have been developed. In the context of the so-called *Smart Roads* [11] these new technologies and possibilities increase the efficiency of road infrastructure maintenance, not only for damage detection but also to improve viability, combining their efforts with other smart methods such as the intelligent traffic light systems [12].

This work presents a novel, non-invasive and automatic method to identify different kinds of road surface degradation, by studying the different acoustic features of three different kind of distress. The acoustic signal is measured thanks to an edge sensor mounted on the vehicle which is able to measure the pressure inside the tyre cavity. The method is particularly useful, since it recognises the state of degradation of road surfaces automatically using AI techniques. Moreover, the acquisition data system is cheaper compared to other systems [13]. The study also involved an in-depth analysis of the spectral features of the acoustic signal coming from different types of distresses, which was carried out in order to gain accurate information about the physical processes at the foundation of tyre/road acoustic interaction.

2. State of the art

Over the years, several acoustic methods have been developed to identify road surface wearing. These methods mostly rely on noise measurements outside or inside the vehicle. Therefore, the signal measured by the sensors is related to both the pavement status and to the type of road surface [14].

Measurements performed outside the vehicle rely on the evaluation of Tyre/Road Noise (TRN), caused by the tyre/road interaction during the motion of the vehicle. Initially, TRN was studied for its role on Road Tra c Noise (RTN), which has been linked to sleep disturbances, cardiovascular disease, hypertension, myocardial infarction [15, 16, 17, 18, 19]. TRN is widely studied in literature and many methods have been developed, such as Close-Proximity (CPX) method [20] or the various Pass-By methods (SPB [21], CB [22], CPB [23]).

Moreover, in the last decade, the interest in the use of AI algorithms to automate the process of road surface classification has increased more and more. In particular, several works were developed with the aim of identifying the pavement type and to classify its degradation [24, 25]. This task can be carried out by using microphones placed close to the contact patch of the tyre, in order to directly measure the TRN [26].

Beside internal/external vehicle noise measurements, another classification technique that was recently proposed relies on the measurement of TCN generated by tyre/road interaction. TCN studies were initially carried out by Bschorr [27] in the context of the study of passenger's acoustic comfort.

Indeed, as the tyre rolls on the pavement, the moving tyre-contact patch compresses the air inside the cavity, thus providing a source for the acoustic pressure field. Due to the size of the tyre cavity, this field allows for the formation of standing waves. Moreover, natural frequencies of the tyre/wheel system include all deformation axes, the effect of the rim, and the suspension system, as studied by Mohamed [28]. For example, a typical Power Spectral Density (PSD) measured with a tyre cavity microphone is shown in figure 1. Standing modes can be easily identified in the low-frequency region, while at frequencies higher than 1 kHz modes are overlapped and no longer recognizable. Lastly, a preliminary model for the classification of road pavements using TCN measurements and machine learning techniques was presented by Krauss and Masino [10, 29].

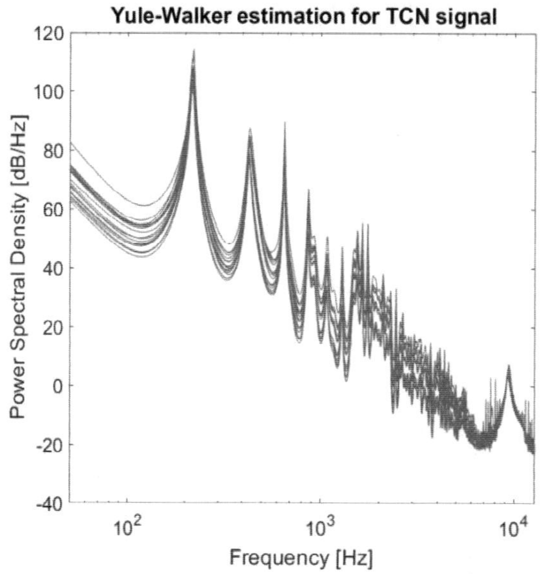

Figure 1. Resonance modes of TCN evaluated through Yule-Walker regression analysis.

In this scenario, this work focuses on the use of an edge computing platform that processes audio signals measured by a TCN microphone, in order to produce a classification of the status of the road surface. In particular, this works shows the results of the analysis performed so far.

Differently from other authors [29, 10], these analyses focus on highlighting specific acoustics features of road pavements related to their maintenance status, capable of classifying the different distresses present on the road surface, rather than on the correlation with CPX measurements or asphalt type. This was possible thanks to a preliminary analysis of the spectral components of the signal aimed at identifying the response of the cavity to the stresses induced by the different types of distress. The TCN signal was compared with accelerometric data measured by accelerometers placed on board. The features analysis carried out represents therefore the first step to develop a multi-layer neural network classifier of pavement distresses.

3. Experimental Methods

The sensor used for these analyses was developed by the Karlsruhe Institute of Technology (KIT) [13] and is composed by:

- A electret microphone CMA-4544pf-w, placed inside the tyre and connected through a SMA connector to an external Bluetooth device;
- A Bluetooth module, composed by a tl072 preamplifier and a BC127 Bluetooh module with a DAC converter;
- An LM540-054 Bluethooth device, placed on board and connected through USB port to a Raspberry Pi 2 Model B V1.1;
- The Raspberry Pi that receives data and saves it in 16 bits .wav format.

The accelerometer signal data acquisition system is instead composed of:

- A piezoeletric accelerometer B&K 4507B, placed inside the vehicle in proximity of the SRTT tyre;
- Apollo light version Apollo lt 8C data acquisition board.

The whole set up is placed on the mobile laboratory, i.e. the vehicle used for measurements. As the vehicle travels on the road, the system gathers data on-the-go and saves it in a local SD card. The vehicle used is a Mercedes-Benz Vito: this choice was performed since it can mount the Standard Reference Test Tyre (SRTT) used for CPX measurements [21] (see figure 2). Indeed, TCN measurements were seldom performed together with CPX noise measurements.

The acoustic signal measured was analyzed in post-processing. The measurement campaign was carried out near the city of Pisa, Italy, and all the pavement analyzed in this work were dense-graded pavements. Moreover, in order to minimize the effects of the vehicle speed on TCN signal, the speed was always been maintained approximately around v = 40 km/h. Acoustic data was used to analyze three different categories:

- Good condition: a signal of a pavement without distress;
- Type-1 distress: superficial and extended distresses which affect a wide part of the road;
- Type-2 distress: superficial distresses which have smaller size than Type-1 distresses, i.e. they have size of maximum 1 m; they are localized on the road.

Figure 2. Sensor set on the mobile laboratory.

Spectral analysis of noise signals measured while travelling on roads belonging to the three categories was used to derive a set of parameters that could be used to perform the classification task. Moreover, accelerometric data was used to gain a deeper insight on the origin of modes present in the noise signal very low frequency region, but was not used to perform the classification.

All post-processing tools were developed using MATLAB 2021b.

4. Results

4.1. Signal Analysis and comparison with accelerometric data

At first, the analysis of transient signals such as Type-2 distresses was performed using spectral analysis in one-third octave bands. One-third octave band levels were obtained through FFT calculation.

A qualitative analysis yields that the signal from Type-2 distresses shows typically an excitation of both tyre cavity eigenmodes and other low frequencies (i.e. f < 100Hz) and high-frequency bands (i.e. f > 2kHz), as shown in figure 3. Low-frequency modes cannot be explained by taking into account only the tyre [28] and, for this reason, other physical sources were investigated, such as the suspensions and wheel axle using the accelerometric data.

In order to provide a more in-depth analysis of the low-frequency modes, TCN signals were compared to the accelerometric data provided by the accelerometer. Using one-third-octave band analysis, the correlation between accelerometric data and TCN was calculated. In particular, as shown in figure 4, the correlation between low-frequency TCN and accelerometer bands results high, since it lies between 0.61 and 0.75.

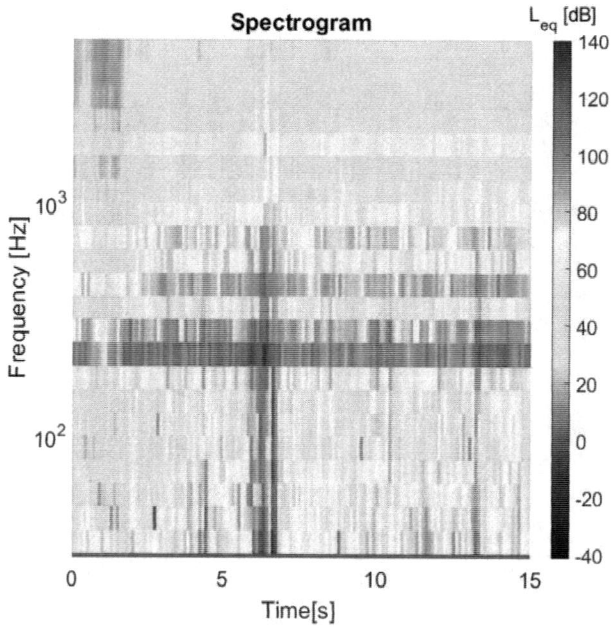

Figure 3. Example of one-third octave bands analysis for a Type-2 distress signal. Both low-frequency and high frequency modes are excited: low-frequency modes refer to long-lasting phenomena such as vertical oscillations of the vehicle, while the tyre interaction with the road surface distress excites high frequency modes.

4.2. Parameters for NN classification

Type-1 distresses provide a different excitation compared to Type-2 ones. Indeed, one-third octave bands analysis of the TCN signal on Type-1 distresses shows that the most excited modes are the tyre cavity eigenmodes, instead of the suspension modes which are relevant for Type-2. An example of the signal generated by this kind of distress is shown in figure 5. The differences that arise between the spectral analysis of the two distresses could be explained by taking into account the spatial characteristics of the distresses: Type-2 distresses are highly localized, and provide an impulse-like excitation of the tyre, while, on the on the other hand, Type-1 distresses increase the sound energy inside the cavity.

Several features for distinguishing the different signal produced by each distress were identified, by using a mix of broadband and frquency-dependent features. Broadband parameters are:

- $L_{p,tot}$: signal Sound Pressure Level (SPL);
- $L_{p,(fd-fu)Hz}$: signal SPL within the frequency range (f_d-f_u);

While frequency-dependent features used in this work are:

- $R_I = L_{p,(200\text{-}240)\ Hz}/L_{p,(270\text{-}400)\ Hz}$: ratio between the L_p evaluated in the tyre first resonance and the L_p evaluated between the first and the second tyre resonance;
- $R_{II} = L_{p,(400\text{-}460)\ Hz}/L_{p,(480\text{-}620)\ Hz}$: ratio between the L_p evaluated in the tyre second resonance and the L_p evaluated between the second and the third tyre resonance;
- $R_{II} = L_{p,(630\text{-}660)\ Hz}/L_{p,(700\text{-}830)\ Hz}$: ratio between the L_p evaluated in the tyre third resonance and the L_p evaluated between the third and the fourth tyre resonance;
- $HF = L_{p,(2,5\text{-}8)\ kHz}=L_{p,(1,5\text{-}2)\ kHz}$: it indicates the L_p evaluated at high frequencies, i.e. 2,5-8 kHz;
- $LF = L_{p,(20\text{-}100)\ Hz}$: it indicates the L_p evaluated at low frequencies, i.e. $f < 100$ Hz.

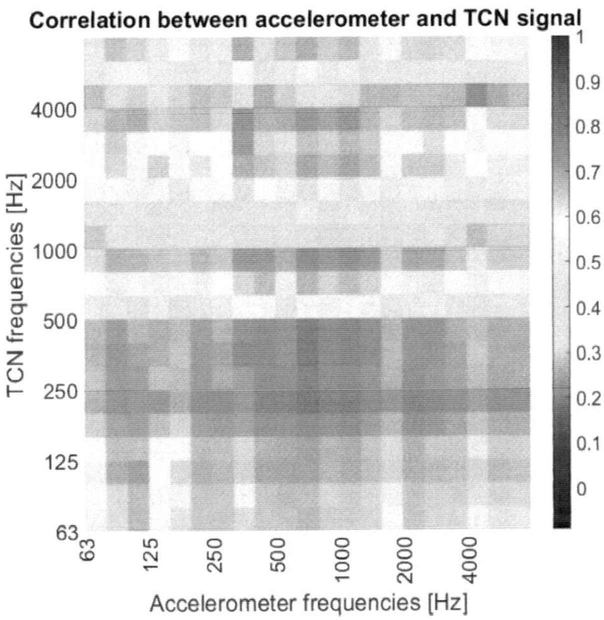

Figure 4. Correlation between accelerometer and TCN signal, shown in one-third octave bands.

A measurement campaign was carried out for the purpose of acquiring data to study the possibility of evaluating the feasibility of a NN classifier based on these features. TCN signals were divided into samples of long 0.4 s each. In this way, the total number of samples from the measurement campaign for the Good condition class is $N_{GC} = 1770$. The sample size of Type-1 class is $N_{CR} = 2136$ while $N_P = 41$ samples belonged to Type-2 class. In table 1, 2 and 3 the correlations between the parameter chosen for the three different classes are shown.

The low correlation found between the features used for Good Condition (table 1) and Type-1 distresses (table 2) shows that the features are indeed suitable for a NN classifier, while for the Type-2 distresses (table 3) R_I , R_{II} , R_{III} and HF are strongly correlated. For this reason, the parameters chosen to train the NN were only $L_{p,tot}$, R_I and LF.

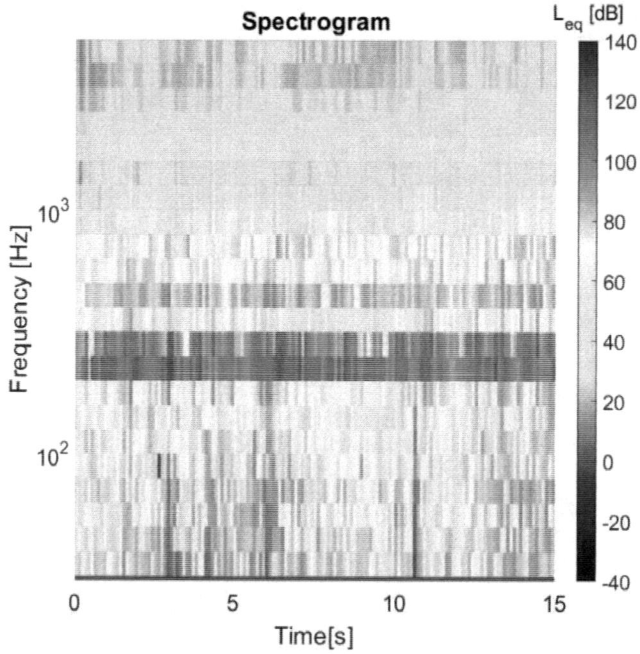

Figure 5. Example of one-third octave bands analysis for a signal caused by a Type-1 distress: low-frequency modes are less excited compared to Type-2; for Type-1, the interaction excites mainly the tyre cavity resonance modes.

Table 1. Features correlation for the signal of Good condition class.

	$L_{p,tot}$	R_I	R_{II}	R_{III}	HF	LF
$L_{p,tot}$	1	0.04	-0.18	-0.27	0.08	0.72
R_I	0.04	1	0.26	0.41	-0.01	-0.20
R_{II}	-0.18	0.26	1	0.32	-0.01	-0.19
R_{III}	-0.27	0.41	0.32	1	-0.18	-0.24
HF	0.08	-0.01	-0.01	-0.18	1	-0.03
LF	0.72	-0.20	-0.19	-0.24	-0.03	1

4.3. NN development and performances

By using the aforementioned parameters, a multi-layer NN with 2 hidden layers and 100 neurons per layer was chosen. This choice was deemed suitable since the number of input parameters of the model is small and this type of net performs well on edge devices [30]. However, the small sample size for Type-2 class could affect the net's recognition ability: for this reason, the sample size was increased to 205 samples by adding White Gaussian Noise to the original samples [31], with a signal-to-noise ratio SNR = 20. 70% of the sample size was used for the training phase, while for validation 15% of the samples was used and the remaining 15% was used for the testing phase.

The net performance was evaluated by testing it on a road classified previously by means of visual inspection. Table 4 shows the performances for training, validation and testing.

Table 2. Features correlation for the signal of Type-1 class.

	$L_{p,tot}$	R_I	R_{II}	R_{III}	HF	LF
$L_{p,tot}$	1	-0.36	-0.36	-0.39	0.26	0.78
R_I	-0.36	1	0.66	0.70	-0.48	-0.44
R_{II}	-0.36	0.66	1	0.66	-0.54	-0.35
R_{III}	-0.39	0.70	0.66	1	-0.64	-0.38
HF	0.26	-0.48	-0.54	-0.64	1	0.23
LF	0.78	-0.44	-0.35	-0.38	0.23	1

Table 3. Features correlation for the signal of Type-2 class.

	$L_{p,tot}$	R_I	R_{II}	R_{III}	HF	LF
$L_{p,tot}$	1	-0.75	-0.86	-0.88	0.90	0.44
R_I	-0.75	1	0.88	0.84	-0.75	-0.16
R_{II}	-0.86	-0.88	1	0.92	-0.86	-0.24
R_{III}	-0.88	0.84	0.92	1	-0.85	-0.30
HF	0.90	-0.75	-0.86	-0.85	1	0.25
LF	0.44	-0.16	-0.24	-0.30	0.25	1

Table 4. NN classifier's performances of success on training, validation and testing dataset.

Class	Training	Validation	Test
Good condition	93.9%	95.5%	94.0%
Type-1	96.7%	96.5%	93.1%
Type-2	78.9%	90.0%	75.8%

The accuracy, i.e. the overall percentage of success, is shown in table 5. The values obtained are:

- 80.0% for Type-2;
- 94.1% for Type-1;
- 96.2% for Good condition;

In table 5 the metrics used to evaluate the NN performances are reported, i.e. accuracy, precision, recall and F1-score.

Table 5. NN classifier's performances.

Class	Accuracy	Precision	Recall	F1-score
Good condition	94.1%	94.9%	94.1%	94.5%
Type-1	96.2%	93.8%	96.1%	95.0%
Type-2	80.0%	98.8%	80.0%	88.5%

5. Discussion

The first part of the work consisted in analyzing the acoustic signal produced by different kind of distresses inside the tyre cavity, using also data from an on-board accelerometer. Comparing the signal measured by the microphone and the signal measured by the accelerometer, the presence of low-frequency modes inside the tyre finds a reasonable explanation. Indeed, since the modes are visible in accelerometric signals, they could be caused by the vehicle suspensions.

Then, the features of different distresses were analyzed by means of a spectral analysis performed in one-third-octave bands. While low frequencies take into account phenomena with a longer time scale, high-frequency modes are useful for locating short-time events, such as Type-2 distresses. Indeed, due to their nature, they act as an impulsive force on the system and are thus capable of exciting the whole spectrum, rather than providing energy only to the natural frequencies of the tyre.

On the other hand, analysis of the correlation between $L_{p,tot}$, R_I , R_{II} , R_{III} , HF and LF shows that $L_{p,tot}$, R_I and LF are sufficient to train the net.

The classifier shows a good percentage of success in identifying Type-1 distresses; however the identification of Type-2 distresses is still non adequate. This is consistent with previous analyses [26] and might be due to the smaller amount of samples of Type-2 distresses compared to other features. The performance of our classifier and the choice of adopted features show nonetheless that different road pavement distresses can be recognized. However, a larger data-set with more distress classes and more samples per class could pave the way towards a more complete efficient classifier. The sample size remains however a critical issue for training NNs and, therefore, on-site measurements are fundamental when defining the dataset.

6. Conclusions

In this work, AI technologies were used to classify three different types of road pavement distresses. The goal was to design a system capable of classifying automatically the type of road pavement distress. This tool could improve the time efficiency of the damage recognition and could thus reduce the cost of of pavement maintenance [32] besides providing means to support decision-making for planning road maintenance.

In this stage, the main focus was to study how to classify good road conditions and two different kind of class of distress: distresses extended over the road and distresses localized in a short region, called respectively Type-1 and Type-2 distresses. A preliminary analysis of the spectral features of each kind of distress and the study of the correlation of the Type-2 distresses with the signal taken from accelerometers allowed to identify the proper features in order to train a NN classifier. Network performance, evaluated by using accuracy, precision, recall and F1-score, is appropriate for a valid classification of types of distress. Therefore, the system shows potential for the identification of features associated with pavement defects. Indeed, most of the problems found so far regard the Type-2 identification: this could be due to the smaller sample size in the data-set. This limitation should be overcome with more measurements. Another the major limit of this work is that the net classifies during post-processing analysis, so different classifying algorithms such as k-Nearest Neighbors (KNN) or Convolutional Neural Networks (CNN) could be tested in the future in order to build a net which gives a real-time classification during the vehicle motion. Moreover, adding more types of distress to the classifier could be useful for obtaining more detailed and precise information about its structural conditions. At the same time repeated measurements over time could lead to a study of the temporal evolution of distress, proving a time scale for each distress that can contribute to the decision-making for planning road maintenance. This may be possible by trying to build different neural network models such as Long-Short Term Memory (LSTM) networks, that take into account data from different time-scales.

References

[1] Schnebele E, Tanyu B, Cervone G and Waters N 2015 *European Transport Research Review* **7** 1-19
[2] Wang W, Yan X, Huang H, Chu X and Abdel-Aty M 2011 *Transportation Research Part C: Emerging Technologies* **19** 682-694
[3] Saeed F, Rahman M, Chamberlain D and Collins P 2019 *Construction and Building Materials* **196** 530-538
[4] Ragnoli A, De Blasiis M R and Di Benedetto A 2018 *Infrastructures* **3** 58
[5] Dong Z, Ye S, Gao Y, Fang G, Zhang X, Xue Z and Zhang T 2016 *Sensors* **16** 2067
[6] Abbondati F, Biancardo S A, Veropalumbo R and Dell'Acqua G 2021 *Measurement* **171** 108763
[7] Alessandroni G, Klopfenstein L, Delpriori S, Dromedari M, Luchetti G, Paolini B, Seraghiti A, Lattanzi E, Freschi V, Carini A et al. 2014 *Proceedings of the UBICOMM* 210-215
[8] Wang W, Wang M, Li H, Zhao H, Wang K, He C, Wang J, Zheng S and Chen J *2019 Journal of Traffic and Transportation Engineering (English Edition)* **6** 535-556
[9] Barrile V, Bernardo E, Fotia A, Candela G and Bilotta G 2020 *Transactions on Environment and Development* **16** 649-659
[10] Masino J, Foitzik M J, Frey M and Gauterin F 2017 *The Journal of the Acoustical Society of America* **141** 4220-4229
[11] Cafiso S, Di Graziano A, Fedele R, Marchetta V and Praticò F 2020 *International Journal of Pavement Research and Technology* **13** 573-580
[12] Prasad V, Ali U, Assaf M H, Das S R and Biswas S N 2020 *Transactions on Systems* **19** 20-25
[13] Masino J, Daubner B, Frey M and Gauterin F 2016 *2016 Annual IEEE systems conference (SysCon)* (IEEE) pp 1-5
[14] Li T 2018 *Measurement* **128** 325-351
[15] Ouis D 2001 *Journal of environmental psychology* **21** 101-120
[16] Muzet A 2007 *Sleep medicine reviews* **11** 135-142
[17] Babisch W, Swart W, Houthuijs D, Selander J, Bluhm G, Pershagen G, Dimakopoulou K, Haralabidis A S, Katsouyanni K, Davou E et al. 2012 *The Journal of the Acoustical Society of America* **132** 3788-3808
[18] Bluhm G L, Berglind N, Nordling E and Rosenlund M 2007 *Occupational and environmental medicine* **64** 122-126
[19] Selander J, Nilsson M E, Bluhm G, Rosenlund M, Lindqvist M, Nise G and Pershagen G 2009 *Epidemiology* **20** 272-279
[20] 2018 ISO 11819-2 Acoustics | Measurement of the in uence of road surfaces on traffic noise | Part
1: Statistical Pass-By method Standard International Organization for Standardization Geneva,
CH
[21] 2017 ISO 11819-2 Acoustics | Measurement of the in uence of road surfaces on traffic noise Part 2: The Close-Proximity method Standard International Organization for Standardization Geneva, CH
[22] 2019 ISO13325:2019 Tyres { Coast-by methods for measurement of tyre-to-road sound emission. Standard International Organization for Standardization Geneva, CH
[23] 2000 NF S31-119-2 - Acoustique - Caracterisation in situ des qualites acoustiques des revetements de chaussees - Mesurages acoustiques au passage - Partie 2: procedure vehicule maitrisé. Standard ANFOR La Plaine Saint-Denis, FR

[24] Ambrosini L, Gabrielli L, Vesperini F, Squartini S and Cattani L 2018 *Audio Engineering Society Convention 144* (Audio Engineering Society)

[25] Paulo J P, Coelho J B and Figueiredo M A 2010 *The Journal of the Acoustical Society of America* **128** 1747-1754

[26] Ramos-Romero C, Leon-R os P, Al-Hadithi B, Sigcha L, De Arcas G and Asensio C 2019 *Measurement* **146** 718-727

[27] Bschorr O, Mittmann J and Wolf A 1981 *Theoretische und experimentelle Untersuchungen zur Abstrahlung von Reifenlarm: Zwischenbericht; Contract 522-7291-TV-79-713* (Zentrale Berichtsstelle, Technisch-wissenschaftliche Information)

[28] Mohamed Z and Wang X 2015 *Mechanical Systems and Signal Processing* **50** 498{509

[29] Krauss O and Gauterin F 2012 *EURONOISE Prague 2012-Conference Proceedings*

[30] Merenda M, Porcaro C and Iero D 2020 *Sensors* **20** 2533

[31] Murphey Y L, Guo H and Feldkamp L A 2004 *Applied Intelligence* **21** 117-128

[32] Ahead R R 2009 *Publication RRA-1. AASHTO, US Department of Transportation*

Suitability of 3D printing materials for printing anthropomorphic phantoms: A simulation study

Nikolay Dukov[1], Elica Encheva[1,2] and Kristina Bliznakova[1,a]

[1] Medical University "Prof. Dr. Paraskev Stoyanov", Varna, Bulgaria
[2] University Hospital "St. Marina" , Varna, Bulgaria

[a]E-mail: kristina.bliznakova@mu-varna.bg

Abstract. The use of anthropomorphic phantoms finds an increased application in the last years. From their use in virtual studies to assessment of new imaging technologies and reconstruction algorithms. The current study proposes a resource saving approach, in terms of time and money for assessing the creation of an anthropomorphic model by means of 3D printing. A four component object consisting of segmentations of adipose, gland, skin and lesion tissues was created from an MRI image set. From the resulting object two computational breast models were created – one with the elemental composition of the real breast tissues and a second one with the elemental composition of the low cost 3D printing materials planned to be used for the phantom manufacturing. Then, an in-house developed software was used to generate mammography images, which were subjected to both visual assessment and profile comparison. In this study we used nylon and acrylonitrile butadiene styrene 3D printing materials to represent the x-ray properties of the different breast tissues. The results showed that these 3D materials well represent the x-ray absorption characteristics of both glandular, adipose and skin tissues, while further investigation on suitable materials for representing the lesion characteristics is needed.

1. Introduction

Anthropomorphic phantoms are a valuable tool, which has been increasingly used during the last years in carrying out virtual studies dedicated to assess the efficiency of a given treatment, the development and optimization of new medical technologies, as well as assessment of new algorithms for image reconstruction. In the field of breast imaging, anthropomorphic phantoms may be used for quality control procedures, as well as for optimisation of existing, emerging and performance testing of novel imaging techniques. In general, the anthropomorphic breast phantoms may be classified into computational and physical phantoms [1, 2]. The computational phantoms are created by applying a wide range of techniques such as constructive solid geometry, tomographic, non-uniform rational B-splines, polygon-mesh, and hybrid techniques to model the breast as accurate as possible, with all anatomical variations which arise during its development as well as between individuals [3-7]. Physical anthropomorphic phantoms could be produced by molding techniques, 3D printing and low cost paper-based approaches. Among them, 3D printing technologies provide an excellent opportunity to create realistic models of the breast by using a number of printing materials with physical and x-ray characteristics similar to these of human tissues. Although, there is a substantial advancement in 3D technologies and the creation of new 3D printing materials, the number of 3D printing materials which satisfy the requirements for radiological similarity with breast tissue data is still limited, especially for representing lesion tissue [2]. Furthermore, the process of creating a physical x-ray anthropomorphic phantom is lengthily, even with 3D printing technologies. Traditionally, this is an iterative process that involves 3D printing with a given material/materials, scanning at an x-ray facility and evaluating the results for suitability. However, making use of software tools for virtual studies to assess the appropriateness of a given 3D printing material for the physical production of x-ray anthropomorphic phantoms can greatly reduce the time and costs involved.

The purpose of this study is to define whether the available low cost 3D printing materials used with fused deposition modelling (FDM) printing technique, are suitable for the production of a four component breast anthropomorphic phantom dedicated to x-ray imaging studies. The study is entirely original and will result in a developed methodology for studying different 3D printing materials, to be used with other 3D printing materials and anthropomorphic phantoms.

2. Materials and methods

An overview of the proposed approach for exploring the possibilities for production of anthropomorphic breast phantoms with low cost 3D printing materials is shown in figure 1. Shortly described, the source of the anthropomorphic phantom is from MRI breast imaging. The tomographic data are then segmented into four breast tissues: adipose, gland, skin and lesion. The obtained segmentation data is then combined into one three-dimensional object. Based on that object two matrices are created with: a) a matrix with the elemental composition of the real breast tissues and b) a matrix with the elemental composition of the 3D printing materials that are planned to be used in printing the physical anthropomorphic breast model. Finally, a comparison of the obtained simulated mammography images is performed.

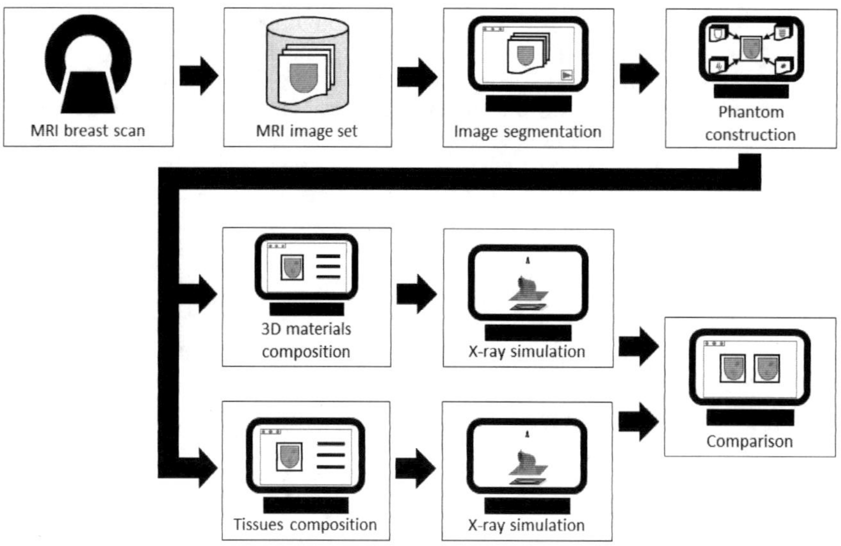

Figure 1. Overview of the approach to define whether specific 3D printing materials are suitable for the production of four component anthropomorphic phantom dedicated to x-ray breast imaging.

2.1. Patient data

For the creation of the breast phantom a set of breast MRI images from T1-weighted scan were used. The image set consists of images with voxel size of 0.70mm x 0.70mm x 0.80 mm of real patient diagnosed with breast cancer. The scanning was carried out with GE Signa HDxt 1.5 T MRI scanner and with contrast agent present, while the scanning sequence was axial multi-phase VIBRANT (3-phase).

This retrospective study was approved by the Ethics Committee of Medical University of Varna (Approval number 102/22.04.2021).

2.2. Segmentation and phantom construction

In order to separate the four tissue types of interest for x-ray imaging: adipose, gland, skin and lesion, a segmentation procedure was developed in Matlab R2020a and performed on the MRI image set. The procedure is semi-automatic and starts with the DICOM images from the set, followed by a set of thresholding and morphological operations, and a reviewing process from a medical doctor with expertise in the field of medical imaging for removing residual artifacts. After a region of interest is chosen (ROI), the skin is segmented followed by the glandular tissue and the tumor tissue. Then the three components are assembled together into one object.

With the purpose of segmenting the adipose tissue a mask covering the whole breast is first generated. This mask is then subtracted with a mask representing the previously created object from the segmentations of the skin, gland and lesion tissues. This results in an "adipose mask", which is used for the segmentation of the adipose tissue.

Finally, the adipose tissue is added to the previously created three component object, now resulting in a four component object comprised of adipose, gland, skin and lesion tissues.

2.3. Computational models

Two computational breast models were created based on the four component object: one model, corresponding to the elemental composition of the skin, gland, adipose and tumor tissues (CBM1) and a second model with suggested 3D printing material substitutes for the four tissues (CBM2). The elemental composition for CBM1 and CBM2 are shown in table 1. Data for the elemental composition of the real breast tissues and the 3D printing materials were taken from [8]. The 3D printing materials were acrylonitrile butadiene styrene (ABS) and nylon for approximating the radiological properties of adipose and the rest of the three tissues, respectively. Their choice is based on the investigation of the radiological properties of the different materials for the purposes of creating 3D anthropomorphic phantoms [8].

Table 1. Elemental composition and materials used for the two breast computational models CBM1 and CBM2.

	CBM1	CBM2	
Tissue	Elemental composition	3D printing material	Elemental composition
Adipose	H: 0.112	ABS	H: 0.075
	C: 0.619		C: 0.855
	N: 0.017		N: 0.053
	O: 0.251		O: 0.016
	P: 0.001		S: 0.001
Gland	H: 0.102	Nylon	H: 0.099
	C: 0.184		C: 0.626
	N: 0.032		N: 0.119
	O: 0.677		O: 0.156
	P: 0.005		
Skin	H: 0.098	Nylon	H: 0.099
	C: 0.178		C: 0.626
	N: 0.050		N: 0.119
	O: 0.667		O: 0.156
	P: 0.007		
Lesion	H: 0.102	Nylon	H: 0.099
	C: 0.184		C: 0.626

N: 0.032 O: 0.677 P: 0.005	N: 0.119 O: 0.156

As can be seen from table 1, for the purpose of mimicking the different breast tissues, for the breast model CBM2, two materials are consider – Nylon and ABS, which are available for fuse deposition modelling printers (FDM). Thus, resulting in a simulation of a manufactured breast model of a printing procedure with two materials.

2.4. Simulation

For the purposes of the simulation study a dedicated in-house developed software was used [9, 10]. This software tool allows simulation of several x-ray imaging acquisition geometries such as radiography, limited tomography, as well as circular and helical computed tomography. The tool is comprised of two basic modules: a) Module for computational modelling of phantoms, which includes solid geometry and voxel based approaches; b) Module for modelling radiological images, which includes different geometries such as isocentric and partial isocentric rotational geometries, as well as radiography and mammography. The main parameters of the simulated geometry are the distances from the source to the center of rotation and to the detector (SID, SDD), as well as the length of the gantry acquisition arc and the angle step, including imaging and beam parameters.

The two created models were used with an x-ray simulations of a mammography setup, with the modelled geometry shown in figure 2. Specifically, in our simulation, the x-ray tube and the detector were placed on a distance SDD set to 800 mm, while the distances SID was 650 mm. The simulated projection images were modelled with a pixel size of 100 μm and the incident fan beam energy was set to 19 keV. Noise modelling was not concerned in this simulation study.

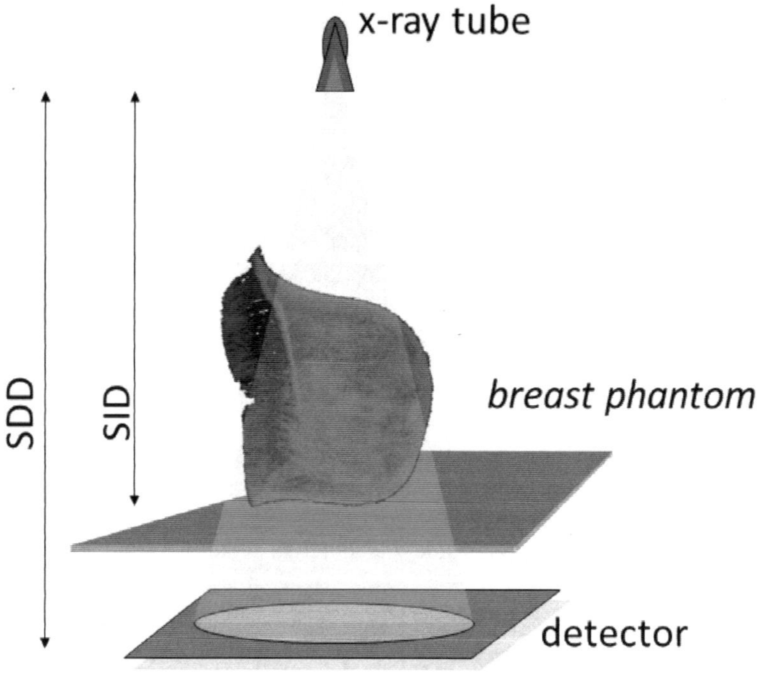

Figure 2. Modelled geometry for the simulation of x-ray images. SDD is the source to detector distance, while SID is the source to isocenter distance.

The generated x-ray images are line-integral images, presenting the $\mu(x, y, z)dl$ in the Beer's law:

$$I = I_0 * \exp\left(-\int_{ll} \mu(x, y, z) dl \right) \quad (1)$$

where $\mu(x,y,z)$ is the spatially dependent linear attenuation coefficient, l is the path length through the object and I_0 is the intensity of radiation at the source segment that emits to the area of the detector, taken equal for all viewing angles.

2.5. Comparison

The comparison of the resulting simulations of models CBM1 and CBM2 was performed by means of comparison of line profiles taken through the generated planar mammography images as well as by means of visual assessment.

3. Results

Figure 3 shows the created software breast model from the breast MRI data. Figure 3a shows a selected slice from the originally created 3D breast model (before any tissue assignment), while figure 3b shows the three-dimensional visualisation of the whole original computational breast phantom.

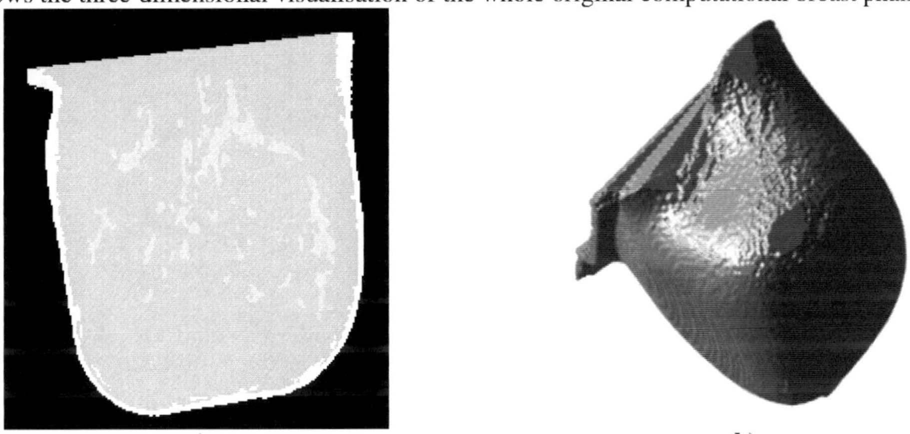

a) **b)**

Figure 3. Created software breast phantom from MRI data – a) a slice from the breast phantom, b) three dimension visualisation.

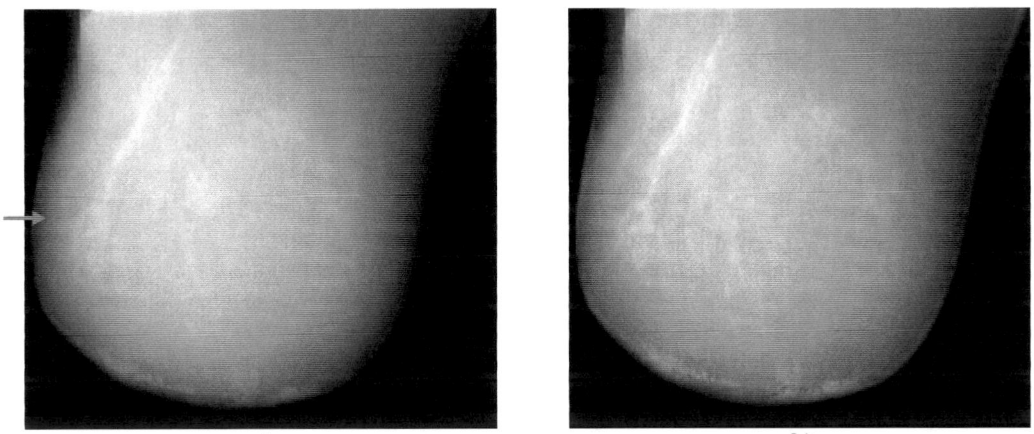

a) **b)**

Figure 4. Results from the x-ray simulations of the created software breast phantom from MRI data – a) x-ray simulation with elemental composition corresponding to adipose, gland, skin and lesion tissues (CBM1), b) x-ray simulation with elemental composition corresponding to 3D printing material substitutes for the four tissues: adipose, gland, skin and lesion tissues (CBM2). The red arrow indicates the horizontal position of the plot profiles shown in figure 5.

Figure 4 visually depicts the results from the x-ray simulations of the two breast matrices – elemental composition of breast tissues (CBM1) and elemental composition of suggested 3D printing material substitutes (CBM2).

Figure 5 shows the line profiles taken through the x-ray simulated projections depicted in figure 4. The line profiles in both images are taken horizontally and at the same position. In figure 4, the red arrow on the left next to figure 4a, indicates the position at which the horizontal profile lines were drawn. The visual representation of the numerical data obtained from the line profiles of both x-ray projections are combined into a single figure (figure 5), for better visualization assessment and comparison.

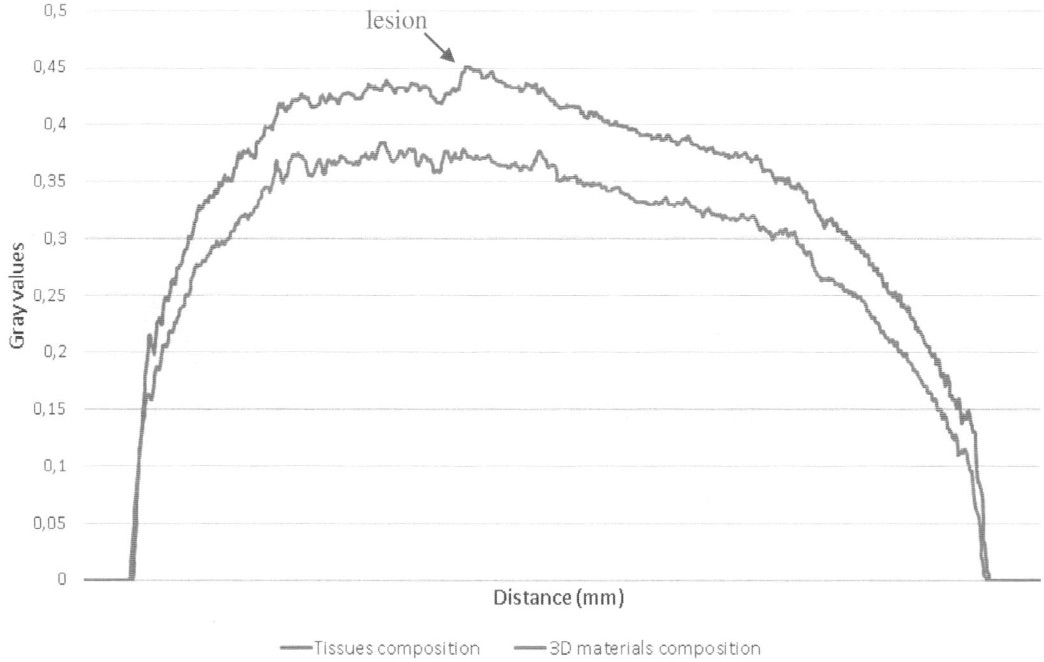

Figure 5. Profiles, taken horizontally from end to end in the middle of the images shown in figure 4.

4. Discussion

This work tackles the issue of creating anthropomorphic breast phantoms for x-ray imaging studies with widely available and affordable 3D printing materials for FDM printers. As such task is cumbersome, time consuming and expensive we opt for a computerized solution by utilizing computational models of the breast model candidate for 3D printing and x-ray simulation techniques. The proposed approach can quickly show if the 3D materials to be used in printing physical models for radiological purpose are suitable and thus this will result in reduced costs.

The chosen breast MRI image set is from a patient with rather large mammary glands, which provides a good opportunity, due to the size, for assessment of the x-ray characteristics of the tissues of interest and the considered tissue substitutes.

We opted for a semi-automatic segmentation of the tissues, as even if the segmentation procedure was fully automated, close collaboration with a medical doctor with expertise in the field is necessary as well as thorough evaluation, which is not the focus of this study. The segmentation and construction procedures described in section 2.2 resulted in a breast model with similar shape to a real breast shape, as seen from figure 3. Furthermore, the distribution of the breast glandular tissue as well as the skin after segmentation appear similar to the distribution of these tissues in real patient images [11-14].

The x-ray simulation of the first computational breast model (CBM1), displayed in figure 4a, showed that the lesion is well differentiated from the rest of the tissues. The rest of the tissues (adipose, gland and skin) also show expectable visual characteristics and the projection image is considered to be realistic. Performing x-ray simulation on the second computational model (CBM2), which considers 3D printing materials (shown in figure 4b) shows that in the projected image the tumor is not well differentiated from the rest of the tissues. The rest of the tissues (adipose, gland and skin), however were evaluated to appear realistically on the simulated x-ray image.

The comparison of profiles (shown in figure 5), taken across the simulated planar images (figure 4) shows a relatively good match with a maximum difference of 15% between the line-integral images. Although, the first model (CBM1) exhibits slightly higher grey values compared to the second (CBM2), the profile patterns of both models follow the same trend. The comparison also indicates that in the projection image of the breast phantom with the 3D printed materials (CBM2), the lesion structure is not well represented. This can be observed on the graph where an increase of grey values to 0.45 for the tissue composition (CBM1) in figure 5 (the red arrow in figure 5 indicates the position) stemming from the influence of the tumor formation. In the case of 3D printing materials composition (CBM2) however, similar increase in that position is not prominent, indicating that the material used for the tumor formation does not exhibit desirable characteristics. Thus, another 3D printed material is needed to adequately represent the radiological properties of the breast lesions. Such a material might be polylactic acid or nylon mixed with a 3D material with higher density.

5. Conclusions

The current study tested through modelling and simulations, the performance of two widely available 3D printed materials for FDM: ABS and nylon to “manufacture” breast phantoms for x-ray imaging purposes. The results showed that the use of nylon and ABS is appropriate for the development of anthropomorphic models and that these 3D printing substitutes well represent the absorption characteristics of glandular and adipose tissues, while in the case of tumor representation, there is a need for further exploration of new materials with suitable such characteristics. Through a simulation study of 3D materials and their comparison by means of subjective and objective evaluation with the x-ray behaviour of real breast tissues, this approach turns out to reduce the cost and time related to manufacturing of anthropomorphic x-ray breast phantoms. Future work concerns testing of more 3D materials and manufacturing the physical version of these models.

References

[1] Glick, S.J., Ikejimba, L.C.: Advances in digital and physical anthropomorphic breast phantoms for x-ray imaging. Medical Physics. **45**(10), 870-885 (2018)

[2] Bliznakova, K.: The advent of anthropomorphic three-dimensional breast phantoms for X-ray imaging. Physica Medica. **79**, 145-161 (2020)

[3] Xu, X.G.: An exponential growth of computational phantom research in radiation protection, imaging, and radiotherapy: A review of the fifty-year history. Physics in Medicine and Biology. **59** (18), 233-302 (2014)

[4] Kainz, W., Neufeld, E., Bolch, W.E., Graff, C.G., Kim, C.H., Kuster, N., Lloyd, B., Morrison, T., Segars, P., Yeom, Y.S., Zankl, M., Xu, X.G., Tsui, B.M.W.: Advances in computational human phantoms and their applications in biomedical engineering - A topical review. IEEE Transactions on Radiation and Plasma Medical Sciences. **3** (1), art. no. 8556501, 1-23 (2019)

[5] Bliznakova, K., Buliev I., Bliznakov, Z.: Anthropomorphic Phantoms in Image Quality and Patient Dose Optimization: A EUTEMPE network, Publisher: Institute of Physics Publishing (2018)

[6] Mamdouh, R., El-Bakry, H. M., Riad, A., El-Khamisy, N.: Converting 2D-Medical Image Files "DICOM" into 3D- Models, based on Image Processing, and Analysing their Results with Python Programming. WSEAS Transactions on Computers. **19**, 10-20 (2020)
[7] Dib, R.: Active Contours with Term of Smoothing: Application in Medical Imaging. WSEAS Transactions on Signal Processing. **16**, 108-117 (2020)
[8] Ivanov, D., et al.: Suitability of low density materials for 3D printing of physical breast phantoms. Phys Med Biol. **63**(17), 175020 (2018)
[9] Bliznakova, K., I. Buliev, and Z. Bliznakov, Anthropomorphic Phantoms in Image Quality and Patient Dose Optimization, in A EUTEMPE Network book. 2018, IOP Publishing.
[10] Bliznakova, K., et al.: Experimental validation of a radiographic simulation code using breast phantom for X-ray imaging. Comput Biol Med. **40**(2), 208-214 (2010)
[11] Dukov, N., et al.: Models of breast lesions based on three-dimensional X-ray breast images. Physica Medica. **57**, 80-87 (2019)
[12] Bliznakova, K., et al.: Development of breast lesions models database. Physica Medica. **64,** 293-303 (2019)
[13] Sarno, A., et al.: Dataset of patient-derived digital breast phantoms for in silico studies in breast computed tomography, digital breast tomosynthesis, and digital mammography. Med Phys (2021)
[14] Sarno, A., et al.: Advanced Monte Carlo application for in-silico clinical trials in X-ray breast imaging. SPIE. (2020)

Acknowledgements
This research is supported by the Bulgarian National Science Fund under grant agreement DN17/2.

Computational simulations and assessment of two approaches for x-ray phase contrast imaging

Kristina Bliznakova[1,a], Zhivko Bliznakov[1] and Nikolay Dukov[1]

[1] Medical University "Prof. Dr. Paraskev Stoyanov", Varna, Bulgaria

[a]E-mail: kristina.bliznakova@mu-varna.bg

Abstract. X-ray phase-contrast imaging is a high-resolution imaging that permits an increase of the perceptibility of the details in three-dimensional objects, such as human tissues compared to conventional absorption imaging. There are different approaches for implementing phase-contrast imaging and their introduction into clinical practice requires advanced computational tools. A long-term goal of our research is the development of computational models of breast phase-contrast imaging. The aim of this study is to develop a software module for implementing grating-based phase-contrast imaging. For this purpose, an existing in-house software application for x-ray imaging with a function to model and simulate propagation-based phase-contrast x-ray images has been extended to include a model of grating-based imaging. To test the new functionality, four computational phantoms reflecting features, which can be screened in the real breast tissue and which differ in their complexity, were designed. Planar x-ray images in absorption, propagation-based and grating-based modes were generated and compared. Results showed improved visual appearance of the simulated objects in images obtained by simulating grating-based imaging setup. The developed subroutine is planned to be experimentally validated at synchrotron facility. The new software functionality will be exploited in studies related to new x-ray imaging techniques for breast screening and diagnosing.

1. Introduction

Breast cancer is the most common incident and mortality form of cancer for women. From the estimated 3.91 million new cases of cancer in Europe in 2018, 28.2% are due to breast cancer, while from the estimated 1.93 million deaths from cancer, 16.2% are caused by breast cancer [1]. Nowadays, mammography continues to be the 'gold technique' in screening and diagnosing the breast. Mammography images present a 2D distribution of the intensity of the x-rays, which are transmitted through the woman breast and then reach the detector. The technique is based on the x-ray absorption properties of the tissues in the breast. In case of dense breasts, however, tumour tissues could be hidden by the dense breast tissue; thus the mammography technique in this particular situation may result in superposition of large areas of the glandular tissue and the cancerous ones on the 2D detector, which will influence the diagnose of these structures [2]. Therefore, new and more efficient x-ray imaging techniques dedicated to screening and diagnosing the breast tissue, such as phase-contrast mammography [3, 4], breast computed tomography [5, 6] and contrast-enhanced mammography [7], are continuously under development, extensive testing and optimisation.

Phase-contrast mammography is an emerging alternative approach to the absorption based mammography. The base of the x-ray phase-contrast imaging is both, (a) the attenuation of the x-rays, and (b) the change in the phase of the x-rays, which is related to diffraction and refraction effects when x-rays propagate in the breast tissue [8, 9]. The amount of phase change in a tissue is associated with the local value of its wavelength-dependent refractive index n, expressed as

$$n(\lambda) = 1 - \delta + \iota\beta \tag{1}$$

where $n(\lambda)$ is the complex refractive index of the tissue at wavelength λ, β is the absorption index, δ is the decrement (with respect to unity) of the real part of the refractive index n and $\iota^2 = -1$ [8]. The coefficient β reflects the x-ray attenuation in a tissue thickness s, while the coefficient $\delta > 0$ determines the total phase change

$$\Phi = -\frac{2\pi}{\lambda}\int_0^s \delta(z)dz \tag{2}$$

integrated along the direction of propagation, z, of the x-ray beam.

During the years, different techniques have been proposed for detecting phase-contrast effects: propagation-based approach, diffraction-enhanced imaging, crystal interferometry, grating interferometry, the edge-illumination approach, and speckle-tracking [9, 10]. Amongst the popular approaches that have been implemented for clinical work, is the so-called in-line or propagation-based phase-contrast imaging approach that already have clinical implementation for breast cancer imaging [11]. In this approach, there are no optical elements, such as gratings and apertures, placed in the beam path during acquisition of x-ray images. In this mode, intensity changes in the Fresnel near field at the surface of a fine-pitch area detector positioned at a distance from the object (as shown in figure 1a), are related to the transverse Laplacian of the phase changes in the object. With this arrangement, the visual appearance of phase-contrast effects in the final x-ray image presents a characteristic contrast enhancement at the interfaces between structures with different x-ray refractive indices [12]. This mode has been computationally modelled and used in successful simulations of phase-contrast mammography and breast tomosynthesis by our team [12-16]. In addition, this approach was implemented at ELETTRA synchrotron facility via the mammography SYRMEP project [17, 18]. One of the limitations of the propagation based approach is the clinical implementation, which is done only at difficult to access synchrotron facilities. An encouraging approach is developing of grating interferometry (with a typical design shown in figure 1b), which may turn to be applicable for imaging of the whole breast. This approach is implemented by using phase and analyser gratings. The most flexible approach to carry out development and optimisation work is the approach of modelling the components of the imaging system and subsequently imaging simulation.

There is a limited number of computational studies in the field of phase-contrast grating imaging reported in the literature [19-24]. All these studies are based on the use of simple objects, such as spheres and cylinders, and aim at proof of concept and optimisation of dose and imaging parameters. To the best of our knowledge, studies in the field of mammography have not been reported. To construct a model of a breast imaging technique based on gratings and advance in this field, a realistic three-dimensional (3D) model of the female breast with accurately modelled anatomical details and a computational model of the grating-based imaging itself, are needed.

A long-term goal of our research is the development of computational models of phase-contrast imaging for screening and diagnosing the breast. The research is entirely original, aiming at a complete computational platform with anthropomorphic breast models and modelled different approaches for phase-contrast imaging allowing comparison of techniques and further optimisation. Our team has the expertise in creating realistic computational breast phantoms [25, 26], which turns out to be the main tool in studying novel imaging techniques, amongst which is also the phase-contrast imaging. The propagation-based phase-contrast approach was already successfully modelled for mammography 2D and 3D imaging [12-16]. The model of the phase-contrast based on gratings is still to be developed. The first step in achieving this goal is the development of a dedicated software, capable of modelling the phase-contrast grating imaging. The aim of this study is to develop a software module for implementing grating-based x-ray phase-contrast imaging and to assess the results of applying this technique in comparison to both, absorption and propagation-based x-ray imaging approaches.

Journal of Physics: Conference Series **2162** (2022) 012013 doi:10.1088/1742-6596/2162/1/012013

2. Materials and methods

2.1. Simulated setup

Figure 1 shows the simulated phase-contrast setups for generation of x-ray images. The propagation-based mode, shown in figure 1a, is modelled based on the Fresnel-Kirchhoff diffraction theory [8] and is described in details in [12]. The model considers pure coherent radiation of wavelength λ. The x-rays are emerging from a point-like monochromatic source, located at a distance SID (source-to-isocenter distance) from the object plane. The object is modelled in respect to the isocenter point, while the detector is located at a distance SDD (source-to-detector distance) from the source. The propagation-based model was validated to correctly reproduce phase-contrast images against images of simple and complex physical phantoms obtained from experimental work carried out at the European Synchrotron Radiation Facility (ESRF), Grenoble, France. In propagation mode, the detector is placed away from the object in order to register the phase shift, while no optics are used.

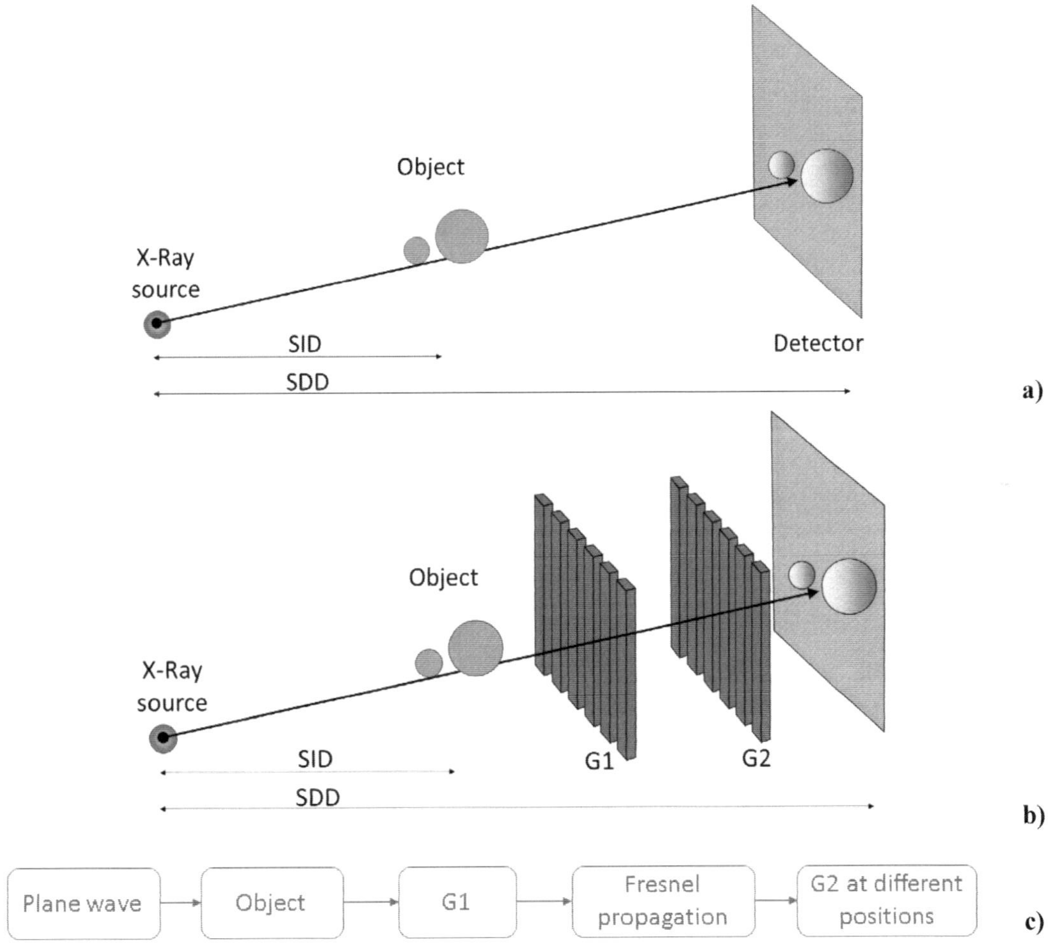

Figure 1. Simulated phase contrast setup. a) free propagation setup, b) grating-based setup, c) main steps in implementing the grating-based phase-contrast imaging.

The modelled grating-based phase-contrast setup is demonstrated in figure 1b. It is based on a coherent x-ray source that emits an x-ray beam towards the detector [27, 28]. The first grating, G1, is the phase grating with a phase shift of π that generates an interference pattern, known as Talbot carpet registered at a distance

$$d_m = (m-1)\frac{g_1^2}{4\lambda} \tag{3}$$

In this equation, g_1 is the grating pitch, while m is a positive integer value. The second grating, G2, is the analyser grating, placed just in front of the detector at a distance d_m from the phase grating G1. The pitch g_2 of the analyser grating matches the periodicity of the undistorted interference pattern at the corresponding detector position.

In the computational implementation of the grating-based phase-contrast imaging, G2 is always placed at a distance at which the Talbot carpet is obtained. The main steps in implementing this imaging are shown in figure 1c. The plane wave is propagated through the object and the first grating G1 by applying the x-ray tracing algorithm [12] followed by propagation of the field, implemented by using the Fresnel diffraction integral. Finally, the propagation of the wave through the G2 is performed. For the simulations performed in this study, the phase grid G1 was designed to have a pitch of 2.5 μm, while the analyser grid G2 had a pitch of 1.25 μm.

2.2. *Computational phantoms*

The computational phantoms were designed with the in-house developed software application *XRayImagingSimulator* [12]. This simulator has three main modules: a module for creation of solid-based 3D phantoms, a module that creates x-ray images, when simulated x-rays with a given energy (in keV), following transmission through the computational phantoms, reach the detector, and a module for visualization of both, created computational phantoms and generated x-ray images.

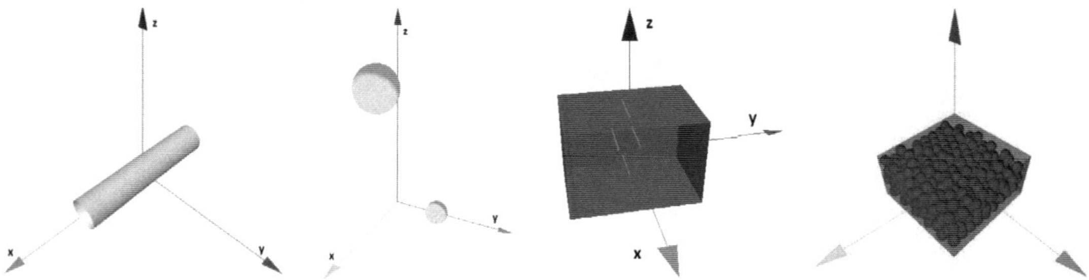

a) **b)** **c)** **d)**

Figure 2. Computational phantoms used in the simulations: a) a nylon wire with a diameter of 2 mm; b) two water spheres located at different z heights and with diameters of 2 mm and 1 mm, respectively; c) a PMMA container with a thickness of 40 mm and four air cylinders with a diameter of 1 mm, d) a paraffin container with a thickness of 5 mm and an area 10 mm x 10 mm and 473 water spheres with a diameter of 1 mm.

Four computational phantoms, shown in figure 2, were created in order to test and validate the software code. The first phantom, shown in figure 2a, consists of a nylon wire 2 mm in a diameter placed in the air. The second phantom, shown in figure 2b, is represented by two water spheres with diameters of 2 mm and 1 mm placed at different positions in air, while the third phantom contains four cylinders filled with air, having diameters of 1 mm, arranged at different planes, as shown in figure 2c. Finally, the complex phantom, phantom 4 shown in figure 2d, is represented by a paraffin container filled with 473 water spheres with a diameter of 1 mm. The phantoms in figure 2 are shown together with the coordinate system of the modelled x-ray imaging geometry, depicted in figure 1. The center of the x-ray imaging system is the isocenter of the system. In +z direction, a point-like x-ray source is modelled. In –z direction, the flat panel detector is modelled as a parallelepiped, absorbing all the x-rays reaching its surface.

Three acquisition setups were simulated by implementing planar acquisition x-ray geometry, i.e. one x-ray image is generated. The acquisition geometry is described with the two distances SID and

SDD (shown in figure 1), which were set to 1000 mm and 1300 mm, respectively. The high-resolution projection images are of a pixel size of 5 μm and an image size between 1500 pixels x 1500 pixels and 12000 pixels x 12000 pixels, depending on the computational phantom size. Incident x-ray beam energy was set to 20 keV. Detector noise was not considered in this simulation. Registered x-ray images present line integrals. The propagation of the x-rays until the G1 are based on the Fresnel–Kirchhoff diffraction theory, computationally described in [12]. The propagation through the gratings G1 and G2 is implemented by the approach discussed in section 2.1.

3. Results and discussions

Figure 3 shows side by side the simulated x-ray projection images of the first three computational phantoms for the simulated absorption, propagation-based and phase-contrast x-ray imaging acquisition geometries. Although simple in design, these phantoms are of imaging interest, i.e. optimisation of acquisition distances, x-ray energy, tissue composition of objects, gratings material. These computational objects, modelled in with shapes and sizes, as shown in figure 2, could approximate well normal and abnormal tissues in real breast. Rows 1 and 2 correspond to the first computational phantom (the nylon wire, shown in figure 2a), rows 3 and 4 correspond to the second computational phantom (the two water spheres, shown in figure 2b), while rows 5 and 6 correspond to the third computational phantom (the PMMA container with air channels, shown in figure 2c). For each phantom, there are three x-ray projection images simulated in absorption mode (figure 3a), propagation mode (figure 3b) and grating-based approach (figure 3c). The absorption mode is implemented by applying the geometry, shown in figure 1a with the detector, placed behind the object.

The plotted profiles, shown below the corresponding x-ray projection images, are taken through the center of the projection images. The first column, figure 3a, shows the simulated projection images when the detector was placed just behind the phantoms (i.e. no phase-contrast effects are observed), while second and third columns, figure 3b,c, show the simulated projection images when the detector was placed at 300 mm away from the phantoms. For all phantoms, the visual comparison of the projection images shows that both, the propagation-based and the grating-based acquisition geometries result in improved object edges compared to the absorption image (first column in figure 3). This observation is well supported by the comparison of the line profiles, which show quantitatively the edge effect due to the phase effect. Specifically, the objective evaluation, which included calculation of the edge enhancement in phase-contrast images (figure 3b,c) by using edge-enhancement index, EEI [29]:

$$EEI = \frac{(P-T)/(P+T)}{(H-L)/(H+L)} \tag{4}$$

where P and T are the peak and trough intensity values at the edge in the plot graphs in figure 3b,c, while H and L are values taken in the high- and low-intensity regions next to the edge [29]. The results for the EEI shows that for *the nylon wire* the edge enhancement due to the phase effect is approximately 5 times higher for grating-based acquisition geometry than for propagation-based imaging. Similarly, the edge enhancement for the second and the third phantoms, i.e. the *water spheres* and the *PMMA container with four air channels*, in case of grating-based acquisition imaging were approximately 4 and 3 times, respectively compared to the propagation-based approach, demonstrated in figure 3b. Finally, the resulting phase-contrast images with gratings visually match with these reported in the study of Fu *et al* [23], based on numerical calculations for a phantom composed of spheres and one cylinder.

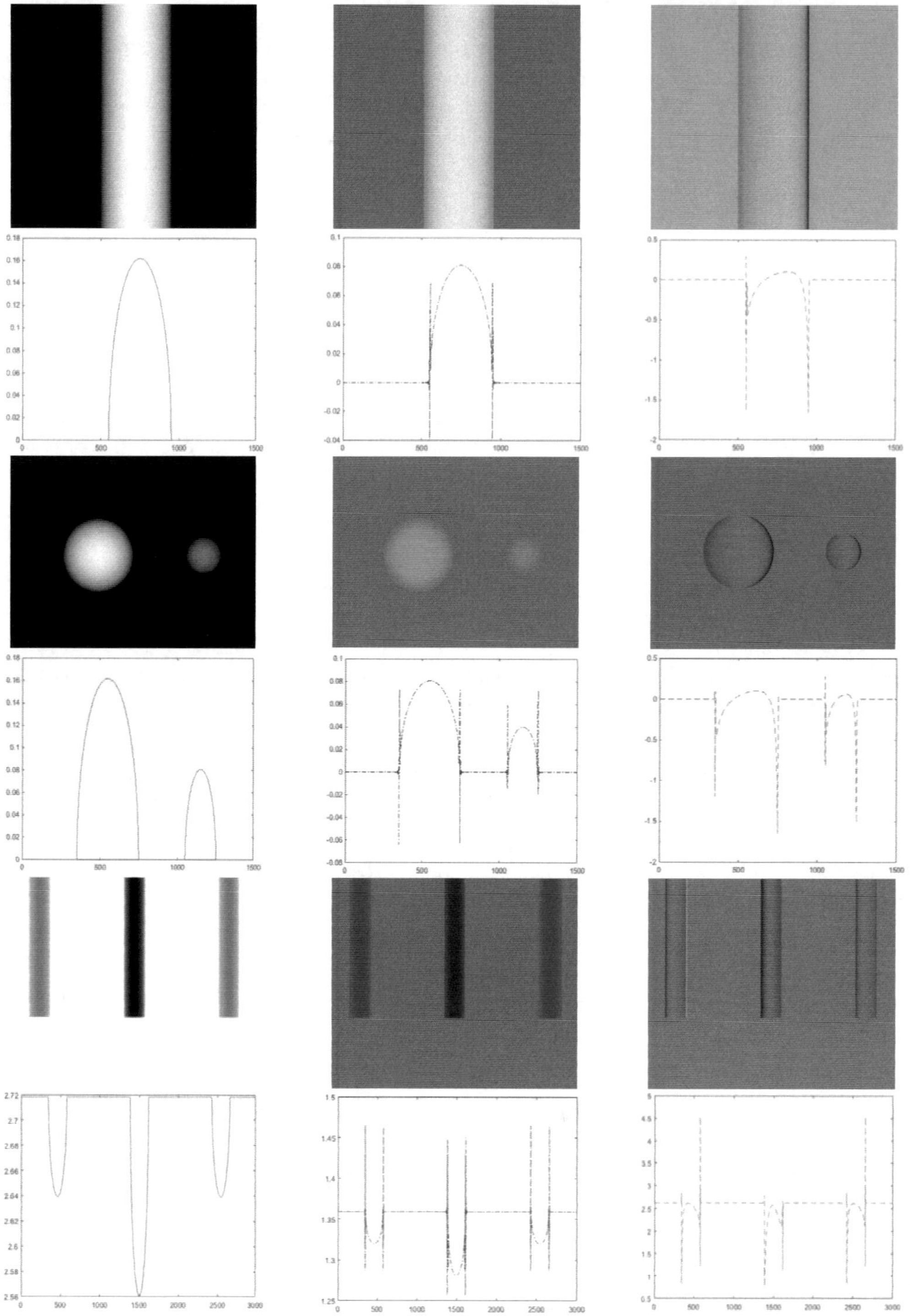

A.

B.

C.

a) b) c)

Figure 3. Comparison of a) – c) projection images of the three phantoms: A. cylinder, B. phantom with two spheres, C. PMMA container with four air channels obtained through simulation in absorption mode a), propagation mode b), and grating-based approach c). The plotted profiles (in a. u.) through the center of the projection images are displayed bellow the corresponding x-ray projection images.

To speed the process of producing the images, the software code for producing x-ray projection images based on grating approach was rewritten for parallel programming using the *omp* library. While for the third phantom, the time needed to obtain an image with resolution of 5 μm and size of 1500 pixels x 1500 pixels was 5 hours, the image obtained with the parallel code on a PC with 24 threads was only few seconds.

Further, this edge-enhancement is well demonstrated in the simulated phase-contrast images from the texture phantom, shown in figure 2d. This complex phantom is initially used to test the newly developed software module to produce grating-based phase-contrast images and to evaluate the visibility of the texture on these images. Figure 4 shows side by side the simulated projection images of the fourth computational phantom for the simulated acquisition geometries. Similarly to the presentation approach in figure 3, the first row in figure 4 shows the three projection images simulated in absorption mode (figure 4a), propagation mode (figure 4b) and grating-based approach (figure 4c). The second row in this figure shows regions of interest taken from the whole x-ray projection images.

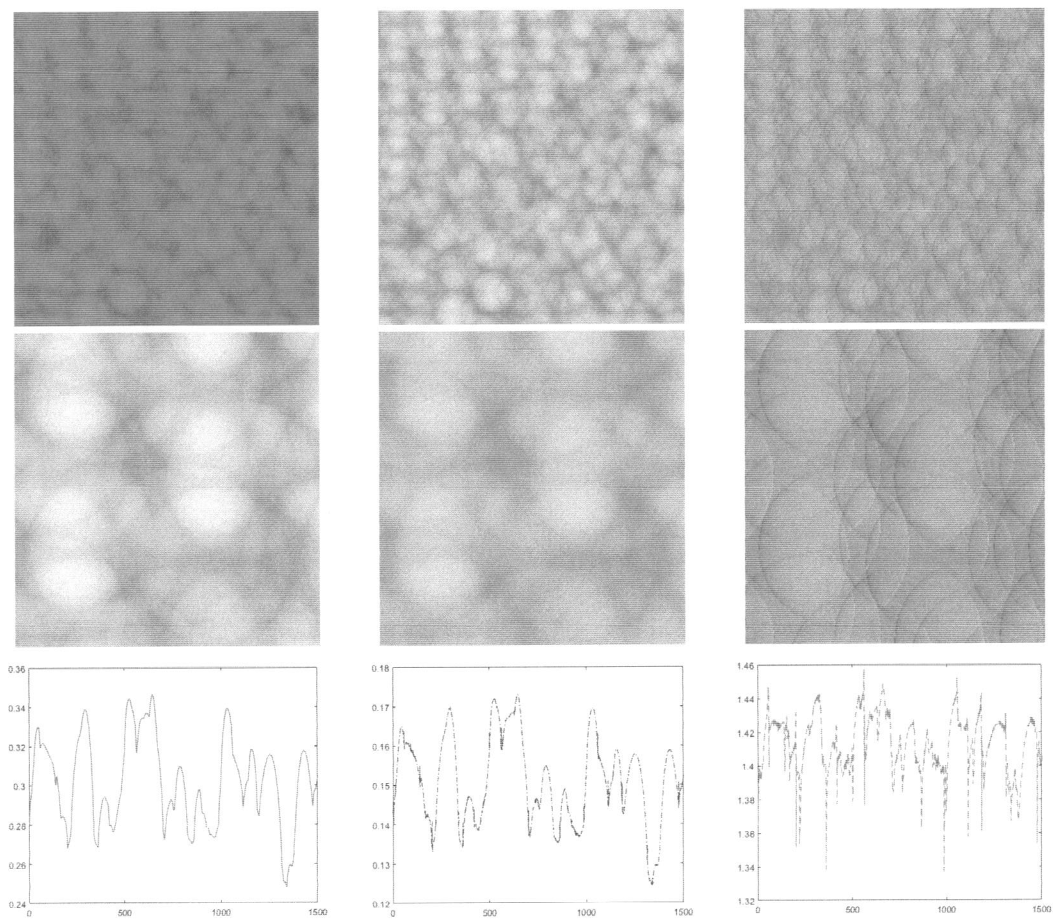

a) **b)** **c)**

Figure 4. Comparison of a) – c) projection images of phantom 4 obtained through simulations in absorption mode a), propagation mode b), and grating-based approach c). The second row of images is a zoomed in version of the first row, for better visualization. The plotted profiles (in a. u.) through the center of the projection images are displayed bellow the corresponding x-ray projection images.

The visual comparison of x-ray projection images showed well differentiated projected spheres on the detector plane for the grating-based acquisition geometry in respect to the other two simulated x-ray imaging acquisitions. This improvement is well demonstrated by the peaks in the plotted profiles, shown below the corresponding x-ray projection images, the latter taken though the center of the projection images. The calculated EEI for the spherical edges are 2 to 3 times higher in grating-based images compared to propagation-based images. The result from this experiment showed that texture is well visualised with apparent edge enhancement.

This simulation study showed that both, phase-contrast techniques result in improved edge enhancement effect, as the images obtained through the grating-based approach are superior in comparison to the images obtained in propagation-based setup. These simulation results are to be experimentally validated at synchrotron facility. The experimental procedure requires: (a) physical models of the computational phantoms, shown in figure 2, and (b) experimental acquisition of x-ray images of these physical model at synchrotron facility with the same acquisition setup, shown in figure 1. For the creation of physical models, the 3D printing facilities in our laboratory will be used [30, 31]. They will be placed on a hardware platform at a synchrotron beamline, providing x-rays, suitable for mammography applications. Images are then acquired at different distances and with the addition of gratings. Once the results of the proposed computational approach are validated, it will be used in investigations with anthropomorphic computational breast models.

4. Conclusions

This computational study demonstrated the development of a software module for generating x-ray grating-based phase-contrast images and was tested to generate phase-contrast images in comparison to x-ray simulated images, obtained in absorption and propagation-based acquisition geometries. The simulation results showed that the module can produce phase-contrast images in grating-based acquisition geometry. When compared to absorption and propagation-based geometries, the grating-based x-ray phase-contrast images demonstrate higher edge enhancement for all computational phantoms used in the study, even for the phantom containing highly structured background. The module will allow comparison of different phase-contrast techniques, applied for detection of various types of lesions, which can be found in the breast tissue. A necessary step to be undertaken is the experimental validation of this software module, which validation could be easily performed at a synchrotron facility. Once validated, it will be used in breast imaging studies with anthropomorphic physical and computational breast models.

References

[1] Ferlay J, Colombet M, Soerjomataram I, Dyba T, Randi G, Bettio M, et al. Cancer incidence and mortality patterns in Europe: Estimates for 40 countries and 25 major cancers in 2018. European Journal of Cancer 2018; 103: 356-87

[2] Yeom Y, Chae E, Kim H, Cha J, Shin H and Choi W. Screening mammography for second breast cancers in women with history of early-stage breast cancer: Factors and causes associated with non-detection. BMC Medical Imaging 2019; 19: art. no. 2

[3] Auweter S, Herzen J, Willner M, Grandl S, Scherer K, Bamberg F, el al. X-ray phase-contrast imaging of the breast - Advances towards clinical implementation. British Journal of Radiology 2014; 87: art. no. 20130606

[4] Momose A. X-ray phase imaging reaching clinical uses. Physica Medica 2020; 79: 93-102

[5] Aminololama-Shakeri S, Hargreaves J, Boone J and Lindfors K. Dedicated Breast CT: Screening Technique of the Future. Current Breast Cancer Reports 2016; 8: 242-47

[6] Sarno A, Mettivier G and Russo P. Dedicated breast computed tomography: Basic aspects. Medical Physics 2015; 42: 2786-2804

[7] Suter M, Pesapane F, Agazzi G, Gagliardi T, Nigro O, Bozzini A, et al. Diagnostic accuracy of contrast-enhanced spectral mammography for breast lesions: A systematic review and meta-analysis. Breast 2020; 53: 8-17

[8] Born M. and Wolf E. Principles of optics. 6th Edition, Pergamon; 1980
[9] Taba S, Gureyev T, Alakhras M, Lewis S, Lockie D and Brennan P. X-ray phase-contrast technology in breast imaging: principles, options, and clinical application. American Journal of Roentgenology 2018; 211: 133-45
[10] Tao S, He C, Hao X, Kuang C and Liu X. Principles of different x-ray phase-contrast imaging: A review. Appl. Sci. 2021; 11: 2971
[11] Baran P, Pacile S, Nesterets Y, Mayo S, Dullin C, Dreossi D, et al. Optimization of propagation-based x-ray phase-contrast tomography for breast cancer imaging. Phys Med Biol 2017; 62 2315-32
[12] Bliznakova K, Russo P, Mettivier G, Requardt H, Popov P, Bravin A, et al. A software platform for phase contrast x-ray breast imaging research. Computers in Biology and Medicine 2015; 61: 62-74
[13] Bliznakova K, Kamarianakis Z and Buliev I. Computer-based platform for phase contrast tomosynthesis: targeting an application for breast imaging. In: Kyriacou E, Christofides S, Pattichis C. (eds) XIV Mediterranean Conference on Medical and Biological Engineering and Computing 2016. IFMBE Proceedings, vol 57. Springer, Cham. 2016. 57: 367-71
[14] Bliznakova K, Russo P, Kamarianakis Z, Mettivier G, Requardt H, Bravin A, et al. In-line phase-contrast breast tomosynthesis: A phantom feasibility study at a synchrotron radiation facility. Physics in Medicine and Biology 2016; 61: 6243-63
[15] Mettivier G, Bliznakova K, Sarno A, Lillo, F, Castriconi R and Russo P. In-line phase contrast tomography of the breast with a dedicated micro-CT scanner. 2016 IEEE Nuclear Science Symposium, Medical Imaging Conference and Room-Temperature Semiconductor Detector Workshop (NSS/MIC/RTSD), 2016: 1-3
[16] Bliznakova K, Mettivier G, Russo P and Bliznakov Z. Validation of a software platform for 2D and 3D phase contrast imaging: Preliminary subjective evaluation. Society of Photo-Optical Instrumentation Engineers (SPIE) Conference Series 2020; 11513: 1151312.
[17] Dreossi D, Abrami A, Arfelli F, Bregant R, Casarin K, Chenda V, et al. The mammography project at the SYRMEP beamline. Eur J Radiol 2008; 68: S58-S62
[18] Castelli E, Tonutti M, Arfelli F, Longo R, Quaia E, Rigon L, et al. Mammography with synchrotron radiation: first clinical experience with phase-detection technique. Radiology 2011; 259: 684-94
[19] Bartl P, Durst J, Haas W, Michel T, Ritter A, Weber T, et al. Simulation of x-ray phase-contrast imaging using grating-interferometry. IEEE Nuclear Science Symposium Conference Record, (NSS/MIC) 2009; 3577-80
[20] Wang Z, Huang Z, Zhang L, Chen Z and Kang K. Implement X-ray refraction effect in Geant4 for phase contrast imaging. IEEE Nuclear Science Symposium Conference Record, (NSS/MIC) 2009; 2395-8
[21] Langer M, Cen Z, Rit S and Letang J. Towards Monte Carlo simulation of X-ray phase contrast using GATE. Opt Express 2020; 28: 14522-35
[22] Sanctorum J, De Beenhouwer J and Sijbers J. X-ray phase contrast simulation for grating-based interferometry using GATE. Opt Express 2020; 28: 33390-412
[23] Fu J, Shi X, Guo W and Peng P. Fast X-ray Differential Phase Contrast Imaging with One Exposure and without Movements. Scientific Reports 2019; 9: 1-8
[24] Vignero J, Marshall N, Bliznakova K and Bosmans H. A hybrid simulation framework for computer simulation and modelling studies of grating-based x-ray phase-contrast images Phys Med Biol 2018; 63: 14NT03
[25] Bliznakova K, Bliznakov Z, Bravou V, Kolitsi Z and Pallikarakis N. A three-dimensional breast software phantom for mammography simulation. Phys Med Biol 2003; 48: 3699-719.

[26] Mettivier G, Bliznakova K, Sechopoulos I, Boone J, Di Lillo F, Sarno A, et al. Evaluation of the BreastSimulator software platform for breast tomography. Phys Med Biol 2017; 62: 6446-66
[27] Weitkamp T, Diaz A, David C, Pfeiffer F, Stampanoni M, Cloetens P, et al. X-ray phase imaging with a grating interferometer. Optics Express 2005; 13: 6296-6304
[28] Pfeiffer F, Bech M, Bunk O, Donath T, Henrich B, Kraft P, et al. X-ray dark-field and phase-contrast imaging using a grating interferometer. Journal of Applied Physics 2009; 105: 102006
[29] Donnelly E, Price E and Pickens D. Quantification of the effect of system and object parameters on edge enhancement in phase-contrast radiography. Med Phys 2003; 30: 2888-96
[30] Dukov N, Bliznakova K, Teneva T, Marinov S, Bakic P, Bosmans H, et al. Experimental evaluation of physical breast phantoms for 2D and 3D breast x-ray imaging techniques. IFMBE Proceedings 2011; 80: 544-52
[31] Feradov F, Marinov S, Bliznakova K, Physical Breast Phantom Dedicated for Mammography Studies. IFMBE Proceedings 2020; 76: 344-52

Acknowledgements
This research is supported by the Bulgarian National Science Fund under grant agreement DN17/2.

Open-source software tools for strategic noise mapping: a case study

G Graziuso[1], A B Francavilla[1], S Mancini[2] and C Guarnaccia[1,a]

[1] Department of Civil Engineering, University of Salerno, via Giovanni Paolo II 132, 84084, Fisciano, SA, Italy
[2] Department of Information and Electric Engineering and Applied Mathematics, University of Salerno, via Giovanni Paolo II 132, 84084, Fisciano, SA, Italy

[a]E-mail: cguarnaccia@unisa.it

Abstract. Currently, the European Union has adopted a series of measures to combat noise problems, focusing on some key indicators, such as the daily and night level of noise. Among the possible actions to manage environmental noise, there are the creation of noise maps and the adoption of action plans. Noise maps are, usually, produced by modelling and simulating data relating to traffic and various activities of an area. However, in some cases, these input data are not available or it is not possible to include all noise sources in the models used, causing difficulties in the prediction and analysis of the environmental impacts of noise. In this paper, starting from the analysis of open-source software tools for data collection, noise modelling and mapping, i.e. OpenstreetMap, NoiseModelling and QGIS, based on free and public licenses, a methodology for the creation of noise maps will be presented. Finally, the outputs of the application of the methodology to the city centre of the Municipality of Nocera Inferiore, in south Italy, will be the creation of noise maps characterized by the spatial representation of the main standard noise indicators.

1. Introduction

The European Environment Agency recognizes acoustic noise has one of the main hazards for human health in urban areas. It is estimated that more than 100 million people are daily exposed to dangerous noise levels [1]. In 2002, the EU issued the Environmental Noise Directive 2002/49/EC [2] in which, among the others, the countries are obliged to draw strategic noise maps in each city exceeding 250,000 inhabitants, for all major roads which have more than six million vehicle passages a year, major railways airports within their territories. For the purpose of noise management and reduction [3], beside spot noise level measurements to be collected with proper observation time [4, 5], the adoption of predictive software is allowed.

Beside the several statistical models [6], also more advanced techniques can be implemented [7], such as, for instance, stochastic and Poisson processes [8] and cellular automata [9]. Usually, these models require in input several features of the noise sources occurring in the areas, such as road traffic, that is generally considered as the main source in urban zones [10]. These inputs can be provided by means of video processing, both automatic [11] and manually [12, 13], or by statistical analysis [14]. In some application, crowdsourced data can be used to feed the models [15, 16]. The crowdsourcing process has some criticism related to the quality of the data [17, 18] but, when validated with standard measurement processes or with reliable data sources, it becomes a relevant tool for input collection, that can be implemented in geographic information systems (GIS), and can be provided to policy maker for influencing the mitigation actions design and implementation [19, 20].

In this paper, the importance of Open Data License (ODbL) software tools in noise mapping will be discussed, reporting a methodological approach and a case study example. The different input sources will be presented, both crowdsourced and from municipal technical office support. The cleaning of opensource maps will be presented, as well as the import process in a predictive software for environmental noise levels. Finally, the implementation of noise maps in free and open-source GIS platform. Results will show how a correct methodology, based on free-to-access and opensource software tools, can provide reliable noise maps in urban areas, to be used for controlling the existing sources and for designing proper mitigation actions, as well as new transportation infrastructures and other potential noise sources.

2. Material and methods

2.1. Software tools for noise mapping

The software adopted in this paper (OpenStreetmap, NoiseModelling and QuantumGis) for the data acquisition, noise level calculation and creation of noise maps are all open-source and freeware, characterized by Open Data and General Public Licenses.

Open Source Software (OSS), generally, are characterized by licenses with a wide range of capabilities. Their applicability in producing and representing geographic information allows people to participate in the creation of contents, modify data and information, combine data from different sources and produce new ones. These types of tools, based on Volunteered Geographic Information (VGI) acquisition, can be considered an alternative or a support for the validation of maps produced by official bodies or for the creation of updated versions at no cost. Despite the quality and reliability of this kind of information are discussed in [15, 16], the main advantages of the voluntary geographic data are the potential global coverage, their representation in georeferenced maps and their use by groups of citizens and public administrations for the definition of a spatial strategy or vision and the support to public decisions [21, 22].

The georeferenced maps are downloaded from OpenStreetMap® and distributed according to the "Open Data Commons–Open Database License" (ODbL) from the OpenStreetMap Foundation (OSMF). In particular, the license is the "Creative Common Attribution-ShareAlike 2.0 Generic" (CC BY-SA 2.0). OpenStreetMap is a collaborative project, in which the users share, create, adapt and, generally, contribute to the map data, as well as in the wiki and in the forums of the project. The websites and the infrastructure are maintained by a group of volunteers [23]. Maps created by OpenStreeMap contains several information about roads, paths, railway stations and buildings around the world.

The elements of OSM database consist of points (nodes), lines (ways), polygons, described by at least one attribute, and relations, necessary to define complex objects. OSM attributes are known as tags, characterized by the combination of keys and values: for example, a motorway can be tagged in OSM with "highway" for the key and "motorway" for the value ("highway=motorway"), in addition to optional tags related to its intrinsic features, such as its width, number of lanes or maximum speed [22]. The contributors may use the tags listed in [24] or add new values.

The OSM tools allow a simple and functional data collection with the use of smartphone applications or field papers [25] and data exporting using the "Export" button from the OSM homepage [23], the API Overpass [26] and, finally, from other websites whose sources are linked to the main OSM platform, such as GeoFabrik [27] and Planet OSM [28].

In order to produce noise maps, the free and open-source tool, NoiseModelling [29], can be used. It belongs to an integrated platform for the acquisition, modelling and evaluation of environmental noise, called Noise-Planet project [30], designed by two French research teams: the Environmental Acoustics Laboratory (Eiffel University, former IFSTTAR) for environmental noise research and the DECIDE Team (Lab-STICC - CNRS UMR) for GIScience. In this project, the data are collected from the free and open-source Android NoiseCapture application, shared from the OnoM@p Spatial Data

Infrastructure (SDI) [31, 32] and NoiseModelling, allows the design of noise maps based on the CNOSSOS-EU standard method for the noise emission (only road traffic) and propagation. [33].

This GIS plugin is distributed under a free and copyleft general public license (GPL3). In order to handle a large amount of spatial features, it connects to a spatial database H2GIS or PostGIS, which must be hosted by a GeoServer. The communication between the GeoServer and NoiseModelling can be operated from a user-friendly WPSBuilder (Web Processing Services Builder) web interface, based on the creation of graphical process workflows that can be easily executed and reproduced [34].

To compute a noise map, different layers must be uploaded (i.e.: Buildings, Roads, Ground type, Topography and Receivers). Then, it is possible to run the calculation of the noise levels and export the output table, for example, in *shp* format, that can be read with many GIS tools.

Finally, it is possible to visualize noise maps in a GIS environment. There are several free and open source GIS user interfaces, mostly similar each other. The one used for the applications in this paper is Quantum GIS (QGIS), that is an official project of the Open Source Geospatial Foundation (OSGeo). It runs on Linux, Unix, Mac OSX, Windows and Android and supports numerous vector, raster, and database formats and functionalities. For example, it can display and overlay vectors and rasters of different formats and with different projections, without the need for any format conversion, allows the creation of maps, the analysis of spatial data through the graphical interface that offers a variety of tools, and the publication of the data on the internet using a webserver [35].

2.2. Methodology

In order to create noise maps with the use of open source tools, a methodology, organized in three phases, can be proposed, as shown in figure 1.

Defined the area of interest, the first phase regards the data acquisition, that can be implemented through crowdsourced data collection or official information acquired from public and private entities and technical offices.

Phase 2 is developed in the WPSBuilder platform. It is characterized by two alternative procedures, in relation to the type of available data.

In the first alternative, the database is composed exclusively by the data downloaded from OpenStreetMap platform. For the creation of noise maps, it is necessary to choose the transport map and import it in WPSBuilder, as it contains information related to roads, buildings and ground type. At this point, it is possible to associate the sound power level to the roads, using the Road_Emission_From_Traffic Tool. The output data corresponds to a table called LW_ROADS, that shows the sound power levels on site of a (moving or constant) point source, for different frequency levels and for the different time bands considered. The values obtained are in accordance with the emission laws of the CNOSSOS model [33]. After defining the positions of the points of the receivers, the noise propagation from the source to the receiver is computed using the Noise_Level_From_Source tool. This block generates a receiver grid corresponding to four noise levels: L_{day}, $L_{evening}$, L_{night} and L_{den}. Specifically, they represent the standard noise descriptors for road, rail and industrial noise, expressed in dBA, i.e. the A-weighted long-term average sound level determined over all the day, evening and night periods of a year (L_{den}), and the day-time (L_{day}), evening-time ($L_{evening}$) and night-time (L_{night}) noise indicators determined respectively over the day, evening and night periods of a year. Then, it is possible to create a level interpolation between the points of the receiver grid and the result is a map

In the second procedure of the phase 2 of the methodology, the database is made of official shape files of roads, buildings and ground, suitably modified, and the Digital Elevation Model (DEM), that are, generally, produced by public administrations or other entities. This procedure is similar to the first alternative, except in the first two steps, where the three above mentioned shape files and the DEM are now necessary.

After having exported the results of the phase 2, the third phase consists in their implementation and display in a GIS environment, with the appropriate categorization of the Noise Level and Contouring Noise Maps relating to main noise descriptors.

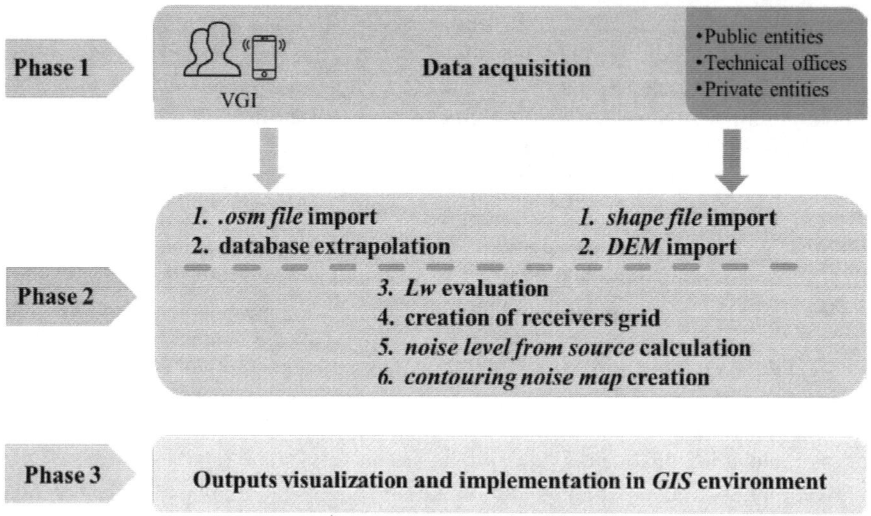

Figure 1. Methodology diagram.

2.3. Case study

The methodology presented in the previous subsection has been applied to the case study of the city of Nocera Inferiore (south Italy). This city is a medium size municipality, with an overall area of about 20.85 km^2 and 44,969 inhabitants, according to 2019 census [36]. The geographical position of this city is strategic, since it is located in between Napoli and Salerno, that are the two main cities of Campania Region (figure 2). This position, as well as the presence in downtown of all the major facilities and services (secondary schools, hospital, court and shops), influences the road traffic volume. In particular, downtown is affected by very large traffic volumes, for both for light and heavy vehicles, compared to the city surface. This consideration, together with the high density of population, suggests focusing on the downtown of the city and makes the case study very interesting for noise mapping purposes. For this reason, the simulations have been restricted to the centre of the city, neglecting what happen in the suburbs, that is mainly countryside with agriculture activities.

The municipality local administration funded a field measurement campaign for traffic flows and volumes in critical points of the city downtown, during specific time window. Results are reported in the simulation model described in the Urban Mobility Plan [37]. A map of the measurement points is reported in figure 3. Since only the major roads have been investigated with traffic measurements, local roads volumes have been fixed with standard and reasonable day, evening and night traffic volumes, according to the class of the roads. These datasets have been used in input of the noise predictive software during simulation 2, together with the information coming from OpenStreetMap about roads.

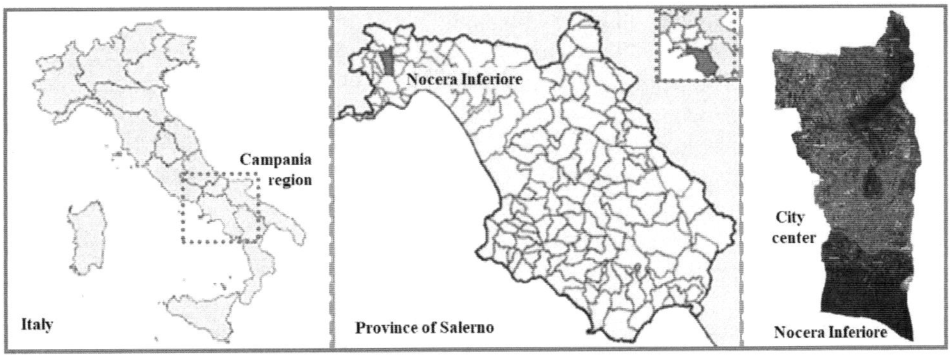

Figure 2. Case study location: city center of Nocera Inferiore, Italy.

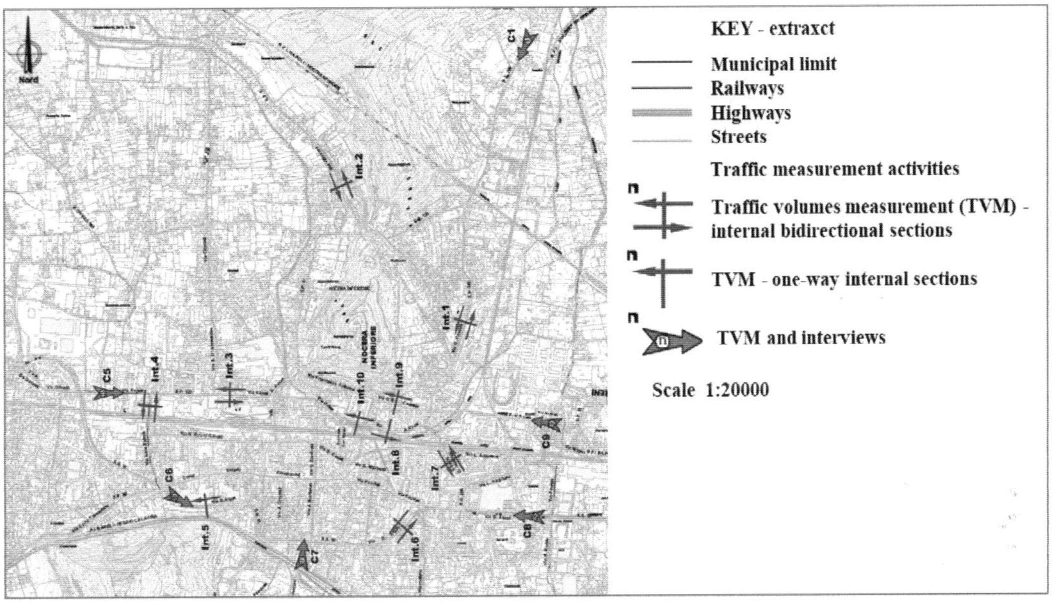

Figure 3. Extract of traffic volumes measurement locations (modified from [35]).

3. Results and discussion

The results of the methodology presented in previous subsections applied to the case study are presented in this section.

The NoiseModelling software automatically produced noise maps of L_{day}, $L_{evening}$, L_{night} and L_{den}. As indicated in the previous section, the emission model of the implemented traffic is the CNOSSOS-EU model [33]. For the sake of brevity, in this paper the authors will report only results related to L_{night} and L_{den} that are selected as main environmental noise descriptors.

As already mentioned in the methodology subsection, the first simulation is obtained feeding the model with the data coming from OpenStreetMap file. Since data are freely uploaded by different registered users, as vehicles drivers, road hauler, pedestrians who are travelling along the area, it contains several information about roads, paths, railway stations and buildings. Anyway, this information is not fully reliable.

Results for L_{night} and L_{den} are reported respectively in figure 4(a) and 4(b). Each map is categorized into 11 classes of different ranges, in dBA, represented on the map with different colors. The lower class refers to values less than 35 dBA and it is graphically represented with the white color. The higher class is represented in blue and includes values greater than 85 dBA. The other 9 intermediate classes have been defined with an equal width of 5 dBA and the range of colors varies from light green to orange and then to red according to the increasing of the value.

The results for L_{night} (Fig 4(a)) show that noise emissions along most of the roads in the study area are higher than the night-time guideline value of 45 dBA, recommended by the WHO [38], and most of residents may be experiencing various adverse health effects, for example sleep disturbance, due to high levels of night-time noise exposure. On the map, it is very easy to appreciate immediately the overcoming of the threshold recommended value. Indeed, great part of the area surrounding the city center (in grey) is depicted in orange and blue. Green areas, with low levels, fall in the suburban part of the city far from the main roads and with less density of buildings.

The results for L_{den} (Fig 4(b)) show how during daytime the general condition gets worse and noise emissions diffusively exceed 55 dBA, that is the EU threshold for excess exposure defined in the Environmental Noise Directive [1]. This indicates that noise emissions in this area are possibly affecting people quality of life.

From the analysis of the results and the comparison between OpenStreetMap data and the information related to critical points of the city reported in the Urban Mobility Plan, various anomalies can be found with reference to the traffic volume in some roads, the heavy vehicles and the characterization of some road pavements. All these differences can influence the results. For this reason, in order to refine results and to achieve a better modelling of the case study, simulation 2 is obtained only implementing the data coming from the official documents recalled in subsection 2.3, i.e. the documents produced during the field measurement campaign of the traffic volumes. Results are reported in figure 5(a) and 5(b). The noise maps are very similar to the ones produced by simulation 1, even if they are more accurate thanks to a more reliable database. It can be appreciated a slight difference in terms of lower values in the second simulation. There is, in fact, a minor density of orange and red areas, due to higher traffic volumes in the first simulation data.

Of course, both these simulations represent an informative tool which can be useful to support the local authority in identifying susceptible areas and to address the noise abatement policies that need to be adopted in a near future, such as noise abatement action plans for the area.

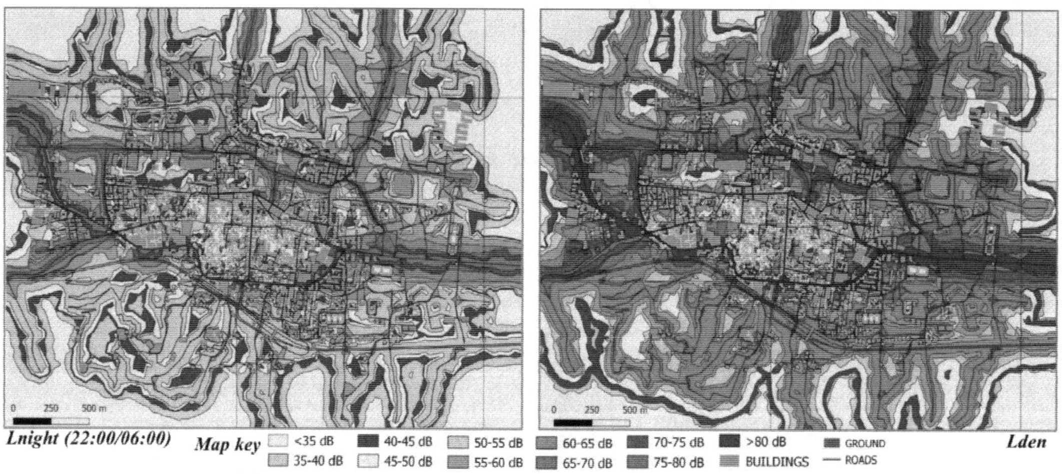

(a) (b)

Figure 4. Simulation 1: (a) L_{night} , (b) L_{den}. Values are reported with A-weighting.

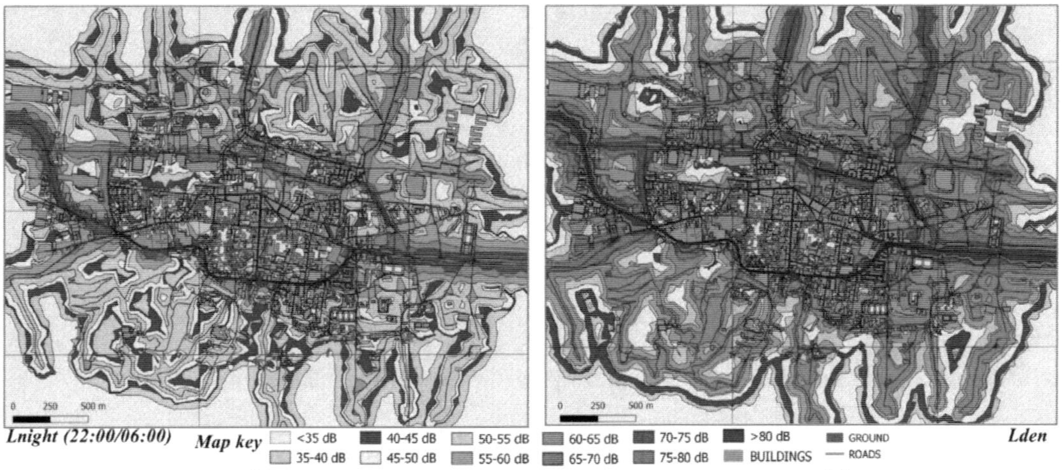

(a) (b)

Figure 5. Simulation 2: (a) L_{night}, (b) L_{den} . Values are reported with A-weighting.

4. Conclusions

Starting from an open-source database and with the use of free and public licensed software tools, a methodology for the generation of noise maps was presented in this paper.

After having analyzed three different tools for data collection, noise modelling and mapping (OpenstreetMap, NoiseModelling and QGIS), it was possible to create different noise maps for the city center of the Municipality of Nocera Inferiore (south Italy).

Two simulations were carried out: the first one was based on voluntary geographic data collected by the OpenStreetMap platform and the second simulation was integrated by official data elaborated from local authorities. Since the results of the first simulation showed some anomalies in different critical points, the traffic flows and volume were compared with the official data of the Urban Mobility Plan in those points and, opportunely handled, became the integrated data of the second simulation.

In both cases, the data were processed using the open source WPSBuilder tool, that permitted the creation of noise maps, which were then implemented in the QGIS environment.

From the analysis of the obtained maps, characterized by the spatial representation of the main standard noise indicators, the most suitable interventions could be defined, such as, for example, the use of sound-absorbing floors or the installation of acoustic barriers in places where sound pressure levels remain high over time.

In addition, this study contributed to define indications and suggestions for both data acquisition and sharing. For example, a greater active involvement of volunteers in data acquisition could open interesting prospects for a large number of subjects, including public authorities, which could benefit from the enrichment of their databases. Furthermore, this could lead to a future scenario in which geographic and non-geographic information is continuously updated at no cost. In this scenario it is possible to witness a complete revolution in the role of the planners, who would be able to constantly analyze the available data. Consequently, a continuous support for design, spatial analysis and decision-making could be implemented with significant innovations in urban and regional planning methodologies.

In conclusion, further useful indications for monitoring the acoustic state of an area can be suggested, i.e. to incentivize the use of voluntary geographical maps in order to have the widest possible coverage of the territories, to ensure the presence of protocols for data creation, to encourage the integration of OSM data with official sources in order to have a reliable database and to encourage public administrations to use open source tools in order to have a continuously updated database available for planning.

References

[1] European Environment Agency 2020 *Environmental noise in Europe — 2020* Report No 22/2019, (online), https://www.eea.europa.eu/publications/environmental-noise-in-europe
[2] Directive 2002/49/EC of the European Parliament and of the Council of 25 June 2002 relating to the assessment and management of environmental noise *Offl J of the Eu Comm (OJEC)* L189/12-25, 18 July 2002
[3] Jakati J S, Kuntoji S S 2021 A Noise Reduction Method Based on Modified LMS Algorithm of Real time Speech Signals. WSEAS Trans Environ Dev 16 (13)
[4] Liguori C, Ruggiero A, Russo D and Sommella P 2017 Influence of measurement time on acoustic noise uncertainty *Proc. I2MTC 2017 - 2017 IEEE International Instrum Meas Tech Conf. (Turin)* 7969850
[5] Liguori C, Ruggiero A, Russo D and Sommella P 2018 Accurate estimation of the environmental noise through sampling approach: Selection of the measurement time *Proc. of IEEE T Instrum Meas* **67**(5) pp 1006-13
[6] Steele C 2001 A critical review of some traffic noise prediction models *Appl. Acoust.* **62** pp 271
[7] Guarnaccia C 2013 Advanced tools for traffic noise modelling and prediction *WSEAS Trans. Syst.* **12**, 121-130
[8] Guarnaccia C, Quartieri J, Tepedino C, Rodrigues ER 2016 A time series analysis and a non-homogeneous Poisson model with multiple change-points applied to acoustic data *Appl. Acoust.* **114** pp 203-212
[9] Quartieri J, Mastorakis N, Iannone G, Guarnaccia C 2010 Cellular automata application to traffic noise control *Proc 12th international conference on Automatic control, modelling & simulation ACMOS10*, May 2010 Pages 299–304
[10] Guzman J A N, Fernandes B R B, Delshad M, Sepehrnoori K, Zapata J F 2020 The Impact of the Ring Road Conclusion to the City of Guimarães, Portugal: Analysis of Variations of Traffic Flows and Accessibilities. *WSEAS Trans Environ Dev* **16** (2)
[11] Guarnaccia C 2020 EAgLE: Equivalent acoustic level estimator proposal *Sensors* **20** pp 1-12
[12] Singh D, Francavilla A B, Mancini S and Guarnaccia C 2021 Application of Machine Learning to Include Honking Effect in Vehicular Traffic Noise Prediction *Applied Sciences* **11** pp 1-16
[13] Pascale A, Fernandes P, Guarnaccia C and Coelho M C 2021 A study on vehicle Noise Emission Modelling: Correlation with air pollutant emissions, impact of kinematic variables and critical hotspots *Science of the Total Environment* **787** pp 1-14
[14] Iannone G, Guarnaccia C, Quartieri J 2013 Speed distribution influence in road traffic noise prediction Env Eng & Manag J **12** pp 493-501
[15] Graziuso G, Mancini S, Francavilla AB, Grimaldi M and Guarnaccia C 2021 Geo-Crowdsourced Sound Level Data in Support of the Community Facilities Planning. A Methodological Proposal *Sustainability* **13**(10) 5486
[16] Graziuso G, Grimaldi M, Mancini S, Quartieri J and Guarnaccia C. Crowdsourcing Data for the Elaboration of Noise Maps: a Methodological Proposal. 2020 *J. Phys.: Conf. Ser.* **1603** 012030
[17] Hashemi P and Abbaspour R A 2015 Assessment of Logical Consistency in OpenStreetMap Based on the Spatial Similarity Concept *In OpenStreetMap in GIScience, Experiences, Research, and Applications (LNCS)*, edited by JJ Arsanjani, A Zipf, P Mooney and M Helbich, 19–36. Berlin Heidelberg: Springer.
[18] Salk C F, Sturn T, See L, Fritz S and Perger C 2016 Assessing quality of volunteer crowdsourcing contributions: lessons from the Cropland Capture game *Int J Digit Earth* **9**(4) pp 410-426
[19] Sebillo M, Vitiello G, Grimaldi M and De Piano A 2020 A Citizen-Centric Approach for the

improvement of territorial services management. *ISPRS Int. J. Geo-Inf.* **9** 223

[20] González-Méndez M, Olaya C, Fasolino I, Grimaldi M and Obregón N 2021 Agent-based modeling for urban development planning based on human needs. Conceptual basis and model formulation *Land Use Policy* **101** 105110

[21] Goodchild M F 2007 Citizens as Sensors: the World of Volunteered Geography *GeoJournal* **69** pp 211-221

[22] Fonte C, Minghini M, Antoniou V, See L, Patriarca J, Brovelli M and Milcinski G 2016 Automated methodology for converting OSM data into a land use/cover map *Proc. 6th Inter. Conf. on Cartography & GIS*, 13-17 June 2016, Albena, Bulgaria, 10 p

[23] https://www.openstreetmap.org last accessed on July 22, 2021.

[24] http://wiki.openstreetmap.org/wiki/Map_Features last accessed on July 27, 2021.

[25] www.fieldpapers.org last accessed on July 30, 2021.

[26] www.wiki.openstreetmap.org/wiki/Overpass_API last accessed on July 30, 2021.

[27] http://download.geofabrik.de last accessed on July 14, 2021.

[28] http://planet.openstreetmap.org last accessed on July 15, 2021.

[29] Bocher E, Guillaume G, Picaut J, Petit G and Fortin N 2019 NoiseModelling: An Open Source GIS Based Tool to Produce Environmental Noise Maps *ISPRS Int. J. Geo-Inf.* **8** 130

[30] Picaut J, Fortin N, Bocher E, Petit G, Aumond P and Guillaume G 2019 An open-science crowdsourcing approach for producing community noise maps using smartphones *Building Environ* **148** pp 20-33

[31] Noise Planet Project website, https://noise-planet.org/index.html, last accessed on July 28, 2021

[32] Bocher E, Petit G, Picaut J, Fortin N and Guillaume G 2017 Collaborative noise data collected from smartphones *Data in Brief* **14** pp 498-503

[33] CNOSSOS-EU (2012) Common Noise Assessment Methods in Europe, (online), https://ec.europa.eu/jrc/en/publication/reference-reports/common-noise-assessment-methods-europecnossos-eu last accessed on July 30, 2021.

[34] https://noisemodelling.readthedocs.io/en/latest/index.html last accessed on July 30, 2021.

[35] https://www.qgis.org/en/site/about/index.html last accessed on July 20, 2021.

[36] http://demo.istat.it/popres/index.php?anno=2020&lingua=ita last accessed on June 12, 2021.

[37] http://old.comune.nocera-inferiore.sa.it/puc/380-piano-urbano-della-mobilita-pum last accessed on June 12, 2021.

[38] World Health Organization 2018 Environmental noise guidelines for the European region *Regional Office for Europe* Copenhagen

Acknowledgments

Authors would like to acknowledge Rosanna Natale for contributing to this research work during her bachelor thesis.

Sustainable building materials based on hemp shives and geopolymer paste

L Vitola[1,a], M Vilnitis[2], I Pundiene[1,3] and D Bajare[1]

[1] Institute of Materials and Structures, Riga Technical University, Kalku 1, LV – 1658, Riga, Latvia
[2] Institute of Construction Technology, Riga Technical University, Kalku 1, LV – 1658, Riga, Latvia
[3] Institute of Building Materials, Vilnius Gediminas Technical University, Sauletekio av. 11, Vilnius, LT-10223, Lithuania

[a]E-mail: laura.vitola_1@rtu.lv

Abstract. European regulations states that the construction industry must encourage the rational and prudent use of non-renewable natural resources, as well as and increase the use of reusable resources and promote the use of renewable natural resources in production. Bio-composites based on natural aggregates such as hemp shives are widely studied and used in building sector due to their extensive properties of a range that primarily is so wide, as hemp shives are increasingly combined with various binders and as their production technology is evolving rapidly. In this study three series of geopolymer paste (GP) samples were studied to obtain the most suitable binder for the production of hemp shive-based bio-composites, as well as a bio-composites were created from the most suitable binder and described their compressive strength and thermal conductivity performance.

1. Introduction

Since 1987. along with publication of report of Brundtland Commission the sustainability has become an important aspect in all areas, as well as in Building sector. In accordance to this report sustainable development is defined as development that meets the needs of the present without compromising the ability of future generations to meet their own needs [1]. To ensure the sustainability in civil engineering sector and the well-being of ecosystem different type of solutions are being sought [2]–[5], including the production of new environmentally friendly building materials using renewable natural resources and industrial by-products and waste [5]–[10].

Several researchers have emphasized the potential of geopolymers in the production of sustainable building materials [5], [11]–[13]. Assi et.al. has compiled the available literature on geopolymers as sustainable materials and a potential alternative to ordinary Portland cement [13]. Geopolymers can be made of commercially available raw materials (for example, calcinated clays and alkali solutions), as well as from lower quality materials, industrial by-products, construction demolition wastes, and other type of wastes. Geopolymers may present lower CO_2 emission (so do impact on environment) and competitive mechanical and durability properties comparing to ordinary Portland cement based materials [13].

During recent decades bio composites have became more and more popular in building industry as primary building materials because they provide an environment that is favourable to the environment and human [14], [15]. There are plenty of variations of different type of plant origin aggregates used for bio-composite production (such as hemp and flex fibres and shives, oat and wheat husk e.c.) [16], [17]. New plant origin aggregates-based bio-composites presents potentially long-term durability, improved physical, mechanical, and thermal properties, and low impact on the environment, as well as recyclability [17]. The use of plant origin aggregates provides sustainable solution for reduction of resource depletion and CO_2 emissions [18].

In this study three series of geopolymer paste (GP) samples were studied to obtain the most suitable binder for the production of hemp shive-based bio-composites, as well as a bio-composites were created from the most suitable binder and described their compressive strength and thermal conductivity performance.

2. Materials and methods

2.1. Raw materials

In this research commercially available metakaolin ASTRA MK-40 (MK) produced by ASTRA (Poland) has been used as aluminum silicate source. The SiO_2 and Al_2O_3 make up, respectively, 57.30% un 35.73% of the MK content. MK has amorphous phase (halo at 15-30°), as well as crystalline phases of quartz (SiO_2), kaolinite ($Al_2(OH)_4Si_2O_5$) and mullite ($Al_6Si_2O_{13}$).

To ensure the geopolymerisation reaction three type of activation solutions have been used – 8M NaOH (A1) solution and two modified 8M NaOH solutions. The 8M NaOH solution was modified by substituting 10% (A2) and 20% (A3) solution mass by sodium metasilicate solution (also known as waterglass).

Two type of hemp shives have been studied; A – commercially available hemp shives grown and processed in Latvia (Zalers) as a by-product of hemp fiber production, B – commercially available hemp shives grown and processed in Lithuania (Naturalus Plostas).

2.2. Testing methods

Flowability of geopolymer pastes was characterized by flow table test in accordance to ASTM C 1437. The density was performed by dividing the sample's mass by the respective volume.

The bulk density of obtained geopolymer pastes were calculated as mass and volume division, for 28-days old samples with dimensions 20 mm × 20 mm × 20 mm. While the bulk density of obtained bio-composites were tested in accordance to EN 1602 plate shaped samples with dimensions 350 mm × 350 mm × 50 mm.

Compressive strength of 6 parallel samples (20 mm × 20 mm × 20 mm) of each geopolymer paste composition using different mixing times were tested on Day 3 and Day 28. Samples were tested by using a Zwick Z100 universal testing system (ZwickRoell, Kennesaw, GA, USA) with a testing speed of 0.5 mm/min. Compressive strength was calculated from force applied to the sample's specific area.

Compressive stress at 10% deformation rates was tested according to the EN 826 method by using a Zwick Z100 universal testing system. The maximum loading force was 10 kN, the loading accuracy was ± 0.5% and the loading speed accuracy was ± 0.05%. Six specimens were tested for each composition with the dimensions 100 mm × 100 mm × 70 mm. Prior to conducting the test, the specimens were conditioned for six hours at 23°C ± 5°C. The specimens were then aligned on the bottom support and subjected to an initial loading pressure of 250 ± 10 Pa. The loading speed during the tests was 0.7 mm/min.

Thermal conductivity of bio-composites has been determined using Fox 600 heat flow measuring device. Samples with dimensions 350 mm × 350 mm × 50 mm were fixed horizontally. During the test temperature of lower surface was +20°C and 0°C of upper surface (mean temperature +10°C). Heat transmission coefficient was determined considering constant heat flow between two sides of sample.

2.3. Samples` compositions and preparation process

2.3.1. Composition and preparation of geopolymer pastes.

In this study 3 series (3 mixtures of each series) of geopolymer paste (GP) have been studied (Table 1), metakaolin (MK) was activated with three different alkaline solutions (A0, A10 and A20, respectively) using 3 different MK/activation solution ratios – 0.5, 0.6 and 0.7. Activation solutions were prepared and kept at room temperature (20 ± 2ºC) for several hours before preparation of GP. Powdered MK was dosed and mixed together with certain amount of chosen activation solution in accordance to table 1. for 1, 3, 5 and 7 minutes. Immediately after mixing moulds (20 mm × 20 mm × 20 mm) were filled with prepared paste and parallelly the flowability of paste was tested. Specimens were put into sealed bags and kept at room temperature (20 ± 2ºC) for 1 h, and then curried at 85ºC for 24 h. After demoulding samples were kept at room temperature (20 ± 2ºC) until Day 3, when part of samples was tested and the rest of samples were after-cured in air and in water media at room temperature (20 ± 2ºC) until Day 28.

The percentage of oxides in compositions was calculated using data given by producers of raw materials.

Table 1. Compositions of studied GP and ratios of main oxides in GP.

Name of composition	Mixture design, mass part				L/S	Main oxides, %			Ratios of main oxides in pastes, % in weight		
	MK	A0	A10	A20		SiO_2	Al_2O_3	Na_2O	SiO_2 / Al_2O_3	SiO_2 / Na_2O	Al_2O_3 / Na_2O
MK-0-0.5	1.0	0.5			0.5	38.2	23.8	13.4			
MK-0-0.6	1.0	0.6			0.6	35.8	22.3	12.5	1.6	2.9	1.8
MK-0-0.7	1.0	0.7			0.7	33.7	21.0	11.8			
MK-10-0.5	1.0		0.5		0.5	40.2	23.8	14.0			
MK-10-0.6	1.0		0.6		0.6	37.7	22.3	13.1	1.7	2.9	1.7
MK-10-0.7	1.0		0.7		0.7	35.4	21.0	12.4			
MK-20-0.5	1.0			0.5	0.5	42.1	23.8	14.7			
MK-20-0.6	1.0			0.6	0.6	39.5	22.3	13.8	1.8	2.9	1.6
MK-20-0.7	1.0			0.7	0.7	37.2	21.0	13.0			

2.3.2. Composition and preparation of bio-composites

The most appropriate GP were chosen for production of hemp-based bio-composites. First paste was mixed in accordance to the procedure mentioned above, then GP were covered with plastic bag and left for 1 h. Exact amount of GP (2.0, 3.0 or 4.0 mass parts) were added to the wet hemp shives (1.0 mass part of dry hemp shives mixed with 1.0 mass part of water) and mixed manually until the homogenous structure were obtained (for ~ 5 min). The obtained fresh material was inserted, pressed into moulds and covered with PVC film, after which the sample was compressed to the required height (50 mm) with the help of threads and spacers. The material was pressed under pressure for 30 minutes and then placed in an oven at 85°C for 24 hours.

3. Results and discussions

3.1. Assessment of GP suitability for bio-composites production

As can see in figure 1, spread diameter of GP increases by increasing the solid/liquid ratio. Also, type of used activation solution and mixing time have some impact on fluidity of GP. As activation solutions A10 and A20 have been obtained by modifying 8M NaOH solution by sodium metasilicate solution, both A10 and A20 have higher viscosity comparing to activation solution A0. That promotes the different fluidity behaviors between pastes with the same L/S (liquid to solid) ratio, but different activation solution used. Additional metasilicate solution in activation solution decreases fluidity of fresh paste. Role of mixing time in fluidity behavior of studied GP depends on L/S ratio (table 1.), higher the ratio is, more noteworthy the impact of mixing time becomes. As presented in figure 1, studied GP presents spread diameter in a range from 9.85 to 22.10 cm, depending on the composition.

The most fluid composition is MK-0-0.7 (19.55 – 22.10 cm), but the less fluid – MK-20-0.5 (9.85 – 10.75 cm).

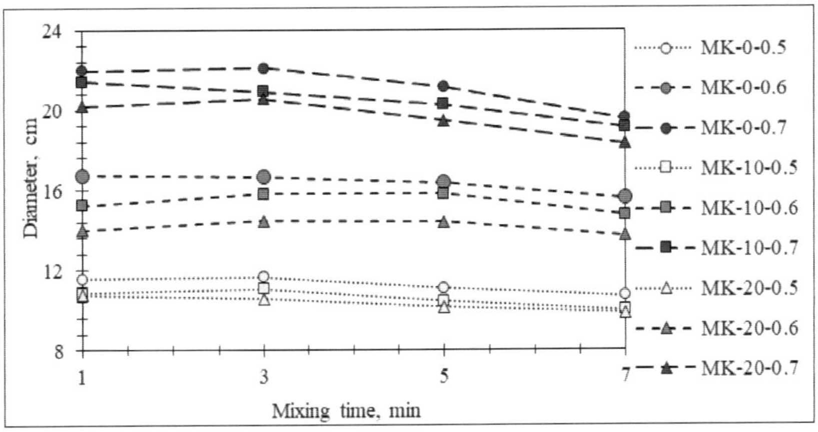

Figure 1. Spread diameter of GP depending on mixing time.

As can see in figure 2 (c). GP made with activation solution A10 (8M NaOH of which 10% replaced by sodium metasilicate solution) presents lower compressive strength comparing to GP obtained with activation solution A0. Respectively, on Day 3 compressive strength is in range from 6.0 to 12.1 MPa, while on Day 28 7.6-14.9 MPa (curried in the air) and 8.2-15.6 MPa (curried in the water).

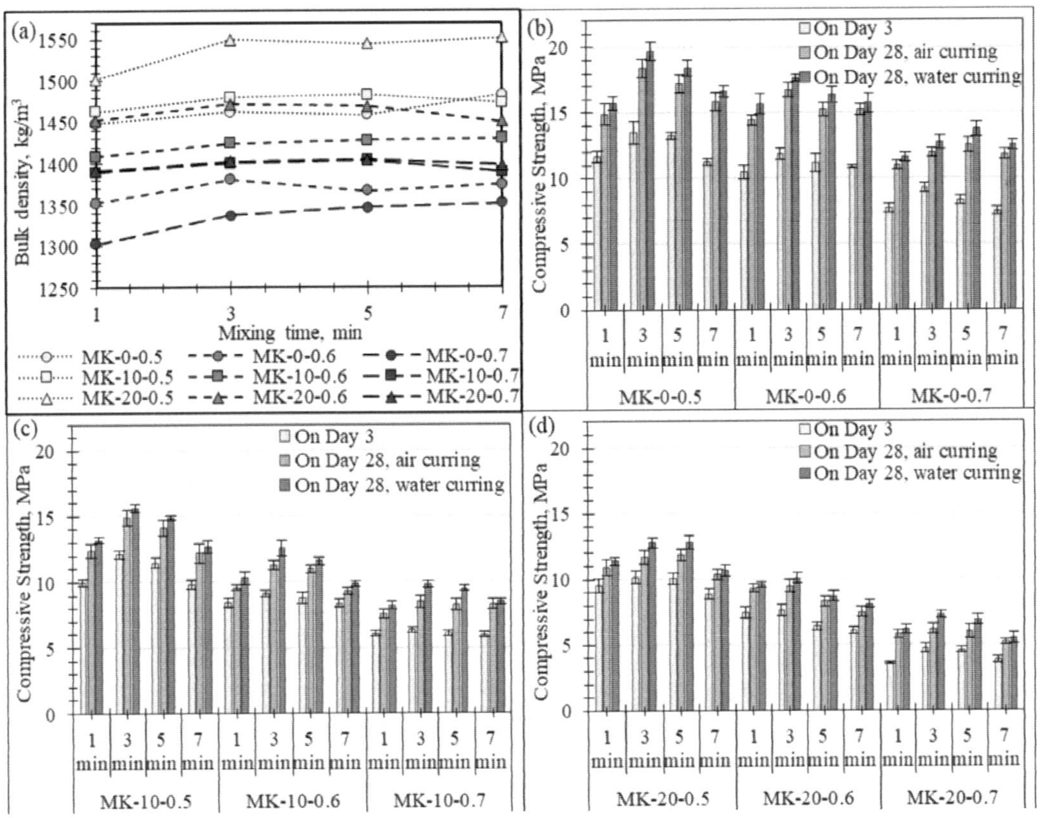

Figure 2. Properties of hardened geopolymer pastes; (a) bulk density, (b) compressive strength of

studied GP obtained with activation solution A0, (c) compressive strength of studied GP obtained with activation solution A10, (d) compressive strength of studied GP obtained with activation solution A20.

Obtained GP present bulk density in range from 1303 ± 19 to 1552 ± 14 kg/m^3 depending on the composition and fresh paste`s mixing time. The bulk density of GP is presented in figure 2 (a), as can see in the graph, density depends on L/S ratio (table 1) and as well as on the used activation solution. Meanwhile the effect of mixing time on bulk density does not have the same tendency for all GP.

Results of compressive strength tests are given in figure 2 (b, c, d). Each composition was tested on the Day 3 and on Day 28 (after air curing and after water curing) (table 1). GP with compressive strength in range from 3.6 to 13.5 MPa on Day 3 have been obtained.

GP containing activation solution A0 (8M NaOH solution, respectively) presents compressive strength on the Day 3 in range from 7.5 to 13.5 MPa depending of L/S ration and mixing time of fresh paste. While on the Day 28 the compressive strength increases up to 19.7 MPa for samples with water curing and up to 18.4 MPa for samples curried in the air (figure 2 (b)).

GP made with activation solution A20 (8M NaOH of which 20% replaced by sodium metasilicate solution) presents compressive strength in the range from 3.6 to 10.2 MPa on Day 3 (figure 2 (d)) depending on the L/S and mixing time of the fresh paste. Until Day 28 compressive strength increases up to 5.2 – 11.9 MPa in case of air curing and up to 5.5 - 12.8 MPa in case of water curing.

Comparing all three series it can be argued that addition of sodium metasilicate to the 8M NaOH solution decreases the compressive strength on Day 3 for up to 22.7% if 10% of 8M NaOH replaced by sodium metasilicate solution and for even up to 53.4% if 20% of 8M NaOH replaced by sodium metasilicate solution. On Day 28 similar behavior has been detected. Sodium metasilicate solution consists of Na_2O, reactive (amorphous) SiO_2 and H_2O, all these three chemical compounds are needed to provide geopolymerisation reaction. As can see in table 1 different L/S ratios do not change the main oxide ratios but impacts the total main oxide amount in the composition. Higher amount of reactive silica and sodium in the composition of geopolymer pastes might cause the formation of the reaction by-products – zeolites [19], [20]. Structure of GP can be damaged during the formation of the zeolites crystals material structure gets damaged. Also, as zeolites and geopolymer gel need the same main oxides and zeolites can format faster than geopolymer gel, less main oxides are free for geopolymerisation reaction.

In accordance to results GP needs time to gain the maximal compressive strength, in further researches development of compressive strength of geopolymer pastes in time period longer than 28 day should be studied deeper.

The relationship between the spread diameter of the fresh GP and compressive strength on Day 3 is given in figure 3. The relationship between compressive strength and fluidity obviously depends on L/S ratio and type of used activation solution.

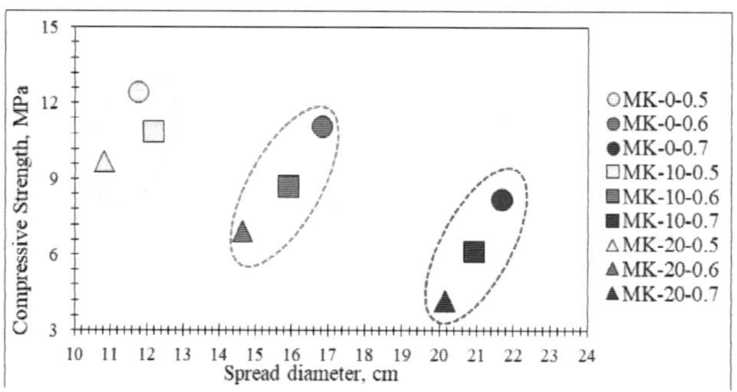

Figure 3. Relationship between average flowability and average compressive strength of geopolymer pastes on Day 3.

From the point of fluidity (figure 1) and bulk density (figure 2 (a)) results fresh GP should be mixed at least for 3 min. As the highest compressive strength is presented by GP made with activation solution A0 (8M NaOH) and higher fluidity, which is important for workability during the production of bio-composites, is presented by GP with S/L ratio 0.7, the GP MK-0-0.7 is chosen for the further study of bio-composites.

3.2. Hemp shive based bio-composites bounded by geopolymer paste

The results of bulk density and compressive strength of obtained bio-composites are given in the figure 4 (a). In this study six different bio-composites have been made, using two type of hemp shives and three geopolymer paste/hemp shive ratios. Obtained bio-composites presents bulk density from 261 - 403 kg/m^3. As can see in figure 4 (a) bulk density depends on GP/hemp shives ratio, while type of hemp shives do not impact the bulk density noteworthy. Increasing the GP/hemp shives ratio twice (from 2.0 to 4.0) the bulk density increases for 54.4% in case of hemp shives A and for 52.3% in case of hemp shives B.

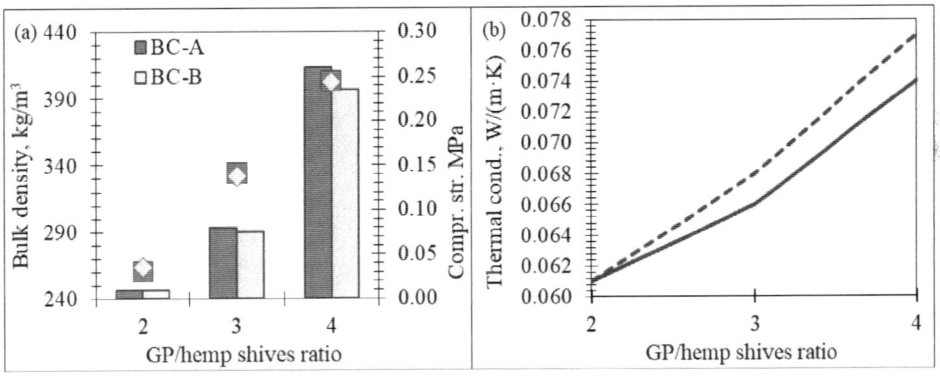

Figure 4. Properties of obtained bio-composites bounded by GP depending on GP/hemp shives ratio; (a) bulk density and compressive strength, (b) thermal conductivity.

Obtained bio-composites present the compressive strength up to 0.48 MPa on Day 28. As can see in figure 4 (a) bio-composites with GP/hemp shives ratio 2.0 presents the compressive strength lower than 0.10 MPa. As expected the increase of GP/hemp shives ratio increases the compressive strength of the obtained bio-composite. The hemp shives A promotes slightly higher compressive strength comparing to hemp shives B.

The thermal conductivity of obtained bio-composites depending on the GP/hemp shives ratio is given in figure 4 (b). Obtained bio-composites bounded by GP MK-0-0.7 presents thermal conductivity in the range from 0.061 to 0.077 W/(m·K) As can see in the graph lower the geopolymer paste/hemp shive ratio, lower the difference between bio-composites with different hemp shives (hemp shive A and hemp shives B, respectively). Meanwhile using the highest GP/hemp shives ratio (4.0) the difference is 0.003 (0.074 of BC-A-4 and 0.077 of BC-B-4), that fall within the permissible error of the measuring equipment.

In accordance to results the properties of bio-composites bounded by GP depends on the amount of geopolymer paste used in the production of bio-composites, while noteworthy the impact of the type of used hemp shives (type A or type B) have not been detected.

4. Conclusions

The following conclusions have been drawn from this study:

1. the optimal SiO_2/Al_2O_3 ratio and liquid to solid ratio for fresh geopolymer paste to ensure flowability (spread diameter >180 mm) and compressive strength on Day 3 (>7 MPa) should be, respectively, 1.6 and 0.7;

2. the 3 min long mixing by mechanical hand mixer is the optimal time for the geopolymer paste preparation from the point of fluidity of fresh geopolymer paste and compressive strength of hardened geopolymer paste;
3. to ensure compressive strength of bio-composites bounded by geopolymer paste higher than 0.05 MPa the geopolymer paste / hemp shive ratio should be at least 3.0, but to ensure higher than 0.25 MPa – at least 4.0, respectively;
4. type of used hemp shives does not impact the properties of obtained bio-composites bounded by geopolymer paste noteworthy;
5. using hemp shives and a metakaolin-based geopolymer binder, it is possible to obtain a self-supporting building material with bulk density 260-400 kg/m^3, compressive strength up to 0.26 MPa and thermal conductivity 0.061-0.077 W/(m·K).

References

[1] UNICEF-ONU-UNESCO, "Our Common Future - Brudtland Report," *Brudtland Report*, p. 383, 1987.

[2] R. Cardoso, A. Paiva, J. Pinto, and J. C. Lanzinha, "Structural and Material Characterization of a Haussmann Building Complex at La Madeleine, Paris. The First Step Before Sustainable Rehabilitation and Strengthening".

[3] M. H. Roushdy, "Recycling of Cullet, Waste Clay Bricks and Wastes Resulted from Wheat and Sugarcane Cultivations in the Manufacture of Fired Clay Bricks", doi: 10.37394/232015.2021.17.19`M.

[4] K. W. Chau, "Incorporation of sustainability concepts into a civil engineering curriculum," *Journal of Professional Issues in Engineering Education and Practice*, vol. 133, no. 3, pp. 188–191, 2007, doi: 10.1061/(ASCE)1052-3928(2007)133:3(188).

[5] K. A. Komnitsas, "Potential of geopolymer technology towards green buildings and sustainable cities," *Procedia Engineering*, vol. 21, pp. 1023–1032, Jan. 2011, doi: 10.1016/J.PROENG.2011.11.2108.

[6] K. Ruģele, Ģ. Būmanis, L. Mežule, T. Juhna, and D. Bajare, "Application of industrial wastes in renewable energy production," *undefined*, 2015.

[7] D. Vaičiukynienė, D. Nizevičienė, A. Kielė, E. Janavičius, and D. Pupeikis, "Effect of phosphogypsum on the stability upon firing treatment of alkali-activated slag," *Construction and Building Materials*, vol. 184, pp. 485–491, Sep. 2018, doi: 10.1016/J.CONBUILDMAT.2018.06.213.

[8] D. Vaičiukyniene, A. Kantautas, V. Vaitkevičius, L. Jakevičius, Ž. Rudžionis, and M. Paškevičius, "Effects of ultrasonic treatment on zeolite NaA synthesized from by-product silica," *Ultrasonics Sonochemistry*, vol. 27, pp. 515–521, Nov. 2015, doi: 10.1016/J.ULTSONCH.2015.06.001.

[9] R. Gailitis, A. Sprince, T. Kozlovskis, L. Radina, L. Pakrastins, and N. Vatin, "Long-Term Properties of Different Fiber Reinforcement Effect on Fly Ash-Based Geopolymer Composite," *Crystals 2021, Vol. 11, Page 760*, vol. 11, no. 7, p. 760, Jun. 2021, doi: 10.3390/CRYST11070760.

[10] R. Gailitis, K. Korniejenko, A. Sprince, and L. Pakrastins, "Comparison of the long-term properties of foamed concrete and geopolymer concrete in compression," *AIP Conference Proceedings*, vol. 2239, no. 1, p. 020012, May 2020, doi: 10.1063/5.0007787.

[11] B. Panda, S. C. Paul, L. J. Hui, Y. W. D. Tay, and M. J. Tan, "Additive manufacturing of geopolymer for sustainable built environment," *Journal of Cleaner Production*, vol. 167, pp. 281–288, Nov. 2017, doi: 10.1016/J.JCLEPRO.2017.08.165.

[12] A. Hassan, M. Arif, and M. Shariq, "Use of geopolymer concrete for a cleaner and sustainable environment – A review of mechanical properties and microstructure," *Journal of Cleaner Production*, vol. 223, pp. 704–728, Jun. 2019, doi: 10.1016/J.JCLEPRO.2019.03.051.

[13] L. N. Assi, K. Carter, E. Deaver, and P. Ziehl, "Review of availability of source materials for geopolymer/sustainable concrete," *Journal of Cleaner Production*, vol. 263, p. 121477, Aug. 2020, doi: 10.1016/J.JCLEPRO.2020.121477.
[14] N. N. Shalobyta, V. V. Tur, T. P. Shalobyta, and V. I. Rakhuba, "ENERGY EFFICIENT COMPOSITES USING NATURAL ORGANIC MATERIALS," in *E3S Web of Conferences*, 2019, vol. 136. doi: 10.1051/e3sconf/201913602027.
[15] F. Asdrubali, F. D'Alessandro, and S. Schiavoni, "A review of unconventional sustainable building insulation materials," *Sustainable Materials and Technologies*, vol. 4, pp. 1–17, Jul. 2015, doi: 10.1016/j.susmat.2015.05.002.
[16] M. Lagouin, C. Magniont, P. Sénéchal, P. Moonen, J. E. Aubert, and A. Laborel-préneron, "Influence of types of binder and plant aggregates on hygrothermal and mechanical properties of vegetal concretes," *Construction and Building Materials*, vol. 222, pp. 852–871, 2019, doi: 10.1016/j.conbuildmat.2019.06.004.
[17] G. Bumanis, L. Vitola, I. Pundiene, M. Sinka, and D. Bajare, "Gypsum, geopolymers, and starch-alternative binders for bio-based building materials: A review and life-cycle assessment," *Sustainability (Switzerland)*, vol. 12, no. 14, Jul. 2020, doi: 10.3390/SU12145666.
[18] A. Laborel-Préneron, C. Magniont, and J. E. Aubert, "Characterization of Barley Straw, Hemp Shiv and Corn Cob as Resources for Bioaggregate Based Building Materials," *Waste and Biomass Valorization*, 2018, doi: 10.1007/s12649-017-9895-z.
[19] J. Davidovits, "PROPERTIES OF GEOPOLYMER CEMENTS", Accessed: Aug. 04, 2021. [Online]. Available: www.geopolymer.org
[20] A. de Rossi *et al.*, "In-situ synthesis of zeolites by geopolymerization of biomass fly ash and metakaolin," *Materials Letters*, vol. 236, pp. 644–648, Feb. 2019, doi: 10.1016/J.MATLET.2018.11.016.

Acknowledgements
Publication is created with support of European Regional Development Fund project "A new concept for low-energy eco-friendly house", Grant Agreement No 1.1.1.1/19/A/017.

Visualization of the Plasma Frequency by means of a Particle Simulation using a Normalized Periodic Model

Joschka Grunwald, Paolo Mercorelli

Institute of Product and Process Innovation, Leuphana University of Lueneburg, Lueneburg, Germany

E-mail: joschkagrunwald@gmail.com, mercorelli@uni.leuphana.de

Abstract. In plasmas the atoms are dissociated into ions and free electrons. Due to the high mobility of the electrons, plasmas are a vibrating medium. As a result, a plasma frequency can be determined, which is an important parameter for characterizing a material. In this study, the plasma frequency is to be visualized. For this purpose, a mathematical model will be set up to describe the electron kinetics. The electrons are briefly deflected with a force against the static ion background in order to examine the interaction between the particles in the form of electrostatic fields. The discretized equations are implemented in a numerical particle simulation to visualize the movement of the electrons. The Particle-In-Cell (PIC) method is used for this. A spatial mathematical grid is created on which the simulated particles are distributed in the cells. The equations of motion and field are then solved on this grid at different times.

Keywords: Plasma Frequency, Particle-In-Cell Method, Periodic Normalized Plasma Model, Particle Simulation, Poisson Equantion, Leapfrog Method

1. Introduction

The matter existing on our earth occurs mainly in the solid, liquid or gaseous aggregate states. These states are extremely rare in the visible universe. More than 99% of matter is in the plasma state, which is also known as the fourth aggregate state [1, cf.: P.227]. The form in which a material exists depends on its thermal energy. If sufficient thermal energy is supplied to a gas, a phase transition occurs. The kinetic energy at the atomic level is so enormous that the binding energy between the nucleus and the electrons is overcome. The atoms dissociate into ions and electrons. This process is called ionization [2, cf.: P.1f.]. Due to the unbound particles, an ionized gas has completely new properties and is therefore no longer called a gas but a plasma.

The free electrons in a plasma create collective phenomena such as the dynamics of space charge fluctuations [3, cf.: P.1]. These charge oscillations have the consequence that electromagnetic radiation with a frequency lower than the natural frequency of the plasma is shielded and reflected on the surface. Due to the lower mass and thus higher mobility of the electrons, they indicate the effective shielding frequency. This shielding frequency or natural frequency is also referred to as the plasma frequency and is given by

$$\omega_{p_e} = \sqrt{\frac{e^2 \rho_e}{\epsilon_0 m_e}} \quad . \tag{1}$$

The elementary charge is represented by e, ρ_e indicates the electron density, m_e the electron mass and ϵ_0 the electric field constant [4, P.529f.].

In this study, the plasma frequency is visualized by means of a Particle-in-Cell (PIC) algorithm, see Fig. 1, which is known for such simulations [5, P.79]. The PIC algorithm is a traditional, simple and valued method for the simulation of plasma phenomena [6]. The kinetics of the electrons are calculated for a one-dimensional periodic plasma model. In a plasma, the positively charged ions and the negatively charged electrons create electrical fields. The particles move according to the forces created by their own and external fields. By the first Maxwell equation, which is also known as Gauss's law, the electric fields can be calculated based on the charge density. The charge density in turn results from the particle positions. The motion of the particles is deduced from the second Newtonian axiom. For this purpose, the forces that result from the electric fields and act on the individual particles are determined. The plasma particles generate the fields, which in turn represent the basis of the particle movements. The PIC method also acts in this way. The electric fields of the charged particles are calculated. Then the electrons are moved slightly and the fields are calculated again. By repeating this computation cycle, the movements of the electrons are simulated step by step.

Since it is not practical to determine the fields or the forces between the charged particles directly using Coulomb's law, a spatial mathematical grid with equidistant grid points is introduced. The charge densities ρ_j of the particles are accumulated at the grid points using an interpolation scheme, see Fig.1, so that Gaussian law can be used to solve the field equations at these discrete points. Then the calculated electric fields E_j are interpolated back, whereby the forces F_i acting on the particles can be calculated. In the following, the interpolations are referred to as weightings. They are to be integrated into the computation cycle already described, as shown in Fig. 1. For a clear differentiation of the various calculation variables, particle-related values are provided with the index i and grid-related values with the index j. The individual computing cycles are also referred to as time steps Δt and can be carried out as often as required. The simulation thus moves in discrete steps z, with $z \in \mathbb{N}$, through time t. The simulation program shows the positions of the electrons between the time steps. In this way, the movement of the electrons is gradually represent and thus the plasma frequency is visualized.

The main nomenclature

z: number of simulated time steps Δt
X_j: $j - th$ grid point
E_j: electric field at the $j - th$ grid point
ρ_j: charge density at the $j - th$ grid point
Φ_j: electric potential at the $j - th$ grid point
Δx: distance between the grid points and width of a super particle
x_i: $i - th$ super particle
F_i: Force that acts on the $i - th$ super particle
q_s: charge of a super particle

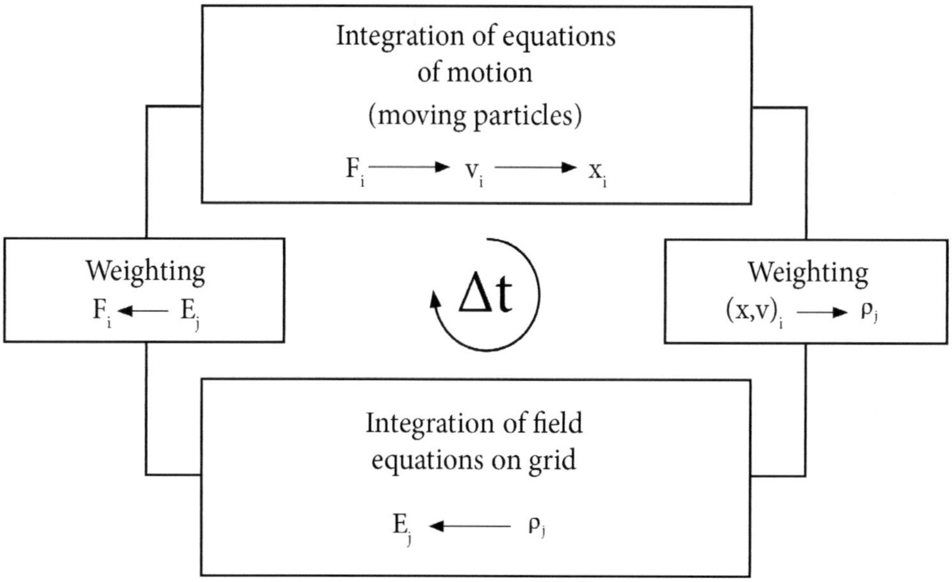

Figure 1: Computing Cycle

2. Model Description

The following model serves as the basis for visualizing the plasma frequency. A layer of width L is considered from a homogeneous, in total neutral plasma. In order to initialize the particle kinetics, the plasma model must be brought out of its equilibrium position. The model is therefore briefly exposed to an external electric field $-E_0$. Due to their higher mobility, the electrons are deflected. The deflaction δx of the electrons is proportional to the strength of $-E_0$. The significantly more massive and therefore more inert ions move only slightly. For a single time step they are regarded stationary. Therefore the ions are only considered as a positively charged, neutralizing background. As shown in Fig. 2a, the deflection of the electrons creates two oppositely charged spatial zones at the edges. After removing the external electric field $-E_0$, an electric field E_0 remains in the interior of the plasma due to these space charge zones, see Fig.2b. The electric field E_0 generates a force on the plasma particles that accelerates them towards their origin. Due to their inertia, the particles move beyond their original position. An opposing electric field arises. The model therefore represents a harmonic oscillator [3, cf.: P.11]. The frequency of the electron oscillation is the plasma frequency.

In the following, the kinetics of the electrons on the basis of the self-generated electrical fields are discussed. The creation of magnetic fields and the collision of particles are neglected. In addition, a reduction in the computational effort is generated by combining the individual electrons into super particles. The super particles can be understood as uniformly charged finite electron clouds. The position of a super particle is represented by the center of mass and the velocity by averaging the individual values. A super particle therefore represents a group of electrons and can be interpreted as a small part of the phase space. So a super particle represents a constant number of electrons.

In the following, the described model is simulated in one-dimensional space with periodic bound-

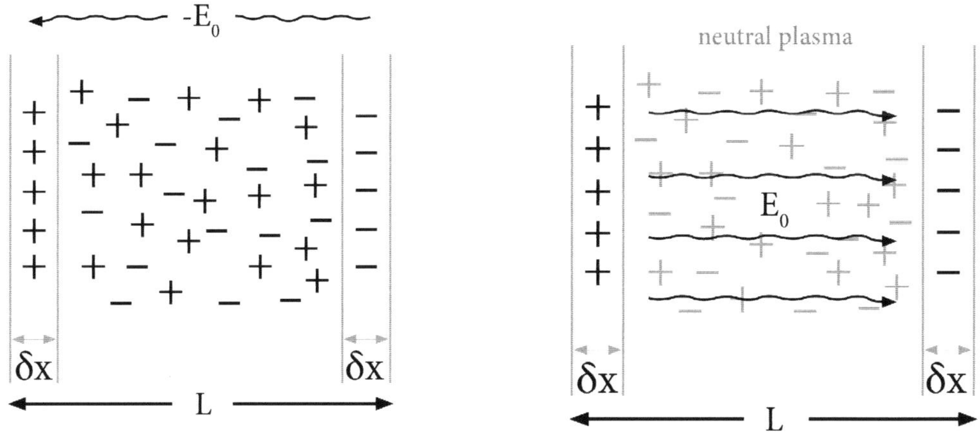

(a) Deflection of the electrons in a neutral plasma

(b) The resulting electrical field in a neutral plasma

Figure 2: The effect of an external electric field on a neutral plasma

ary conditions and thus the plasma frequency is visualized. No specific plasma matter is assumed here. A standardized, generally valid representation of the plasma frequency is generated.

3. Particle-in-Cell Methode

Figure 1 shows the basic computing cycle of the PIC method. In this paragraph the individual calculations for a simulated time step Δt are presented. After the weightings and thus the interpolations of the charge density ρ_j and the force F_i have been presented, the next two paragraph describe the numerical methods for calculating the electric field $E(x_i(t))$ and the new positions $x_i(t)$ as well as velocities $v_i(t)$. At the end of a paragraph, an implementation option in the Matlab programming language is given.

3.1. Weightings

In the calculation cycle, two calculations are titled as weightings. On the one hand, the force $F_i(z)$ acting on the particle, which is required for the integration of the equations of motion, is calculated. On the other hand, the charge density $\rho_j(z)$, which is required for the integration of the field equations, is determined. These weightings imply a form of interpolation through which generates a connection between the particle quantity and the spatial grid. In order to prevent a particle from accelerating itself, the same interpolation method is used for the weighting of the force $F_i(z)$ and the charge density $\rho_j(z)$ [7, cf.: P.129] .

The super particle model serves as the basis for weighting the simulated particles. The super particles are viewed as finite electron clouds with a uniform density. As shown in Fig. 3, the electron clouds have a width of Δx around the position of their center of mass $x_i(z)$. Assuming the $i-th$ super particle is in the interval between the $j-th$ and the $(j+1)-th$ grid point, so $X_j < x_i(t) < X_{j+1}$, then the $i-th$ super particle creates a charge density $\rho_{j,i}(z)$ at the grid point X_j

$$\rho_{j,i}(z) = \frac{\Delta x - x_i(z) - X_j}{\Delta x}\frac{q_s}{\Delta x} = \frac{X_{j+1} - x_i(z)}{\Delta x}\frac{q_s}{\Delta x} \quad (2)$$

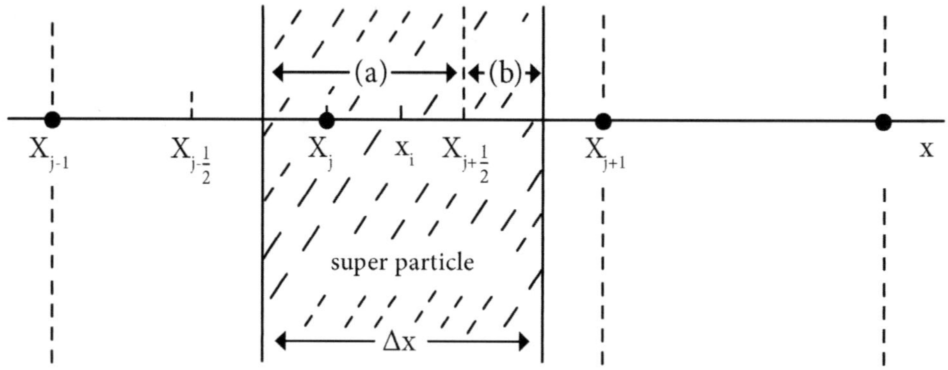

Figure 3: PIC Weighting of the Super Particle x_i

and correspondingly for the charge density $\rho_{j+1,i}(z)$ in the grid point X_{j+1}

$$\rho_{j+1,i}(z) = \frac{x_i(z) - X_j}{\Delta x} \frac{q_s}{\Delta x} \quad , \tag{3}$$

where qs is the charge of a super particle. Δx represents both the distance between two grid points and the width of a super particle. If an equivalent weighting is carried out for all super particles, the charge density $\rho_j(z)$ for the grid point X_j can be summed up by the proportional charge densities of the super particles $\rho_{j,i}(z)$.

$$\rho_j(z) = \sum_{i=1}^{N_{j,s}} \rho_{j,i}(z) \tag{4}$$

Where $N_{j,s}$ represents the number of super particles in the $j - th$ grid cell. This procedure corresponds to a linear interpolation and is referred to as Particle in Cell (PIC) weighting [8, cf.: P.20f.].

For the weighting of the electric fields $E(x_i(z))$ on the super particles, the procedure is equivalent. The electric field for the super particle at position $x_i(z)$ is given by

$$E(x_i(z)) = \left[\frac{X_{j+1} - x_i(z)}{\Delta x}\right] E_j(z) + \left[\frac{x_i(z) - X_j}{\Delta x}\right] E_{j+1}(z). \tag{5}$$

Here, Δx corresponds to the distance between the grid points X_j and X_{j+1}. $E_j(z)$ or $E_{j+1}(z)$ represent the calculated electric fields in these grid points, see section 3.3). If the electric field $E(x_i(z))$ and the charge q_s of a super particle at position $x_i(z)$ are known

$$F(x_i(z)) = E(x_i(z))q_s \tag{6}$$

can be used to calculate the force F_i acting on the particle $x_i(z)$.

The weighting of the electric fields on the super particles is shown together with the particle movements in paragraph 3.2. The implementation of the particle density in the grid points is implemented as follows.

```
function [] = density(dx, rho0, npart, ngrid)

global x rho_el rho_ion qs;

qs = -rho0 * (ngrid/npart); % normalized electron charge
rho_el = zeros(1,ngrid+1);  % electron density

for i = 1:1:npart
    xp = x(i)/dx;
    j1 = floor(xp);
    j2 = j1 + 1;
    g2 = xp - j1;    % weighting factor
    g1 = 1 - g2;     % weighting factor
    j1 = j1+1;
    j2 = j2+1;
    rho_el(j1)=rho_el(j1)+qs*g1; % accumulated charge density
    rho_el(j2)=rho_el(j2)+qs*g2; % ...in the grid points
end

rho_el(1)=rho_el(1)+rho_el(ngrid+1);
rho_el(ngrid+1)=rho_el(1); % periodic conditions
rho_ion=rho0; % ion density (neutralizing background)
end
```

By applying a for-loop, the weighting factors of each individual super particle are calculated for the corresponding grid points. The charge densities of the grid points are then accumulated. Finally, the periodic boundary conditions are implemented and the ion density is defined.

3.2. Integration of the equation of motion - Leapfrog method

The Leapfrog method is a fast, second-order numerical integration method based on little information [9, cf.: P.2f.]. By integrating Newton's second axiom using the Leapfrog method, the new velocity and the new position of the modeled particles can be calculated from the force F_i. In one-dimensional space

$$F(x_i(z)) = m_s \frac{dv_i(z)}{dt} \tag{7}$$

and

$$v_i(z) = \frac{dx_i(z)}{dt} \tag{8}$$

are the relevant equations for calculating the trajectories of the electrons. $F(x_i)$ represents the force that acts on the particle x_i, m_s indicates the mass of a super particle, v_i represents the velocity of the $i-th$ particle and t the time.

The Leapfrog method can be interpreted as an extension of the Euler method. But the Leapfrog method generates higher accuracy by considering a variable not at the beginning but in the middle of an interval [9, cf.: P.2f.]. Takes into account that the particles move gradually through time with Δt, then

$$x_{z+1} = x_z + \Delta t v_{z+\frac{1}{2}} \qquad (z = 1, 2, ..., n) \tag{9}$$

for the new position and

$$v_{z+\frac{3}{2}} = v_{z+\frac{1}{2}} + \Delta t \frac{F(x_{z+1})}{m_e} \qquad (z = 1, 2, ..., n) \tag{10}$$

for the new velocity are the approximated result from Eq. (7) and (8). The number of time steps is given by z with $z \in \mathbb{N}$ and runs from zero (initial position) to n (end of simulation) with $n \in \mathbb{N}$. The step length is essential for the accuracy of the approximation. The longer the time step Δt is chosen, the more imprecise the process becomes.

Due to the fact that v is postponed over time, x and v are not known at the same time. They must be calculated alternately, as shown in Fig. 4. Furthermore, the Leapfrog method achieves

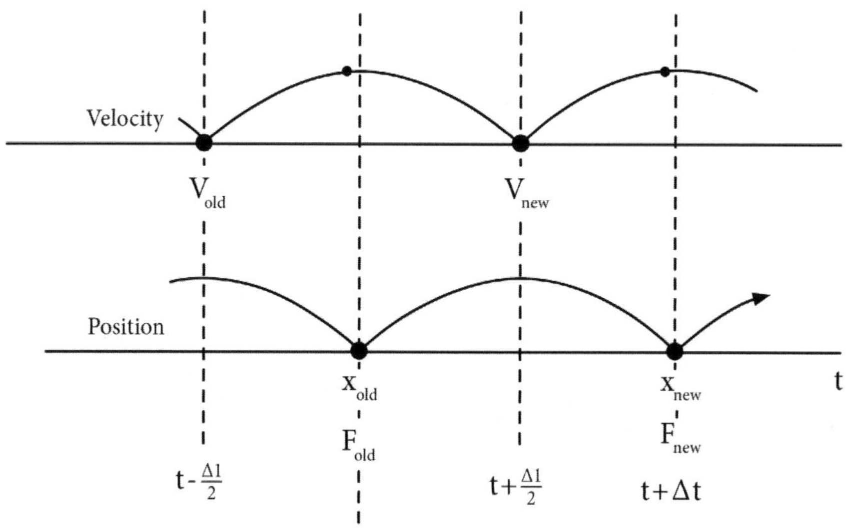

Figure 4: Leapfrog Method

improved stability through the time deflection of v [8, cf.: P.56.]. Compared with alternative second order integration methods it is a stable and relatively precise method for calculating the trajectories of the electrons.

To implement the particle movements, the electrical fields must be weighted. As in paragraph 3.1 a for-loop has been used.

```
function [] = part_move(dt, dx, npart, q_over_m)

global x v E;

for i = 1:1:npart
    xp = x(i)/dx;
    j1 = floor(xp);
    j2 = j1 + 1;
    g2 = xp - j1; % weighting factor
    g1 = 1 - g2;  % weighting factor
```

```
        j1 = j1+1;
        j2 = j2+1;
        E_xi = g1*E(j1) + g2*E(j2); % electric field at x_i
        v(i)=v(i)+q_over_m*E_xi*dt; % calculation of the new velocity
end
x = x + dt*v;  % calculation of the new position
end
```

The new velocity and new positions can then be calculated using the electric fields and the normalized ratio of charge to mass.

3.3. Integration of field equation - Gaussian law

In order to be able to deduce the forces acting on the individual super particles from the charge density $\rho_j(z)$, the electric fields at the grid points must be calculated. With the first Maxwell equation, which is also known as Gauss's law, the divergence of the electric field $E_j(z)$ at the grid points X_j are determined.

$$divE_j(z) = \frac{\rho_j(z)}{\epsilon_0} \quad \text{or} \quad \vec{\nabla}E_j(z) = \frac{\rho_j(z)}{\epsilon_0} \tag{11}$$

The Nabla differential operator $\vec{\nabla}$ represents a vector whose components contain the partial derivative operators [10, cf.: P.54]. Conservative fields can be expressed in terms of the negative gradient of the potential. The electrostatic and thus conservative field considered here is therefore also given by

$$E_j(z) = -grad\Phi_j(z) \quad \text{or} \quad E_j(z) = -\vec{\nabla}\Phi_j(z) \tag{12}$$

The charge density in the grid points is given by $\rho_j(z)$ and the electrical potential by $\Phi_j(z)$. The electric field constant is represented by ϵ_0. A combination of Eq. 11 and 12 yields to

$$div(grad\Phi_j(z)) = -\frac{\rho_j(z)}{\epsilon_0} \quad \text{or} \quad \vec{\nabla}^2\Phi_j(z) = -\frac{\rho_j(z)}{\epsilon_0} \quad , \tag{13}$$

which is called the Poisson Equation [8, cf.: P.16]. The discretization of Eq. (12) and (13) results in

$$E_j(z) = \frac{\Phi_{j-1}(z) - \Phi_{j+1}(z)}{2\Delta x} \tag{14}$$

and

$$\frac{\Phi_{j-1}(z) - 2\Phi_j(z) + \Phi_{j+1}(z)}{(\Delta x)^2} = -\frac{\rho_j(z)}{\epsilon_0} \tag{15}$$

according to the finite difference method [8, cf.: P.16]. Equation (15) can also be written in matrix form.

$$\begin{bmatrix} -2 & 1 & 0 & \dots & 1 \\ 1 & -2 & 1 & \dots & 0 \\ 0 & 1 & \ddots & \ddots & \vdots \\ \vdots & \dots & \ddots & \ddots & \vdots \\ 1 & 0 & \dots & 1 & -2 \end{bmatrix} \begin{bmatrix} \Phi_1(z) \\ \Phi_2(z) \\ \vdots \\ \Phi_{n-1}(z) \\ \Phi_n(z) \end{bmatrix} = \begin{bmatrix} p_1(z) \\ p_2(z) \\ \vdots \\ p_{n-1}(z) \\ p_n(z) \end{bmatrix} \quad \text{with} \quad p_j(z) = -\frac{(\Delta x)^2\rho_j(z)}{\epsilon_0} \quad . \tag{16}$$

The vectors of the matrix in Eq. (16) are linearly dependent. Therefore the additional information that the charge density of the entire system $\langle\rho\rangle$ is

$$\langle\rho\rangle = \sum_{j=1}^{m} \rho_j(z) = 0 \tag{17}$$

must be inserted. Where m indicates the number of grid points. The system of equations can thus be solved uniquely. The Gaussian elimination method is used to calculate the potentials $\Phi_j(z)$. Then the electric fields in the grid points can be determined with the potentials according to Eq. (14).

For the implementation of Poisson's equation, the matrix or the system of equations shown in Eq. (16) is created. The electrical potentials are then determined with a function using the Gaussian elimination method. The electrical fields are then calculated by using a for-loop. Finally, the periodic boundary conditions are implemented.

```
function [] = E_Poisson(dx, ngrid)

global rho_el rho_ion E Phi;
rho = rho_el + rho_ion; % total charge density

D = diag(-2*ones(ngrid,1)); % Matrix
for i = 1:1:ngrid-1
    D(i+1,i)= 1;
    D(i,i+1)= 1;
end
D(ngrid,1)= 1;
D(1,ngrid)= 1;

p = -(dx^2)*rho;
Phi = gaussAlgorithmus(D,p,ngrid); % Gaussian elimination
Phi = Phi(1:ngrid); % el. potentials
E = zeros(1,ngrid+1);
sum_E = 0;

for j = 1:1:ngrid % calculation of the el. fields
      if j==1
          E(j) = 0.5*(Phi(ngrid)- Phi(j+1))/dx;
      elseif j==ngrid
          E(j) = 0.5*(Phi(j-1)- Phi(1))/dx;
      else
      E(j) = 0.5*(Phi(j-1)- Phi(j+1))/dx;
      end
      sum_E = sum_E + E(j);
end
E = E - sum_E/ngrid;
E(ngrid+1) = E(1); % implementation of boundary conditions
end
```

4. Simulation Results

For the visualization and the plasma frequency there is the problem that a one-dimensional model with several thousand super particles cannot be clearly represented. For the visualization of the plasma frequency, the direct representation of the kinetics of the super particles on a line is not functional. For this reason, the movements of the particles are shown using their density functions in the grid points. For each time step, the particle densities are visualized in the grid points, creating a "film" that reflects the particle distribution of the entire system over all time steps.

The relevant complexity of the PIC algorithm is given by the calclation of the elctrical potentials Φ_j. To calculate Φ_j in accordance with Gaussian elemination method the complexity of the calcutaions is proportional to n^3, where n is the number of the grid points.

In Fig. 5 the electron densities in the grid points are visualized. On a grid of length 2π with 100 grid points, 10^4 super particles were implemented.

Figure 5 shows that the system, after it has been deflected once, follows the oscillations of a harmonic oscillator.

5. Conclusion

This paper deals with the visualization of the plasma frequency using a standardized one-dimensional model. The applied PIC algorithm is an excellent method for simulating the electron kinetics. Using Poisson equation the electric fields are calculated from the electral potentials. Alternatively, the electric fields can also be calculated directly from the electron densities using the Gaussian law and the trapezoidal method as an integration method. The simulation confirms the results to be expected from the model.

References

[1] Heuermann, H.: *Mikrowellentechnik: Feldsimulation, nichtlineare Schaltungstechnik, Komponenten und Subsysteme, Plasmatechnik, Antennen und Ausbreitung.* Springer Fachmedien Wiesbaden, 2020. http://dx.doi.org/10.1007/978-3-658-29023-8. http://dx.doi.org/10.1007/978-3-658-29023-8. – ISBN 9783658290238

[2] Kegel, W.H.: *Plasmaphysik: Eine Einführung.* Springer Berlin Heidelberg, 2013. – ISBN 9783642587719

[3] Stroth, Ulrich: *Plasmaphysik: Phänomene, Grundlagen und Anwendungen.* 2. Springer Spektrum, 2018. http://dx.doi.org/10.1007/978-3-662-55236-0. http://dx.doi.org/10.1007/978-3-662-55236-0. – ISBN 978–3–662–55235–3,978–3–662–55236–0

[4] Grimsehl, E. ; Lösche, A.: *Struktur der Materie.* Vieweg+Teubner Verlag, 1990. – ISBN 9783663014522

[5] Lapenta, Giovanni: Kinetic Plasma Simulation:Particle In Cell Method / Center for mathematical Plasma Astrophysics, KU Leuven,Belgium. 2015. – Forschungsbericht

[6] Philip L. Pritchett, Manfred S. Jörg Büchner B. Jörg Büchner ; Dum, Christian T. (Hrsg.): *Space Plasma Simulation.* 1. Springer-Verlag Berlin Heidelberg, 2003 (Lecture Notes in Physics 615). – ISBN 3540006982,9783540006985

[7] Hockney, R.W. ; Eastwood, J.W.: *Computer Simulation Using Particles.* CRC Press, 1988. – ISBN 9781439822050

[8] Birdsall, A.B C.K; L. C.K; Langdon: *Plasma Physics via Computer Simulation.* CRC Press, 2004 (Series in Plasma Physics and Fluid Dynamics). – ISBN 978–1–4822–6306–0,1482263068,277–277–277–2

[9] Young, Peter: *The leapfrog method and other "symplectic" algorithms for integratingNewton's laws of motion*, April 2014. – Physics 115/242

[10] Schwab, A.J.: *Begriffswelt der Feldtheorie: Elektromagnetische Felder, Maxwell-Gleichungen, Gradient, Rotation, Divergenz.* Springer Berlin Heidelberg, 2019. – ISBN 9783662583920

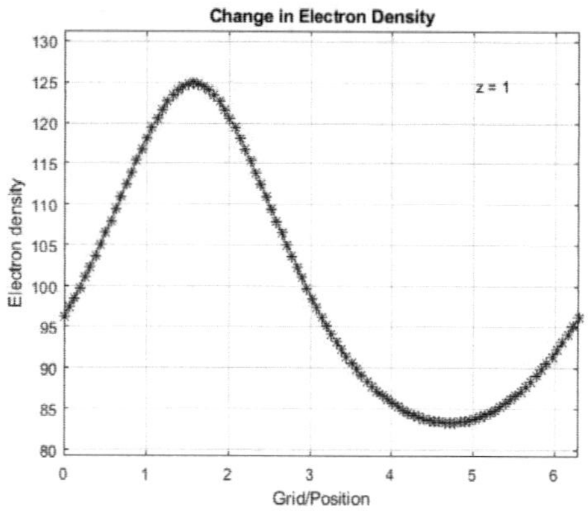

(a) $z = 1$

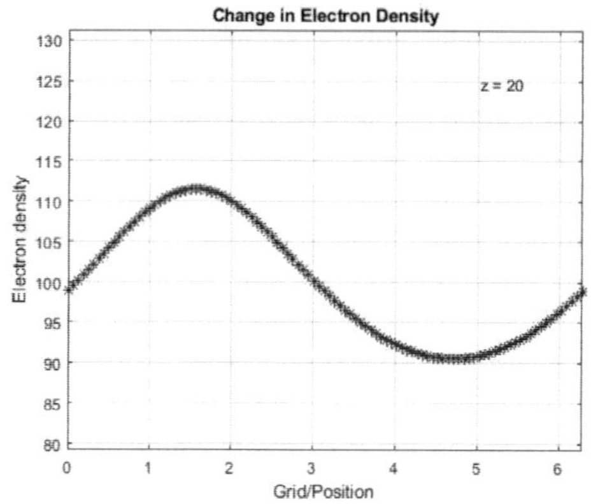

(b) $z = 20$

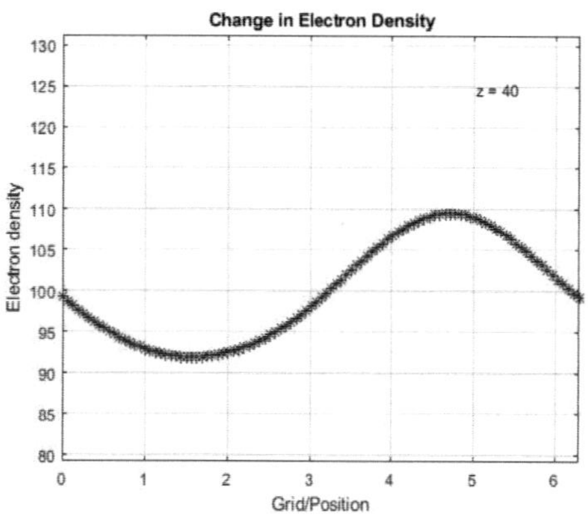

(c) $z = 40$

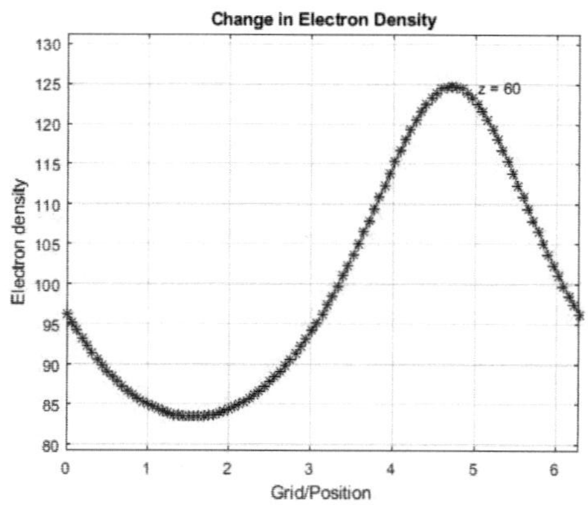

(d) $z = 60$

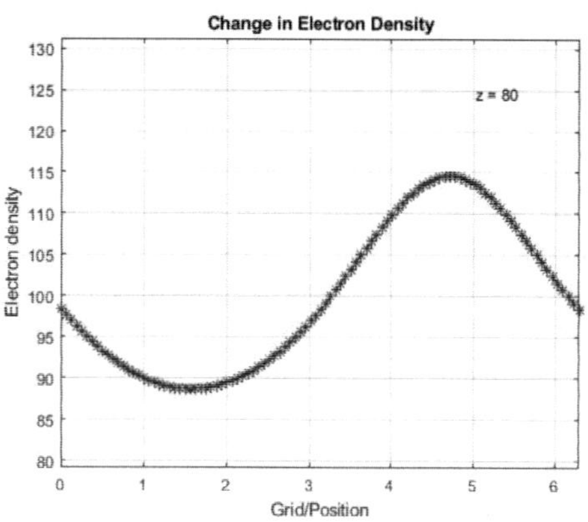

(e) $z = 80$

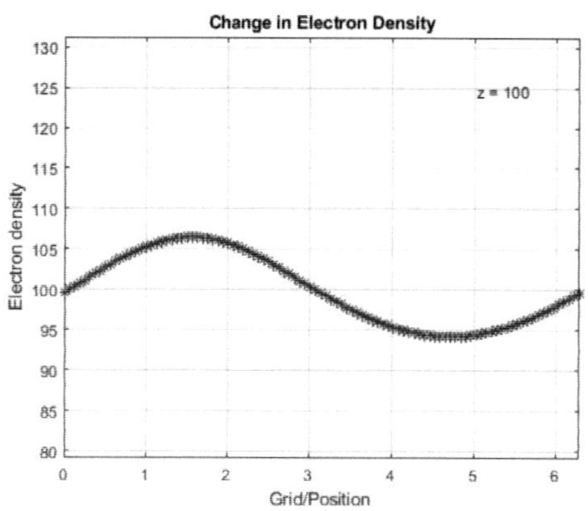

(f) $z = 100$

Figure 5: Electron Density according to different Number of Time Steps

Journal of Physics: Conference Series **2162** (2022) 012017 doi:10.1088/1742-6596/2162/1/012017

Conception and analysis of Cascaded Dual Kalman Filters as virtual sensors for mastication activity of stomatognathic craniomandibular system

Carlotta Bogena[1], Sirarpi Hovsepyan[2], Paolo Mercorelli[3,*]

[123]Leuphana University of Lueneburg, Institute of Product- and Processinnovation, Universitätsallee 1, D-21335 Lueneburg, Germany

*correnspoding author: mercorelli@uni.leuphana.de

Abstract. The presented work shows a possible methodical approach for parameter estimation of a kinematic and dynamic element that characterizes a human mandibular system during the mastication process using position measurement only. The considered parameters are the velocity, friction coefficient, and the mass of the moving part of the mandibular during the mastication activity of a human. Internal or optical motion sensors can still allow imprecision in the measurements. To overcome these, in the present work a system identification algorithm is designed using a combination of three backward cascaded Kalman Filter, which consists of three Extended Kalman Filters. The identification procedure is validated through a matching criterion based on the estimation of the mass, which is assumed to be known in the first stage of the Kalman Filter structure. Three EKFs are tuned as long as the initial value of the mandibular mass is achieved as an estimation of the third one. This is due to the fact that the optimization procedure tries to optimize a non-convex optimization problem that can admit more than one solution. The main contribution of this project is designing state estimation dynamic system, which accurately estimates friction with a linear time varying model. Friction coefficient plays an important role in the early diagnosis of temporomandibular joints disorders, since it is very low under normal condition, and an increase may be associated with abnormalities. Computer simulations show the effectiveness of the proposed method to accurately estimate friction dynamics and refrain from complex nonlinearities.

1. Introduction

In the fields of virtual sensing, sensor fusion, intelligent sensing and sensorless control, the Kalman Filter (KF), with its methodological variations, is one of the highly known algorithms used for linear systems. The Extended Kalman filter (EKF) can be used for the identification of parameters of nonlinear systems. Today sensor fusion plays an important role in motion analysis in all types of applications. The recording of motion is often done by inertial or optical motion sensors. However, the accuracy of such tools is often very low, which can result in incorrect data interpretation. Kalman Filter is used as a real time multi-sensor fusion algorithm to overcome these problems, see [1]. Additionally, it serves for estimating the states of a system [2], to increase the accuracy [3] or to make corrections for system parameters [1]. Sequential filter such as KF "provides an excellent framework for estimating and updating the parameters of a system as data becomes available" [4], therefore a dual KF structure is proposed.

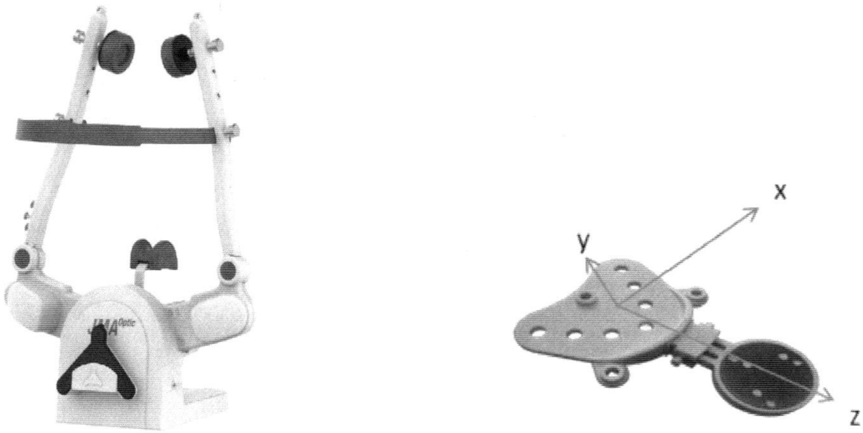

(a) (b)

Figure 1: (a) Instrumental functional analysis for the determination of the horizontal jaw relation with the application of a JMA Optic facebow by Zebris (Zebris Medical GmbH, Isny, Germany). (b) Bite fork with displayed coordinate system (Zebris Medical GmbH, Isny, Germany).

The recording of the lower jaw movements and positions under neuromuscular and kinematic view serves:

- for the qualitative and quantitative determination of the functional state of the the position and movement of the jaw;
- to gain indications for therapeutic measures (function-oriented therapy planning);
- for the documentation of the results due to the course of therapy in order to control and evaluate the initiated therapy (functional therapy documentation and feedback therapy).

Its purpose is the improvement of function in terms of a biomechanical optimisation or prosthetic rehabilitation and/or oral rehabilitation on a function-oriented basis. The data collected from the movement records represent a function-oriented therapeutic approach. They provide patient-oriented values for setting the articulator for the design of individual occlusal surfaces. In the digital workflow, the instrumental motion analysis is an important and necessary supplement in order to map the movement functions of the jaw for the occlusal design of restorations in the CAD/CAM process. Human oral cavity consists of both hard and soft tissues such as chin, teeth, tongue, mucosa and glands as well as the temporomandibular joint (TMJ) which connects the upper and lower jaw bone. The TMJ can be considered as a synovial joint and works in the same way as other synovial joints such as the hip and the knee. The loading and motion conditions in synovial joints are complex therefore "it is important to measure the friction in synovial joints accurately, since the friction level in these natural bearings is generally low and also because of additional difficulties associated with other soft tissues surrounding the joint and mechanical factors that can contribute to the friction measurement" in the context of [5]. The main motivation of this project is to estimate the friction which is generally very low in biological systems under normal conditions, but can drastically increase in deteriorated conditions or due a disease. The main function of the oral cavity as an organ is closely related to speech and food processing. Therefore, friction plays an important role. [6] studies possible effects of high friction and TMJ disc disorders. According to them, for example, due to some changes in the lubrication mechanism the coefficient may rise by a factor of about 4. In [6] it is employed friction as a Coulomb-type friction model, which, however, is considered as a simplification when taking into account characteristics of friction dynamics between a deformable and a rigid articular surfaces. Coulomb-type friction model is inherently nonlinear and relies upon time and velocity dependence of friction [7]. Another well-known dynamical friction estimation model is the LuGre model. In [8], designing a model-based nonlinear controller called Dynamical Adaptive Backstepping- Sliding Mode Control (DAB-SMC) for a human like skeletal-muscle the authors implemented LuGre-based adaptive friction observer. With the latter controller gains more stability in performance, and other metrics, such as position tracking accuracy, are also improved. However, in order LuGre-based friction model, for instance, to incorporate hysteresis behavior of friction in a more complex model structure, the parameter estimation complexity increases due to its high nonlinearity [8]. In this contribution a form of a linear time varying friction model is proposed. In fact, the friction parameter is considered in an EKF and dynamically estimated. This approach is a valid compromise between a structurally simple model and a dynamic estimation of the parameter. This approach can be seen as a linearization of a complex phenomenon such as the mechanical friction effect around the actual working state (position and velocity) of the mandibular system. The effectiveness of the obtained results indicates that the adopted method is a suitable one for the considered application.

"KF is used for estimating an object's position and other parameters of motion based on measurements, which are non linear related to the objects position" [9]. For this reason, the goal was to use the advantages of the KF also for the approximation of a jaw position. The paper presents three EKFs to be tuned as long as the initial value of the mandibular mass is achieved as an estimation of the last EKF. This is due the fact that here in the model it tries to optimize a non-convex optimisation problem which can admit more than one solution. The paper is structured in the following way. The paper starts with the description of the physical model in Section 2. It follows a description of the Kalman Filter background in Section 3 and the Observability of the system in Section 4. The main results are presented in Section 5, with a presentation of the used measurement system and the proposed algorithm to estimate velocity and parameters of the moving mandibular system. A discussion of the found results will be addressed in Section 6. The conclusion closes the paper.

2. Physical model

The concept of using kinematics in translational and rotational capacity is used for the combination of opening-closing and lateral movements of the jaw in [10]. Therefore, the calculated forces, combined with the moment arms and horizontal components can be used for the calculation of the forces that contribute to the lateral movements. The following equations describing the Kinematic Model of the jaw joint system are proposed in [10].
The opening and closing movement M_n can be described as

$$M_{\mathrm{n}} = K_n \; MA_n \; F_n, \tag{1}$$

where $n = LM, RM, LD$ or RD: left masseter (LM), right masseter (RM), left diagastric (LD) and right diagastric (RD) describes the product of the force component in the sagittal plane, MA_{n} describes the moment arm and a subject-specific parameter K_{n} is added to account for inaccuracies. Therefore the total joint moment about the centre of rotation M_{tot} can be determined using

$$M_{\mathrm{Tot}} = M_{\mathrm{LM}} + M_{\mathrm{RM}} + M_{\mathrm{LD}} + M_{\mathrm{RD}} + M_W + M_{\mathrm{Damp}} + O_z, \tag{2}$$

where W is the mandible weight, D is the joint damping and O_z is the offset value that determines resting position in the vertical direction.
The kinematics of the joint can be calculated by

$$d(T_s + \Delta T_s) = d(T_s) + \omega(T_s) + \frac{M_{\mathrm{Tot}}}{2I}\Delta T_s^2, \tag{3}$$

$$\omega(T_s + \Delta T_s) = \omega(T_s) + \frac{M_{\mathrm{Tot}}}{I}\Delta T_s, \tag{4}$$

where $d(T_s + \Delta T_s)$ describes the new angular displacement (in θ-direction) of the mandible, on the previous displacement $d(T_s)$, previous angular velocity $\omega(T_s)$, moment of intertia I and sampling time ΔT_s.

In order to obtain vertical displacement $z(T_s)$ and not polar coordinates, a geometric conversion that uses (R, θ) was used in [10]. The vertical displacement $z(T_s)$ is given by

$$z(T_s) = IP_R\ \sin(IP_\theta + d(T_s)) - IP_R\ \sin(IP_\theta). \tag{5}$$

The transitional component of the movement can be determined similarly, but along the x-axis. The horizontal force components can be calculated using the acute angle between force vector and x-axis, and is described by α. The vertical distance y can be calculated by

$$y = r\ \sin(\theta). \tag{6}$$

α can be determined as follows

$$\alpha = \arctan\left(\frac{r_1 \sin\theta_1 - r_2 \sin\theta_2}{x_2 - x_1}\right), \tag{7}$$

with F_x being the lateral component of the total musculotendon force F

$$F_x = F\ \cos(\alpha). \tag{8}$$

The total force F_{Tot}, can be determined by the sum of the forces in x-direction, with the addition of O_{x} to add lateral force offset and subject specific parameters K_{i} (i=LD, RD, KM, RM) to account for different muscle sizes and strengths

$$M_{\mathrm{Tot}} = K_{\mathrm{LM}}F_{\mathrm{LM}} + K_{\mathrm{RM}}F_{\mathrm{RM}} + K_{\mathrm{LD}}F_{\mathrm{LD}} + K_{\mathrm{RD}}F_{\mathrm{RD}} + F_{\mathrm{Damp}} + O_{\mathrm{x}}. \tag{9}$$

The horizontal displacement $x(T_s + \Delta T_s)$ and $v(T_s + \Delta T_s)$ can be calculated as follows

$$x(T_s + \Delta T_s) = x(T_s) + v(T_s)\Delta T_s + \frac{F_{\mathrm{Tot}}}{2m}\Delta T_s^2, \tag{10}$$

$$v(T_s + \Delta T_s) = v(T_s) + \frac{F_{\mathrm{Tot}}}{m}\Delta T_s. \tag{11}$$

The horizontal motion also requires its own damping components F_{Damp} due to the passive viscous and elastic damping properties of the masticatory system that only affect horizontal movement

$$F_{\mathrm{Damp}} = -\beta\ v, \tag{12}$$

where β is the damping coefficient and v the horizontal velocity.

The adopted model is based on the following mathematical systems:

$$\begin{cases} \dot{z}_{\mathrm{mp}}(t) = z_{\mathrm{mv}}(t) \\ \dot{z}_{\mathrm{mv}}(t) = -\frac{k_v(t)}{m}\ z_{\mathrm{mv}}(t) + \frac{u(t)}{m} \\ \dot{k}_v(t) = 0, \end{cases} \tag{13}$$

including the three states of the system in which $z_{\mathrm{mp}}(t)$ describes the position of the mandible, $z_{\mathrm{mv}}(t)$ the velocity of the mandible, $\dot{z}_{\mathrm{mv}}(t)$ the acceleration of the mandible, and $u(t)$ is total force. The parameter that is being estimated by the system varies. In this specific case the parameter k_v, which describes the Friction coefficient is being estimated.We get $\dot{k_v}(t) = 0$ as it is set as *stochastic constant*. The Jacobian matrix of the system is as follows

$$J = \begin{bmatrix} 0 & 1 & 0 \\ 0 & -\frac{k_v(t)}{m} & -\frac{z_{\text{mv}}(t)}{m} \\ 0 & 0 & 0 \end{bmatrix}. \tag{14}$$

Estimating the Mass m the system can be described as follows

$$\begin{cases} \dot{z}_{\text{mp}}(t) = z_{\text{mv}}(t) \\ \dot{z}_{\text{mv}}(t) = -\frac{k_v(t)}{m} z_{\text{mv}}(t) + \frac{u(t)}{m} \\ \dot{m}(t) = 0, \end{cases} \tag{15}$$

therefore the Jacobian matrix can be defined as

$$J = \begin{bmatrix} 0 & 1 & 0 \\ 0 & -\frac{k_v(t)}{m} & -\frac{-k_v(t)z_{\text{mv}}(t)+u(t)}{m(t)^2} \\ 0 & 0 & 0 \end{bmatrix}. \tag{16}$$

3. Kalman filter background

"The Kalman Filter is a set of mathematical equations that provide an efficient computational (recursive) means to estimate the states of a process, in a way that minimizes the mean of the squared error." [11]

Based on [11] the process of a non-linear system can be described with the following stochastic difference equation

$$x_k = f(x_{k-1}, u_{k-1}, w_{k-1}), \tag{17}$$

with a measurement defined as

$$z_k = h(x_k, v_k), \tag{18}$$

in which w_k and v_k both represent random variables describing the process and measurement noise respectively. Therefore, the a priori estimation of the state is defined as

$$\hat{x}_k^- = f(\hat{x}_{k-1}, u_{k-1}, 0), \tag{19}$$

as well as the a priori error covariance matrix P_k^-, which is as follows

$$P_k^- = J_k P_{k-1} J_k^T + W_k Q_{k-1} W_k^T. \tag{20}$$

The matrix Q_k is the process noise covariance and both J_k and W_k are the process Jacobians at the current step k. Therefore P_{k-1} represents the a posteriori error covariance matrix at step $k-1$ (previous step). Secondly the EKF measurement update equations include the Kalman gain K which is as follows

$$K_k = P_k^- H_k^T (H_k P_k^- H_k^T + V_k R_k V_k^T)^{-1}, \tag{21}$$

with H_k and V_k as the measurement Jacobians and R_k as the measurement noise covariance all changing at each time step k with the sampling time T_s. The system which is being considered in this paper neglects both W_k and V_k, since the noises are assumed to have the same effect on all states and measured variables and are therefore not considered further. The a posteriori estimation is as follows

$$\hat{x}_k = \hat{x}_k^- + K_k(z_k - h(\hat{x}_k^-, 0)), \tag{22}$$

where $z_k(k) = h(\hat{x}_k^-, 0)$. The a posteriori estimation of the error covariance matrix is as follows

$$P_k = (I - K_k H_k) P_k^-. \tag{23}$$

In which H is the Jacobian matrix of the output measurement.

4. Observability of the system

The analysis of the observability is done for each singular system. For that purpose, the idea is to consider separately in a cascaded form a procedure for the estimation of the mastication parameters. Considering the full system with four states implies the following

$$\begin{cases} \dot{z}_{\mathrm{mp}}(t) = z_{\mathrm{mv}}(t) \\ \dot{z}_{\mathrm{mv}}(t) = -\frac{k_v(t)}{m} z_{\mathrm{mv}}(t) + \frac{u(t)}{m} \\ \dot{k}_s(t) = 0 \\ \dot{m}(t) = 0. \end{cases} \tag{24}$$

The observability analysis can start with a observable determination matrix θ as follows:

$$\theta = \begin{bmatrix} \theta_0 & \theta_1 & \theta_{n-1} \end{bmatrix}^{\mathrm{T}}, \tag{25}$$

with $\dot{X}(t) = f\big(X(t), u(t)\big)$ and $y(t) = h\big(X(t)\big)$ and where $y(t) = z_{\mathrm{mp}}(t)$, in which $X(t) = [z_{\mathrm{mp}}(t) \quad z_{\mathrm{mv}}(t) \quad k_s(t) \quad m(t)]$ only the position is a measured state, then

$$\theta = \begin{bmatrix} \theta_0 & \theta_1 & \dots & \theta_{n-1} \end{bmatrix}^{\mathrm{T}}. \tag{26}$$

It follows that

$$\theta = \begin{bmatrix} \frac{\partial h}{\partial X^{\mathrm{T}}} & \frac{\partial \theta_0 f}{\partial X^{\mathrm{T}}} & \frac{\partial \theta_1 f}{\partial X^{\mathrm{T}}} & \dots & \frac{\partial \theta_{n-1} f}{\partial X^{\mathrm{T}}} \end{bmatrix}^{\mathrm{T}}. \tag{27}$$

To have a sufficient condition for the observability, then $\theta = n$. The observability test for system (24) follows as:

$$\theta_0 = \begin{bmatrix} 1 & 0 & 0 & 0 \end{bmatrix}, \tag{28}$$

and

$$\theta_1 = \frac{\partial \theta_0 f}{\partial X^{\mathrm{T}}} = \frac{\frac{\partial z_{\mathrm{mp}}(t)}{\partial \theta}}{\partial X^{\mathrm{T}}} \begin{bmatrix} 1 & 0 & 0 & 0 \end{bmatrix}. \tag{29}$$

This is already proven that the sufficient condition for the full system is not guaranteed. Considering the following system note for the kinematic Forward Euler model to estimate the state of the force $F_{\mathrm{Tot}}(k-1)$ and acceleration $\gamma(k-1)$:

$$\begin{cases} x(k) = x(k-1) + T_{\mathrm{s}}\gamma(k-1) \\ \gamma(k) = \gamma(k-1) + T_{\mathrm{s}}\frac{F_{\mathrm{Tot}}(k-1)}{m} \\ F_{\mathrm{Tot}}(k) = F_{\mathrm{Tot}}(k-1), \end{cases} \tag{30}$$

Calculating as before, then

$$\theta_0 = \begin{bmatrix} 1 & 0 & 0 \end{bmatrix}, \tag{31}$$

$$\theta_1 = \frac{\partial \theta_0 f}{\partial X^{\mathrm{T}}} = \frac{\partial x(k-1)}{\partial X^{\mathrm{T}}(k-1)} = \begin{bmatrix} 1 & T_{\mathrm{s}} & 0 \end{bmatrix}, \tag{32}$$

and

$$\begin{aligned} \theta_2 = \frac{\partial \theta_1 f}{\partial X^{\mathrm{T}}(k-1)} = \frac{\partial}{\partial X^{\mathrm{T}}(k-1)} \left[x(k) + T_{\mathrm{s}}\gamma(k)\right] = \\ \left[x(k-1) + T_{\mathrm{s}}\gamma(k-1) + T_{\mathrm{s}}\Big(\gamma(k-1) + T_{\mathrm{s}}\frac{F_{\mathrm{Tot}}}{m}\Big)\right] \\ = \begin{bmatrix} 1 & 2T_{\mathrm{s}} & \frac{T_{\mathrm{s}}^2}{m} \end{bmatrix}. \end{aligned} \tag{33}$$

and thus

$$\theta = \begin{bmatrix} 1 & 0 & 0 \\ 1 & T_\text{s} & 0 \\ 1 & 2T_\text{s} & \frac{T_\text{s}^2}{m} \end{bmatrix}. \tag{34}$$

The sufficient condition is guaranteed when Rank $\theta = 3$. Now, considering the system of estimating the Friction coefficient k_v represented with the following Kalman filter:

$$\begin{cases} \dot{z}_\text{mp}(t) = z_\text{mv}(t) \\ \dot{z}_\text{mv}(t) = -\frac{k_v(t)}{m} z_\text{mv}(t) + \frac{u(t)}{m} \\ \dot{k_v}(t) = 0, \end{cases} \tag{35}$$

where $u(t)$ is already estimated F_Tot, $\dot{z}_\text{mv}(t)$ is the already estimated velocity, which we use together with $z_\text{mp}(t)$ as measured states:

$$y(t) = \begin{bmatrix} z_\text{mp}(t) \\ z_\text{mv}(t) \end{bmatrix}. \tag{36}$$

Given two measured states the observable determination matrix (27) can be obtained as:

$$\theta_0 = \begin{bmatrix} 1 & 0 & 0 \\ 0 & 1 & 0 \end{bmatrix}, \tag{37}$$

$$\theta_1 = \frac{\partial \theta_0 f}{\partial X^\text{T}} = \frac{\partial}{X^\text{T}} \begin{bmatrix} z_\text{mv}(t) \\ -\frac{k_v(t)}{m} z_\text{mv}(t) + \frac{u(t)}{m} \end{bmatrix} = \begin{bmatrix} 0 & 1 & 0 \\ 0 & -\frac{k_v(t)}{m} & \frac{u(t)}{m} \end{bmatrix}, \tag{38}$$

and

$$\theta_2 = \frac{\partial \theta_1 f}{\partial X^\text{T}} = \frac{\partial}{\partial X^\text{T}} \begin{bmatrix} -\frac{k_v(t)}{m} z_\text{mv}(t) + \frac{u(t)}{m} \\ \frac{k_v^2(t)}{m^2} z_\text{mv}(t) - \frac{k_v(t)}{m^2} u(t) \end{bmatrix} = \begin{bmatrix} 0 & -\frac{k_v(t)}{m} & \frac{z_\text{mv}(t)}{m} \\ 0 & \frac{k_v^2(t)}{m^2} & \frac{2k_v(t)}{m^2} z_\text{mv}(t) - \frac{u(t)}{m^2} \end{bmatrix}. \tag{39}$$

Hence, above described system is always observable iff $z_\text{mv}(t) \neq 0$ and $u(t) \neq 0$. The system for the estimation of the mass m is defined as follows:

$$\begin{cases} \dot{z}_\text{mp}(t) = z_\text{mv}(t) \\ \dot{z}_\text{mv}(t) = -\frac{k_v(t)}{m} z_\text{mv}(t) + \frac{u(t)}{m} \\ \dot{m}(t) = 0. \end{cases} \tag{40}$$

The same procedure of calculation of the corresponding elements of observability determination matrix follows as:

$$\theta_0 = \begin{bmatrix} 1 & 0 & 0 \\ 0 & 1 & 0 \end{bmatrix}, \tag{41}$$

$$\theta_1 = \frac{\partial \theta_0 f}{\partial X^\text{T}} = \frac{\partial}{X^\text{T}} \begin{bmatrix} z_\text{mv}(t) \\ -\frac{k_v(t)}{m} z_\text{mv}(t) + \frac{u(t)}{m} \end{bmatrix} = \begin{bmatrix} 0 & 1 & 0 \\ 0 & -\frac{k_v(t)}{m} & \frac{k_v(t) z_{\text{mv}-u(t)}}{m^2} \end{bmatrix} \tag{42}$$

and

$$\theta_2 = \frac{\partial \theta_1 f}{\partial X^\text{T}} = \frac{\partial}{\partial X^\text{T}} \begin{bmatrix} -\frac{k_v(t)}{m} z_\text{mv}(t) + \frac{u(t)}{m} \\ \frac{k_v^2(t)}{m^2} z_\text{mv}(t) - \frac{k_v(t)}{m^2} u(t) \end{bmatrix} = \begin{bmatrix} 0 & -\frac{k_v(t)}{m} & \frac{k_v(t) z_{\text{mv}(t)} - u(t)}{m^2} \\ 0 & \frac{k_v^2(t)}{m^2} & \frac{-k_v^2(t) z_{\text{mv}(t)} + k_v(t) u(t)}{m^4} \end{bmatrix}. \tag{43}$$

Based on the rank criteria of observability it is easy to show for θ matrix with above described elements that Rank $\theta = 3$ holds true. Thus, the observability of discussed system (40) is guaranteed.

5. Experimental setup

The JMA-Optic System (Zebris Medical GmbH, "Isny", Germany), shown in Fig. 1a, which is a 3D coordinate measuring system, was used to measure the movement of the mandible. It consists of three parts: a handy stand-alone face bow which contains a receiver and a control unit, the lower jaw sensor which works as the transmitter unit and a charging station. The system works due to infrared signals from the mandible sensor when measuring from the facebow. During movement cameras in the headbox are detecting a change of the image of the mandible sensor. The coordinates of the mandible sensors are being calculated in 3D space.

A patented bite fork, shown in Fig. 1b is used for an exact relationship between the movement data in the measuring system and the position of the upper jaw as well as the surfaces of the teeth scanned by the model and intraoral scanner. It allows an easy transfer of the settings of the jaw to mechanical articulators. The measured position data consist of three fixed points on the bite fork. One point is located in the coordinate origin, the other two are equally spaced to the right and left of it. The position values of the three points were queried at the same frequency during the motion sequence.

A simplified mathematical model of the jaw joint system was designed. It includes parameters such as a combination of all acting forces, described as force d, the acceleration a of the movement of the mandible, a spring coefficient k_s, a friction coefficient k_v and the mass m of the mandible. As it has already been described in the previous sections, it is impossible to determine the exact mass of the lower jaw, therefore it is usually assumed that the entire mass is applied to a centered point and only acts at this point. The measured position data was used to determine the parameters mentioned above. For this the motion sequence along the y-axis (vertical axis) was used. A cascaded KF structure was created for this purpose. As there is no input a dual KF was used. It was not possible to approximate all parameters at the same time, as the system would have been unobservable otherwise. Based on the cascaded structure, the parameters could thus be continuously estimated and at the same time compared and checked against given reference values from the literature. Due to the relative large sample time (0.0167 sec.) with which the position is sampled, the dynamics of a feedforward discretisation appears to be high and, thus, makes it difficult to be tuned.

5.1. *Estimation of friction coefficient*

Figure 2 represents a systematic construction of the structure and indicates also the proposed estimation procedure. The three EKFs are tuned as long as the initial value of the mandibular mass are achieved as an estimation of the third one. The diagram shows that the cascaded structure consists of three individual EKF's, with different known inputs and different output. Each EKF can therefore determine an individual parameter (EKF_1 estimates velocity $z_{\mathrm{mv}}(k)$, whereas EKF_2 estimates k_v). The EKF's are separate from each other, otherwise the system would be unobservable.

Given are the equations of the EKF_2 as already generally described in (13) which approximates k_v. After applying Backward Euler discretisation the system structure can be written as the following, where T_{s} represents a sampling time:

$$\hat{x}^-(k) = \begin{bmatrix} \hat{x_1}(k-1) + T_s \left(\frac{T_s\ u(k)/m}{1+T_s\ \hat{x_3}(k)/m} + \frac{\hat{x_2}(k)}{1+T_s\ \hat{x_3}(k)/m} \right) \\ \frac{T_s\ u(k)/m}{1+T_s\ \hat{x_3}(k)/m} + \frac{\hat{x_2}(k)}{1+T_s\ \hat{x_3}(k)/m} \\ \hat{x_3}(k-1) \end{bmatrix}, \qquad (44)$$

$\hat{x}^-(k)$ describes the a priori estimation of the three states $x_1(k)$, $x_2(k)$ and $x_3(k)$, in which

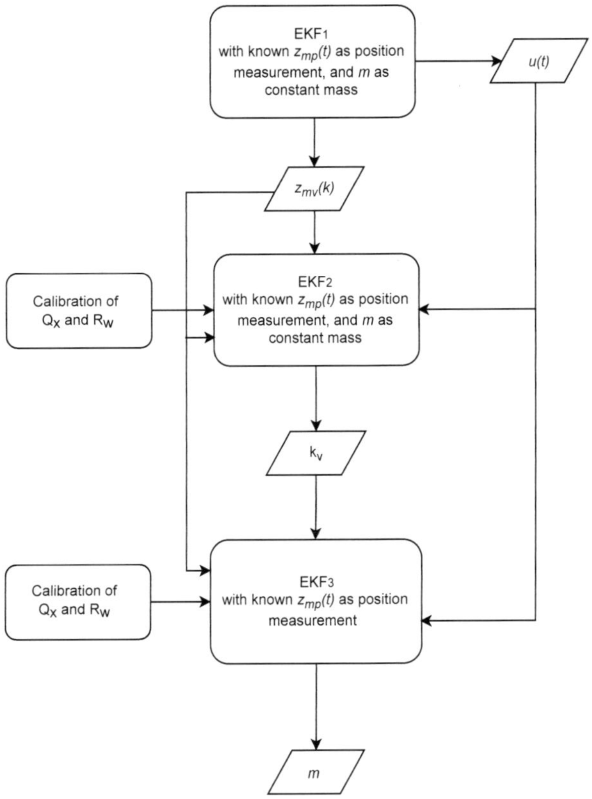

Figure 2: Cascaded dual Kalman Filter structure for parameter approximation

$x_1(k-1)$ is the position, $x_2(k-1)$ the velocity and $x_3(k-1)$ the friction coefficient k_v.

$$H = \begin{bmatrix} 1 & 0 & 0 \\ 0 & 1 & 0 \end{bmatrix}, \tag{45}$$

where H describes the output Jacobian Matrix. The Jacobian matrix of the system is as follows

$$J = \begin{bmatrix} 1 & \frac{T_s}{1+T_s \frac{x_3(k)}{m}} & T_s \left(\frac{-u(k)\frac{T_s^2}{m^2} - \frac{x_2(k-1)\ T_s}{m}}{\frac{(1+x_3\ T_s)^2}{m^2}} \right) \\ 0 & \frac{1}{1+T_s \frac{x_3(k)}{m}} & \frac{-u(k)\frac{T_s^2}{m^2} - \frac{x_2(k-1)\ T_s}{m}}{\frac{(1+x_3\ T_s)^2}{m^2}} \\ 0 & 0 & 1 \end{bmatrix} \tag{46}$$

5.2. Estimation of mass

Given are the equations of the EKF_3 as already described in (15) which estimates m.

$$\hat{x}^-(k) = \begin{bmatrix} x_1(k-1) + T_s \left(\frac{T_s\ u(k)/x_3(k)}{1+T_s\ k_v(k)/x_3(k)} + \frac{x_2(k)}{1+T_s\ k_v(k)/x_3(k)} \right) \\ \frac{T_s\ u(k)/x_3(k)}{1+T_s\ k_v(k)/x_3(k)} + \frac{x_2(k)}{1+T_s\ k_v(k)/x_3(k)} \\ x_3(k-1) \end{bmatrix}, \tag{47}$$

in which $x_1(k-1)$ is the position, $x_2(k-1)$ the velocity and $x_3(k-1)$ the mass m. The Jacobian matrix for it can be described as

$$J = \begin{bmatrix} 1 & \frac{T_s}{1+T_s\ \frac{x_3(k)}{m}} & T_s\left(-\frac{T_s\ u(k)}{(m+T_s\ x_3(k))^2} + \frac{\frac{x_2(k-1)\ T_s\ x_3(k)}{m^2}}{\frac{(1+T_s\ x_3(k))^2}{m^2}}\right) \\ 0 & \frac{1}{1+T_s\ \frac{x_3(k)}{m}} & -\frac{T_s\ u(k)}{(m+T_s\ x_3(k))^2} + \frac{\frac{x_2(k-1)\ T_s\ x_3(k)}{m^2}}{\frac{(1+T_s\ x_3(k))^2}{m^2}} \\ 0 & 0 & 1 \end{bmatrix} \tag{48}$$

while the output Jacobian matrix follows as:

$$H = \begin{bmatrix} 1 & 0 & 0 \\ 0 & 1 & 0 \end{bmatrix}. \tag{49}$$

6. Results

The determination of the parameters was done by means of the cascaded EKF structure, which allowed a continuous approximation and feedback comparison of the parameters. As shown in Fig. 2, the mass in EKF_1 is assigned a fixed value. In the last step of the cascaded structure (EKF_3) the mass is then being estimated as an unknown parameter and compared with the given value. Thus it can be determined whether the estimates of whole cascaded structure are valid. The measured position and the estimated position by EKF are presented in Fig. 3 with a

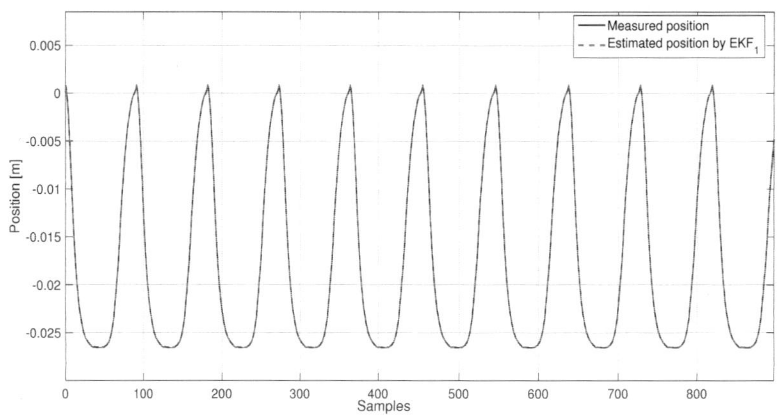

Figure 3: Result of the measured position and the estimated position by EKF

good mashing of the dynamics. Figure 4 shows the measured velocity and the estimated velocity by EKF. Only small deviations are visible in the approximation of measured velocity.

Figure 5 shows the approximation of the friction coefficient k_v. According to reference value [12] estimated k_v of temporomandibular joints varies between approximately 0.021 –0.023 Ns/m. As we can see, for the whole simulation friction value demonstrates steady approximation of these values.

Figure 6 shows the estimation of the mass m which approaches a constant value of approximately 0.25 kg. The estimation of the mass is used as a validation approach. The mass was known in EKF_1 and EKF_2 (Fig. 2) based on reference values given by the literature. The result of the estimation of the mass shows a similar value to the one previously stated. Therefore

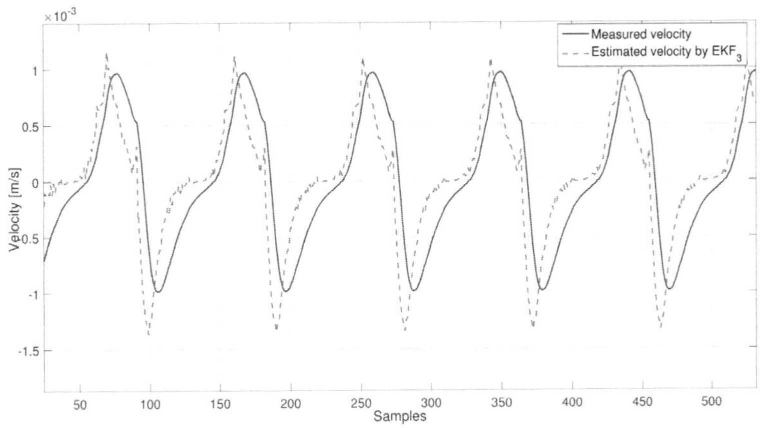

Figure 4: Result of the measured velocity and estimated velocity by EKF

it is emphasized that the other parameters such as k_v, total force as well as the velocity which are being estimated during the course of the cascaded structure approached are reliable values, as the estimation of the mass includes the previously estimated other parameters.

However, the tuning of the KF must be optimal for the generation of such constant and comparable results. Therefore, in each EKF of the estimation of friction coefficient and mass a tuning of Q_x was individually applied in compliance with cascaded structure, and R_w was manually calibrated to achieve the best empirical approximation.

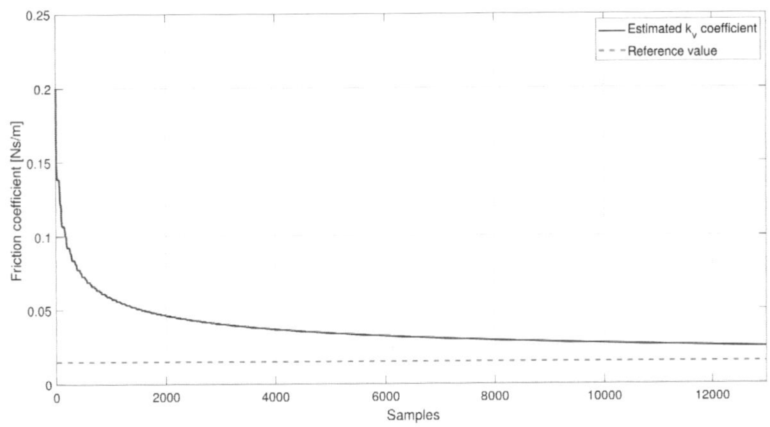

Figure 5: Graphical representation of the friction coefficient

7. Conclusions

The presented topic represents an interesting and important approach for dentists as well as for the industry, which deals with the required measuring systems. Observing friction of temporomandibular joints in the oral cavity can be an important indicator for early disease diagnosis considering crucial functionality of the oral cavity in speech and food processing. Based on demonstrated simulation results the proposed backward cascaded Kalman Filters

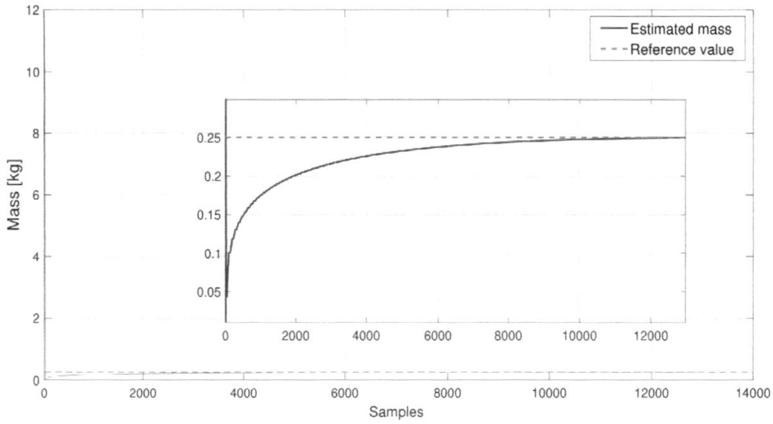

Figure 6: Approximation of the mass of the mandible

which consist of three backward Extended Kalman Filters is a promising to identification algorithm, where under adverse abnormal condition friction coefficient tends to deviate from inherent biological low level. The results showed that for incomplete observable system designed cascaded structure still allows to accurately estimate through a validation process of target mass. The choice of backward discretisation due to the large sampling time of the position effectively restrained the effect of high dynamics.

Further research work will focus on the improvement of the discussed cascaded Kalman Filters structure in terms of reducing the deviation of real states from filtering results while incorporating more measurements. At the same time, for the simulation an adaptive filtering process should be considered to adjust and update the Kalman gain based on adaptive tuning of noise covariance matrices, which will also presumably lead to an improved EKF robustness.

References

[1] Zizzo G and Ren L 2017 *Sensors* **17** 2866 ISSN 1424-8220 URL `http://dx.doi.org/10.3390/s17122866`
[2] Du H, Wang W, Xu C, Xiao R and Sun C 2020 *Sensors* **20** 919 ISSN 1424-8220 URL `http://dx.doi.org/10.3390/s20030919`
[3] Tannous H, Istrate D, Benlarbi-Delai A, Sarrazin J, Gamet D, Ho Ba Tho M and Dao T 2016 *Sensors* **16** 1914 ISSN 1424-8220 URL `http://dx.doi.org/10.3390/s16111914`
[4] Yardim C, Michalopoulou Z H and Gerstoft P 2011 *IEEE Journal of oceanic engineering* **36** 71–89
[5] Jin Z and Dowson D 2013 *Friction* **1** 100–113
[6] Tanaka E, Hirose M, Koolstra J H, van Eijden T M, Iwabuchi Y, Fujita R, Tanaka M and Tanne K 2008 *Journal of oral and maxillofacial surgery* **66** 462–468
[7] Nosonovsky M and Bhushan B 2007 *Materials Science and Engineering: R: Reports* **58** 162–193
[8] Abd Rahman R and Sepehri N 2017 *International Journal of Control* **90** 249–274
[9] Kaniewski P 2020 *Sensors* **20** 1584 ISSN 1424-8220 URL `http://dx.doi.org/10.3390/s20061584`
[10] Xie S S 2016 *Physiological Model of the Masticatory System* (Cham: Springer International Publishing) pp 45–79 ISBN 978-3-319-19896-5 URL `https://doi.org/10.1007/978-3-319-19896-5`$_3$
[11] Welch G, Bishop G *et al.* 1995 An introduction to the kalman filter
[12] Dowson D 2012 *Faraday discussions* **156** 9–30

 2162 (2022) 012018 doi:10.1088/1742-6596/2162/1/012018

A Physical ground for the scaling of Peak Ground Acceleration (PGA) with the integral of squared velocity (I^2_{Vp}) and its potential for Earthquake Early Warning

F Vallianatos[1,2,a] **and I Spingos**[1,2]

[1]Section of Geophysics–Geothermics, Department of Geology and Geoenvironment, National and Kapodistrian University of Athens, 15784, Athens, Greece
[2] Institute of Physics of the Earth's Interior and Geohazards, UNESCO Chair on Solid Earth Physics and Geohazards Risk Reduction, Hellenic Mediterranean University Research Center, 73133, Chania, Greece

[a]E-mail: fvallian@hmu.gr, fvallian@geol.uoa.gr

Abstract. One of the main goals of an Earthquake Early Warning System (EEWS) is to estimate the expected peak ground motion of the destructive S-waves using the first few seconds of P-waves, thus becoming an operational tool for real-time seismic risk management in a short timescale. EEWSs are based on the use of scaling relations between parameters measured on the initial portion of the seismic signal, after the arrival of the P-wave. Scope of the present work is to study the physical basis of the scaling laws observed between the peak ground acceleration (PGA) and the integral of the squared velocity (I^2_{Vp}). Based on Brune's model, which is one of the most widely adopted earthquake source models, we explore the physical principles of the scaling relations between the root mean square (rms) of the velocity acceleration recorded in the first few seconds after P-wave arrival and acceleration (recorded in S-waves) and the I^2_{Vp} . Assuming a relation between the PGA and the rms values estimated, the scaling of PGA with the integral of the squared velocity (I^2_{Vp}) which is calculated directly from the first few seconds-long signal window (*T*) after the P-wave arrival obtained. The latter formulation opens the possibility of using such laws for on-site and inter-site earthquake early warning.

1. Introduction

Advances in signal processing, data communications and real-time seismology have gradually rendered the concept of a reliable Earthquake Early Warning System (EEWS) a possibility [1-17]. Damages from an impending strong motion can be reduced by taking mitigation measures suitable for the given warning time [5, 6]. EEWSs have already been deployed, whether operationally or at a pilot stage, in Japan [1-3, 16], Mexico [18], Taiwan [19], Greece [20, 21], Italy [22-24] and California, USA [25, 26]. A detailed and comparative analysis of the performance of EEWS in Europe is given in [27, 28] while a recent application in the Northern California rail system is presented in [29]

An earthquake generates two fundamental types of body-waves: longitudinal (P) and shear (S) waves. The direct P-waves are weaker in amplitude and have a higher velocity than the S-waves. As a result, the difference in velocity can be used to retrieve information about the earthquake from the first wave arrivals and, consequently, predict the effect of the impending destructive secondary waves [10].

Earthquake early warning systems can mainly be divided into two categories: (a) on-site and (b) regional, even though the combination of both has also been explored [30-34]. The on-site EEWSs use one seismic station to offer insight about individual sites (such as critical infrastructure), e.g. [26, 29]. Regional EEWSs use multiple P arrivals to estimate the epicenter and magnitude of the earthquake and provide information for the affected area, e.g. [21, 22, 25, 33, 34]. This poses fundamental differences in the design, operation, and maintenance of the system. The on-site approach can give a larger warning time, but is susceptible to false triggers and local site effects. Regional EEWSs offer more robust estimations, but are affected by errors in the source estimation and require a large investment in seismic network design and deployment.

The regional EEWS, based on the operation of a regional network that detects earthquake events, locate them and determine its magnitude, using for the analysis the first few seconds of the arrivals of the P waves recorded at the stations [35, 36, 37]. In contrast the on-site system consists of a single or more sensors in the vicinity of target area or sometimes inside the structure under interest to be alerted. In the on-site approach the P-wave recordings to the seismic sensor are used to predict the peak ground motion at the site [16]. The on-site approach could be useful for sites located within the blind zone of a regional EEW system, creating a warning before the arrival of strong shaking S-waves. Worth to mentioned that in a number a applications the two EEW approaches combined [38, 39, 40] in a way that local parameters and predicted ground motions at a regional scale could be capable to provide accurate and even rapid estimations of the parameters of earthquake source along with the size of the expected damage zone [30]].

The regional EEW can be analyzed in terms of the geometry of the source following two main views. The first one assumes a point-source where the earthquake source viewed as a concentrated volume, while in the second approach a finite fault introduced and thus we take into consideration the entire fault area. Most studies have used the "point-source" demonstrating the reliability of this approach for the magnitude estimation of small to moderate events. However, it has been shown that this approach is not always accurate for strong earthquakes with magnitude greater than 7, due to the saturation of the P-wave parameters [16, 41]. In the present work we use the point-source approach and our results are limited to earthquakes with a moderate magnitude (≤ 6.5) that could be considered as point sources.

Scaling laws are of fundamental importance in EEWSs, as they connect early estimated parameters with the strength of impending S-wave. Relations between the integral of squared velocity estimated from the initial P-wave ($\mathrm{I}_{\mathrm{Vp}}^2$) and a metric of the anticipated shaking, such as the macroseismic intensity or the ground's acceleration during a seismic event, have been proposed [31-42].

The integral of squared velocity [42] is defined as:

$$I_{Vp}^2 = \int_{t_p}^{t_p+T} V^2(t)dt \qquad (1)$$

where t_p the P arrival time, T the considered signal window and V(t) the signal in velocity terms [33] estimated from the initial P wave. Therefore, a relation between I_{Vp}^2 and the Peak Ground Acceleration of S-waves (PGA) [33, 34] can be given as:

$$\log(\mathrm{PGA}) = a + b * \log(\mathrm{I}_{\mathrm{Vp}}^2) \qquad (2)$$

which suggests that we can predict the shaking using the initial P waves. Albeit equations similar to (2) have been used to connect I_{Vp}^2 to earthquake strength metrics in a single site, using a sensor near the source to obtain the latter and extrapolating the value of PGA in another target site could provide a significant warning time advantage [34].

In this work, we analyse the physical base of the scaling expression (2), focused on the correlation of I_{Vp}^2 with PGA on-site and inter-site, using Brune's model, one of the most widely adopted earthquake source models [43]. Determination of the strength of shaking from the initial P-wave is an important element for earthquake early warning, since PGA is a quantity commonly used in seismic risk and engineering, e.g. [44].

2. Empirical Scaling laws between *PGA* and I_{Vp}^2

A regional EEWS operates analyzing the information from the seismic network deployed in the vicinity to the epicenter and predicts the regional seismic intensities using a ground motion prediction equation [35]. On the other hand, the on-site system often consists of seismic stations located at particular target sites of interest, providing rapid ground motion estimates, using only information on the characteristics of P waves recorded at one seismic station [10], since a number of empirical relationship used to algorithms that derive ground shaking [45, 46, 47].

The parameter I_{Vp}^2 was first introduced in [42] to estimate the earthquake magnitude, providing good correlation between the two quantities. Recently, a correlation between I_{Vp}^2 and PGA was reported [20, 21]. The relationship between I_{Vp}^2 and PGA could therefore be used to identify, in real-time and before the arrival of S waves, whether a site is going to be adversely affected or not, and, thus, has the potential to become key in the design of on-site or inter-site EEWSs, enabling automatic mitigation measures and assisting civil protection in acting immediately, according to the severity of the situation. We note that the reliability of the scaling between I_{Vp}^2 and PGA has been recently tested and extensively used to establish EEWS [31, 42]

Figure 1 presents PGA as a function of I_{Vp}^2 in the on-site case of some stations in Greece. For all the cases, a correlation as that of equation (2) exists, while b is quite stable close to 0.5. It is noted that while I_{Vp}^2 is computed from measurements in the vertical channel, PGA refers to the maximum observed acceleration between the two horizontal components of the sensor.

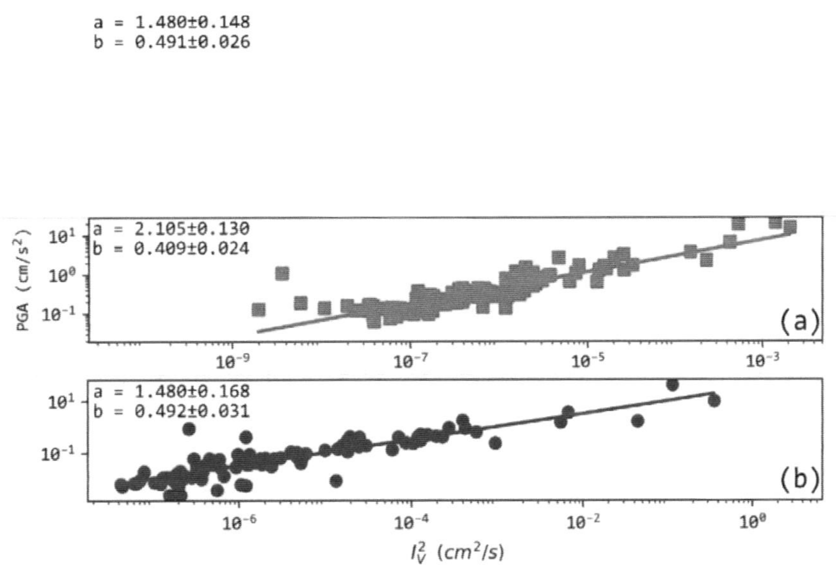

Figure 1. Scaling of PGA with I_{Vp}^2 for stations (a) HT.LIT (red squares) and (b) HA.ATHU (blue dots) using a T=3s time windows after the P-wave arrival. PGA is estimated from the maximum amplitude of the two horizontals after the S-wave arrival, while I_{Vp}^2 is obtained from the vertical channel. Both stations are located in Greece. For details see [33, 34]. Values for the intercept and slope of equation (2) are given, along with their standard errors (top left).

The potential of I_{Vp}^2 as an inter-site tool has been demonstrated in [20, 21], by measuring it at a reference station located closest to earthquake sources (named I_{Vp}^{2close}) and relating it to PGA at target sites (PGA^{ts}) using a scaling relation of the form of equation (2), i.e.,

$$\log(PGA^{ts}) = a + b * \log(I_{Vp}^{2close}).$$

Examples of such relations are given in Figure 2.

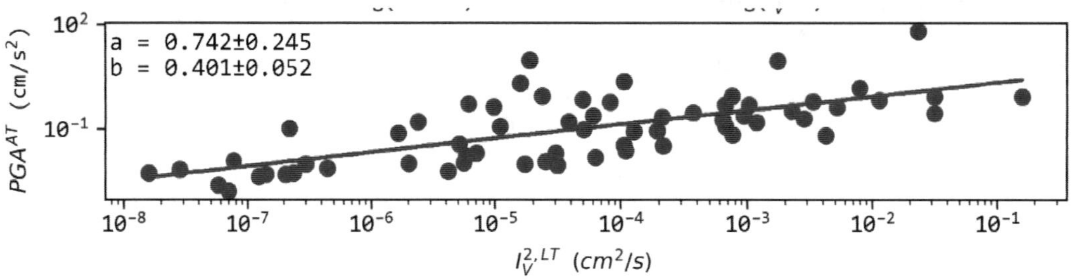

Figure 2. PGA in target site Athens (stations HA.ATHU and HL.ATH) that scaled with I_{Vp}^2 measured at reference stations at Loutraki (HA.LOUT and HP.LTK), using a T=3s time windows after the P-wave arrival. Both sites are located in Greece (see [33]). Values for the intercept and slope of equation (2) are given, along with their standard errors (top left).

From Figures 1 and 2 it is obvious that expression as that of equation (2) could fit the observed data satisfactory. The physical basis of such an expression is still an open question, along with the source and medium parameters that control the fitting coefficients α and b in equation (2). This question will be addressed in the next paragraph.

3. On the Physical basis of PGA- I_{Vp}^2 scaling law

To establish a relation between the strong ground motion parameters and I_{Vp}^2 we will use the root mean square (rms) value of the velocity and acceleration that could be estimated using the first few seconds from the initial part of the recording after the arrival of P-waves.
From the definition of I_{Vp}^2, we can state that:

$$I_{Vp}^2 = \int_{t_p}^{t_p+T} V^2(t)dt = V_{rms_p}^2 \; T \tag{3}$$

where the index rms_p indicates the root mean square (rms) value of the velocity estimated from the initial part of P-waves, and T is the record interval.

The latter expression permits the application of the ideas presented in [48-51], where the rms values of velocity and ground acceleration are related with the source parameters and the attenuation coefficient, using Parseval's theorem, which states that:

$$y_{rms} = \sqrt{\frac{\int_{-\infty}^{\infty}|y(t)|^2 dt}{T}} = \sqrt{\frac{\int_{-\infty}^{\infty}|Y(f)|^2 df}{T}} \tag{4}$$

where y(t) is the ground motion time series, *Y(f)* is the ground motion spectra, and *T* is the record interval. According to Brune's model [43], the far-field ground motions reads as:

$$\frac{d^n}{dt^n}\Omega(f) = (2\pi f)^n \frac{\Omega_o}{1+(\frac{f}{f_o})^2} \tag{5}$$

where f is the frequency and f_0 is the corner frequency. The index $n = 1,2,3$ refers to displacement, velocity, or acceleration spectra, respectively. According Brune's model, the displacement spectra are constant equal to Ω_0 at frequencies well below the f_0 and decay as f^{-2} above it, while the velocity spectra increase proportionally to f below f_0 and decrease as f^{-1} above it. As equation (2) suggests, the acceleration spectra goes as f^2 up to the corner frequency and are flat above it. The spectral parameters Ω_0 and f_0 are related to the seismic moment, M_0, and the stress drop, $\Delta\tau$, as follows [52]:

$$\Omega_o = \frac{M_o U_{\varphi\theta}^{p,s} F_s}{4\pi\rho C^3 R} \quad (6)$$

and

$$f_o = kC_{p,s}(\frac{16}{7}\frac{M_o}{\Delta\tau})^{1/3} \quad (7)$$

Where $U_{\varphi\theta}^{p,s}$ is the radiation pattern of P and S waves, F_s is the free-surface correction factor, R is the hypocentral distance, ρ is the density, k is a constant with different values for P and S waves and C stands for the wave velocity with C_p and C_s for the P and S waves, respectively [53, 54]. The expression (7) is valid for a circular crack embedded within an infinite homogeneous isotropic Poissonian medium. The effect of site and path attenuation can be modelled by multiplying the source spectra with an exponent function [55]:

$$\frac{d^n}{dt^n}\Omega(f) = (2\pi f)^n \frac{\Omega_o}{1+(\frac{f}{f_o})^2} \exp(-\pi\kappa_o f)$$

where κ_o is an attenuation parameter that embodies both anelastic and near surface attenuations. $\kappa_o > 0$ results in a high-frequency decay of the spectra and effectively produces an additional corner frequency, $f_k = 1/(\pi\kappa_0)$.

Using Parseval's theorem, in [48-51] the displacement, velocity, and acceleration ground motion rms are obtained as:

$$(\frac{d^n}{dt^n}D)_{rms} = \Omega_o \sqrt{\frac{2}{T}\int_0^\infty (2\pi f)^n \frac{1}{1+(\frac{f}{f_o})^2} \exp(-2\pi\kappa_o f) df} \quad (8)$$

Since the solutions of the integrals express either the rms of velocity or acceleration, based on the Meijer G-function, and Ci and Si (the cosine and the sine integral functions), respectively, we use some approximate expressions that are valid for moderate and strong earthquake events that match the exact solutions for the rms values of velocity, and acceleration (for details see [48-51]), as follows:

$$V_{rms_p} = 2\pi\Omega_{op}\sqrt{\frac{\pi f_o^3}{2T}} \quad (9)$$

$$A_{rms} = (2\pi)^2\Omega_{os}\ f_o^2 \sqrt{\frac{1}{\pi\kappa_o T}} \quad (10)$$

The latter expressions suggest that for moderate and strong earthquake magnitudes the rms of velocity V_{rms_p} is insensitive to attenuation, while the rms of acceleration, A_{rms}, depends on the attenuation coefficient κ_o.

Combining equations (3), (9) and (10) we obtain:

$$logA_{rms} = \frac{1}{2}logI_{Vp}^2 + log\left(2\frac{U_{\varphi,\theta}^S}{U_{\varphi,\theta}^P}\frac{C_p^3}{C_v^3}\sqrt{\frac{2f_o}{\kappa_o T}}\right) \quad (11)$$

for the on-site scaling

and

$$logA_{rms} = \frac{1}{2}logI_{Vp}^2 + log\left(2\frac{U_{\varphi,\theta}^S}{U_{\varphi,\theta}^P}\frac{R_1}{R_2}\frac{C_p^3}{C_v^3}\sqrt{\frac{2f_o}{\kappa_o T}}\right) \quad (12)$$

for the inter-site scaling

where R_1 the hypocentral distance of the close to epicenter station where first detected the seismic wave, while R_2 the hypocentral distance of the target site.

Taking into account that PGA $= \gamma A_{rms}^{v}$ [48-51], where the local site effects, are taking into consideration in the parameters γ and v, the above expression leads to:

$$logPGA = \frac{v}{2} logI_{Vp}^{2} + vlog\left(2\frac{U_{\varphi,\theta}^{s}}{U_{\varphi,\theta}^{p}}\frac{C_p^3}{C_v^3}\sqrt{\frac{2f_o}{\kappa_o T}}\right) + log\gamma \qquad (13)$$

for the on-site scaling

and

$$logPGA = \frac{v}{2} logI_{Vp}^{2} + vlog\left(2\frac{U_{\varphi,\theta}^{s}}{U_{\varphi,\theta}^{p}}\frac{R_1}{R_2}\frac{C_p^3}{C_v^3}\sqrt{\frac{2f_o}{\kappa_o T}}\right) + log\gamma \qquad (14)$$

for the inter-site scaling .

Expressions (13) and (14) have the form of equation (2) where b=v/2 . The latter suggests that when $v \cong 1$, then $b \cong 0.5$ as in most of the cases observed [31, 33, 34]. Worth to mentioned that the main advantage of equations (11) to (14) is that based on a simple widely adopted physical source model (attenuated omega-square spectra) could reproduced the empirical formulation of equation (2) which presented in a significant number of observations [see 31, 33 and references therein]. As such, it accounts for the three most important source parameters affecting ground motion intensity; the seismic moment (commonly expressed through the earthquake magnitude), the hypocentral distance, and the stress drop, as suggested by equation (7).

4. Concluding remarks

Today, the development of EEWSs represents one of the most useful strategies to mitigate seismic risk in short timescales, and several countries worldwide are promoting and developing such systems. In the context of seismic risk management, they are considered a reasonably cost-effective solution for loss reduction [56]. In this study, we explored the physical basis of the scaling laws, both on-site and between stations located near the earthquake sources and target sites, in order to estimate inter-site relations for I_{Vp}^2 (recorded at the former) and PGA (recorded at the latter). Our results justify the scaling between I_{Vp}^2 and PGA at individual sites, with seemingly identical slope parameters. This scaling bears hope for establishing both on-site and inter-site hazard estimators. The physical ground of the scaling between $\mathrm{I}_{\mathrm{Vp}}^2$ and PGA could be used to improve recent approaches that use $\mathrm{I}_{\mathrm{Vp}}^2$ parameter to estimate the strength of strong ground motion using modern artificial intelligence techniques as that of machine learning [57-60].

Summarizing, in the present work, on-site and inter-site theoretical scaling relations of the far-field ground motions were derived using the omega-squared model. The scaling laws between $\boldsymbol{I_{Vp}^2}$ and PGA are in agreement with recently observed empirical relations and could be used as earthquake early warning parameter. Further work is required to better support this and solve issues such as the actual consequences of local site effects. However, if these matters are resolved and the robustness of both the empirical and theoretical bases is improved, the on-site and inter-site EEWS approaches could be used to issue alerts and, even, estimate potential damages level within very few seconds, faster than the in-operation regional EEWSs.

References

[1] Nakamura Y 1988 On the urgent earthquake detection and alarm system (UrEDAS) *Proc. 9th World Conf. Earthq. Eng.* **7** 673–678

[2] Fujinawa Y. and Noda Y. 2013 Japan's Earthquake Early Warning System on 11 March 2011: Performance, Shortcomings, and Changes. *Earthquake Spectra* **29** (S1), S341–S368. doi:10.1193/1.4000127

[3] Hoshiba M 2014 Review of the Nationwide Earthquake Early Warning in Japan during its First Five Years in *Earthquake Hazard. Risk Disasters*, 509–529. doi:10.1016/b978-0-12-394848-9.00019-5

[4] Erdik M, Fahjan Y, Ozel O, Alcik H, Mert A and Gul M 2003 Istanbul Earthquake Rapid Response and the Early Warning System *Bull. Earthquake. Eng.* **1**, 157–163. doi:10.1023/a:1024813612271

[5] Allen RM, Gasparini P, Kamigaichi O and Böse M 2009 The Status of Earthquake Early Warning Around the World: An Introductory Overview. *Seismological Res. Lett.* **80 (5)** 682–693. doi:10.1785/gssrl.80.5.682

[6] Allen RM and Melgar D 2019 Earthquake Early Warning: Advances, Scientific Challenges, and Societal Needs. *Annu. Rev. Earth Planet. Sci.* **47**, 361–388. doi:10.1146/annurev-earth-053018-060457

[7] Böse M, Ionescu C and Wenzel F 2007 Earthquake Early Warning for Bucharest, Romania: Novel and Revised Scaling Relations. *Geophys. Res. Lett.* **34**, L07302. doi:10.1029/2007GL029396

[8] Kanamori H, Hauksson E and Heaton T 1997 Real–time seismology and earthquake hazard mitigation *Nature* **390** 461–464

[9] Allen RM and Kanamori H 2003 The potential for earthquake early warning in Southern California *Science* **300** 786–789

[10] Kanamori H 2005 Real–time seismology and earthquake damage mitigation. *Annu. Rev. Earth Planet. Sci.* **33** 195–214

[11] Allen RM, Brown H, Hellweg M, Khainovski O, Lombard P and Neuhauser D 2009 Real–time earthquake detection and hazard assessment by ElarmS across California *Geophys. Res. Lett.* **36** L00B08

[12] Hloupis G and Vallianatos F 2013 Wavelet–based rapid estimation of earthquake magnitude oriented to early warning *IEEE Geosci. Remote. Sens. Lett.* **10** 43–47

[13] Hloupis G and Vallianatos F 2015 Wavelet–based methods for rapid calculations of magnitude and epicentral distance: an application to Earthquake Early Warning System *Pure Appl. Geophys.* **172** 2371–2386

[14] Parolai S, Bindi D, Boxberger T, Milkereit C, Fleming K and Pittore M 2015 On–site early warning and rapid damage forecasting using single stations: Outcomes from the REAKT project. *Seismol. Res. Lett.* **86** 1393–1404

[15] Brooks BA, Protti M, Ericksen T, Bunn J, Vega F, Cochran ES, Duncan C, Avery J, Minson SE, Chaves E, Baez JC, Foster J and Glennie CL 2021 Robust Earthquake Early Warning at a fraction of the cost: ASTUTI Costa Rica. *AGU Adv.* **2** e2021AV000407

[16] Colombelli S, Amoroso O, Zollo A and Kanamori H 2012 Test of a Threshold-Based Earthquake Early-Warning Method Using Japanese Data. *Bull. Seismological Soc. America* **102,** 1266–1275. doi:10.1785/0120110149

[17] Nakamura Y, Saita J and Sato T 2011 On an Earthquake Early Warning System (EEW) and its Applications. *Soil Dyn. Earthquake Eng.* **31**, 127–136. doi:10.1016/j.soildyn.2010.04.012

[18] Espinosa–Aranda J, Jiménez A, Ibarrola G, Alcantar F, Aguilar A, Inostroza M and Maldonado S 1995 Mexico City seismic alert system. *Seism. Res. Lett.* **66** 42–53

[19] Wu YM, Lee WHK, Chen CC. Shin TC, Teng TL and Tsai YB 2000 Performance of the Taiwan Rapid Earthquake Information Release System (RTD) during the 1999 Chi–Chi (Taiwan) earthquake *Seismol. Res. Lett.* **71** 338–343

[20] Sokos E, Tselentis GA, Paraskevopoulos P, Serpetsidaki A, Stathopoulos–Vlamis A and Panagis A 2016 Towards earthquake early warning for the Rion–Antirion bridge, Greece *Bull. Earthq. Eng.* **14** 2531–2542

[21] Kapetanidis V, Papadimitriou P and Kaviris G 2019 Earthquake Early Warning application in Central Greece *Bul. Geol. Soc. Greece* **7(95)** 277–278.

[22] Iannacone G, Zollo A, Elia L, Convertito V, Satriano C, Martino C, Festa G, Lancieri M, Bobbio A, Stabile TA, Vassalo M and Emolo A 2010 A prototype system for earthquake early–warning and alert management in southern Italy *Bull. Earthq. Eng.* **8** 1105–1129

[23] Zollo A, Amoroso O, Lancieri M, Wu YM and Kanamori H 2010 A threshold- based earthquake early warning using dense accelerometer networks *Geophys. J. Int.* **183** 963–974

[24] Caruso A, Colombelli S, Elia L, Picozzi M and Zollo A 2017 An On-Site Alert Level Early Warning System for Italy. *J. Geophys. Res. Solid Earth* **122**. doi:10.1002/2016JB013403

[25] Chung AI, Meier MA, Andrews J, Böse M, Crowell BW, McGuire JJ and Smith DE 2020 ShakeAlert earthquake early warning system performance during the 2019 Ridgecrest earthquake sequence *Bull. Seismol. Soc. Am.* **110** 1904–1923

[26] Böse M, Hauksson E, Solanki K, Kanamori H and Heaton TH 2009 Real–time testing of the on–site warning algorithm in southern California and its performance during the July 29 2008 M_w 5.4 Chino Hills earthquake *Geophys. Res. Lett.* **36** L00B03

[27] Clinton J, Zollo A, Marmureanu A, Zulfikar C and Parolai S 2016 State-of-the Art and Future of Earthquake Early Warning in the European Region. *Bull. Earthquake Eng.* **14**, 2441–2458. doi:10.1007/s10518-016-9922-7

[28] Zuccolo E, Cremen G and Galasso C 2021 Comparing the Performance of Regional Earthquake Early Warning Algorithms in Europe. *Front. Earth Sci.* **9,** 686272. doi: 10.3389/feart.2021.686272

[29] Minson SE, Cochran ES, Wu S and Noda S 2021 A Framework for Evaluating Earthquake Early Warning for an Infrastructure Network: An Idealized Case Study of a Northern California Rail System. *Front. Earth Sci.* **9,** 620467. doi: 10.3389/feart.2021.620467

[30] Colombelli S, Caruso A, Zollo A, Festa G and Kanamori H 2015 A P Wave-based, On-site Method for Earthquake Early Warning. *Geophys. Res. Lett.* **42**, 1390–1398. doi:10.1002/2014GL063002

[31] Brondi P, Picozzi M, Emolo A, Zollo A and Mucciarelli M 2015 Predicting the macroseismic intensity from early radiated P wave energy for on–site earthquake early warning in Italy. *J. Geophys. Res. Solid Earth* **120** 7174–7189

[32] Iaccarino AG, Picozzi M, Bindi D and Spallarossa D 2020 Onsite earthquake early warning: Predictive models for acceleration response spectra considering site effects. *Bull. Seismol. Soc. Am.* **110** 1289–1304

[33] Vallianatos F, Karakonstantis A and Sakelariou N 2021 Estimation of Earthquake Early Warning parameters for Eastern Gulf of Corinth and Western Attica Region (Greece). First results *Sensors* 2021 **21** 5084

[34] Spingos I, Vallianatos F and Kaviris G 2021 The scaling of PGA with IV2p and its potential for Earthquake Early Warning in Thessaly (Central Greece). *Bul. Geol. Soc. Greece*, **58**, 177-199 doi: 10.12681/bgsg.27062

[35] Satriano C, Wu YM, Zollo A and Kanamori H 2011 Earthquake Early Warning: Concepts, Methods and Physical Grounds. *Soil Dyn. Earthquake Eng.* **31**, 106–118. doi:10.1016/j.soildyn.2010.07.007

[36] Zollo A, Colombelli S, Elia L, Emolo A, Festa G, Iannaccone G, C. Martino and P. Gasparini 2014 An Integrated Regional and On-Site Earthquake Early Warning System for Southern Italy: Concepts, Methodologies and Performances, in *Early Warning for Geological Disasters, Advanced Technologies in Earth Sciences*, **117–137**. doi:10.1007/978-3-642-12233-0_7

[37] Picozzi, M, Zollo A, Brondi P, Colombelli S, Elia L and Martino C 2015 Exploring the Feasibility of a Nationwide Earthquake Early Warning System in Italy. *J. Geophys. Res. Solid Earth* **120**, 2446–2465. doi:10.1002/2014JB011669

[38] Wang Z and Zhao B 2021 Applicability of Accurate Ground Motion Estimation Using Initial P Wave for Earthquake Early Warning. *Front. Earth Sci.* **9**, 718216. doi: 10.3389/feart.2021.718216

[39] Cremen G, and Galasso C 2020 Earthquake Early Warning: Recent Advances and Perspectives *Earth-Science Rev.* **205**, 103184. doi:10.1016/j.earscirev.2020.103184

[40] Ladina C, Marzorati S, Amato A and Cattaneo M 2021 Feasibility Study of an Earthquake Early Warning System in Eastern Central Italy. *Front. Earth Sci.* **9**, 685751. doi: 10.3389/feart.2021.685751

[41] Velazquez O, Pescaroli G, Cremen G and Galasso C 2020. A Review of the Technical and Socio-Organizational Components of Earthquake Early Warning Systems. *Front. Earth Sci.* **8**, 533498. doi:10.3389/feart.2020.533498

[42] Festa G, Zollo A and Lancieri M 2008 Earthquake magnitude estimation from early radiated energy.*Geophys. Res. Lett.* **35** L22307
[43] Brune JN 1970 Tectonic stress and the spectra of seismic shear waves from earthquakes *J. Geophys. Res.***75**, 4997–5009
[44] Dolce M, Prota A, Borzi B, da Porto F, Lagomarsino S, Magenes G, Moroni C, Penna A, Polese M, Speranza E, Verderame GM and Zuccaro G 2021 Seismic risk assessment of residential buildings in Italy *Bull. Earthq. Eng.* **19** 2999–3032
[45] Wu YM and Kanamori H 2008 Development of an Earthquake Early Warning System Using Real-Time strong Motion Signals. *Sensors* **8 (1),** 1–9. doi:10.3390/s8010001
[46] Wu YM and Kanamori H 2005. Rapid Assessment of Damage Potential of Earthquakes in Taiwan from the Beginning of P Waves. *Bull. Seismological Soc. America* **95 (3),** 1181–1185. doi:10.1785/0120040193
[47] Zollo A, Colombelli S and Emolo A 2016 An Integrated Regional and On-Site Earthquake Early Warning System for Southern Italy: Concepts, Methodologies and Performances in *Early Warning Geol. Disasters, chapter 7*, **117–137**. doi:10.1007/978-3-642-12233-0_7
[48] Lior I, Ziv A and Madariaga R 2016 P–wave attenuation with implications for Earthquake Early Warning *Bull. Seismol. Soc. Am.* **106** 13–22
[49] Ziv A and Lior I 2016 Real–time moment magnitude and stress drop with implications for real–time shaking prediction. *Bull. Seismol. Soc. Am* **106** 2459– 2468
[50] Lior I and Ziv A 2017 The relation between ground acceleration and earthquake source parameters: Theory and observations *Bull. Seismol. Soc. Am* **107** 1012–1018
[51] Lior I and Ziv A 2020 Generic source parameter determination and ground–motion prediction for earthquake early warning. *Bull. Seismol. Soc. Am.* **110** 345–356
[52] Eshelby JD 1957 The determination of the elastic field of an ellipsoidal inclusion, and related problems *Proc. R. Soc. Lond.* A **241** 376– 396
[53] Madariaga R 1976 Dynamics of an expanding circular fault. *Bull. Seismol. Soc. Am.* **65** 163–182
[54] Sato T and Hirasawa T 1973 Body wave spectra from propagating shear cracks *J. Phys. Earth* **21** 415–431
[55] Ktenidou OJ, Cotton F, Abrahamson N and Anderson JG 2014 Taxonomy of κ: A review of definitions and estimation approaches targeted to applications *Seismol. Res. Lett.* **85** 135–146
[56] Valbonesi C 2021 Between Necessity and Legal Responsibility: The Development of EEWS in Italy and its International Framework. *Front. Earth Sci.* **9,** 685153. doi: 10.3389/feart.2021.685153
[57] Zhu J Li S Song J and Wang Y 2021 Magnitude Estimation for Earthquake Early Warning Using a Deep Convolutional Neural Network. *Front. Earth Sci.* **9,** 653226. doi: 10.3389/feart.2021.653226
[58] Ochoa LH, Niño LF and Vargas CA 2017 Fast magnitude determination using a single seismological station record implementing machine learning techniques. *Geod. Geodyn.* **9**, 34–41. doi: 10.1016/j.geog.2017.03.010
[59] Mousavi SM and Beroza GC 2020 A Machine-Learning Approach for Earthquake Magnitude Estimation. *Geophys. Res. Lett.* **47.** doi:10.1029/2019GL085976
[60] Iaccarino AG, Gueguen P, Picozzi M and Ghimire S 2021 Earthquake Early Warning System for Structural Drift Prediction Using Machine Learning and Linear Regressors. *Front. Earth Sci.* **9**, 666444. doi: 10.3389/feart.2021.666444

Vision-Based Deep Learning Algorithm for Detecting Potholes

K Gajjar[1,a], T van Niekerk[2], Thomas Wilm[3] and P Mercorelli[4]

[1] Engineering, Built Environment and IT, Nelson Mandela University, Port Elizabeth, South Africa
[2] Engineering, Built Environment and IT, Nelson Mandela University , Port Elizabeth, South Africa
[3] Volkswagen Group South Africa, Uitenhage, South Africa
[4] Institute of Product and Process Innovation – PPI Leuphana University of Lueneburg, Universitaetsallee, 1 21335 Lueneburg, Germany

[a]E-mail: kanushkagajjar0@gmail.com

Abstract. Potholes on roads pose a major threat to motorists. Driving over a pothole has the potential to cause serious damage to a vehicle, which in turn may result in fatal accidents. Currently, many pothole detection methods exist. However, these methods do not utilize deep learning techniques to detect a pothole in real-time, determine the location thereof and display its location on a map. The success of determining an effective pothole detection method, which includes the aforementioned deep learning techniques, is dependent on acquiring a large amount of data, including images of potholes. Once adequate data had been gathered, the images were processed and annotated. The next step was to determine which deep learning algorithms could be utilized. Three different models, including Faster R-CNN, SSD and YOLOv3 were trained on the custom dataset containing images of potholes to determine which network produces the best results for real-time detection. It was revealed that YOLOv3 produced the most accurate results and performed the best in real-time, with an average detection time of only 0.836 s per image. The final results revealed that a real-time pothole detection system, integrated with a cloud and maps service, can be created to allow drivers to avoid potholes.

1. Introduction

Since South Africa is a developing country, the development and maintenance of roads is of great significance. Well-maintained roads contribute largely towards the country's economy and tarred roads have become an absolute necessity in the 21st century in light of economic demands on government by the citizenry. Potholes can be defined as areas of a road surface that have formed a hole after being cracked and worn away [1]. Potholes start as small cracks, which if not repaired timeously, can increase considerably in size. Flat and smooth surfaces are required to ensure a comfortable drive, however, potholes can result in an unpleasant and potentially dangerous ride. The tyres are likely to get damaged and would require re-alignment. Currently, there is no mechanism to caution drivers of approaching potholes. Consequently, potholes remain a serious hazard. However, technology is available that reduces the impact of a pothole on a vehicle. Variations of this technology have been installed in several vehicles including the S-class Mercedes, Ford Focus, Jaguar Land Rover and Audi A8. The technology in the S-class Mercedes is referred to as Magic Body Control, which is a suspension that can predict surface unevenness and prepare accordingly [2]. This technology does not completely eliminate the effect of a pothole but simply reduces it. However, it is possibly more beneficial for a driver to detect any damage in the road beforehand. Driving over a pothole may have serious implications for a vehicle, for example, a flat, or damage to a tyre, rims, suspension, steering and possibly the body of the car [3]. Complete avoidance of a pothole would eliminate these possibilities. The overall aim of this project is to identify the most suitable deep learning algorithm to detect a pothole whilst driving, determine the GPS location thereof, display the determined position on a map as pin and send the data to a cloud server for information storage and sharing. Various topics relating to solving the proposed problem was explored. From this exploration, it was possible to determine which method would produce the most desirable result. It was revealed that deep learning methods produce desirable results. It should be noted that although vision, vibration, stereo-imaging and laser based techniques can be utilised to solve the pothole detection problem, some of these methods have been utilised for more than a decade. SpotholeAI and CarVi are two applications that were developed in the last year and both utilise deep learning methods to detect potholes. An investigation of various deep learning detection algorithms was conducted. The algorithms selected for this project include Faster R-CNN, SSD and YOLOv3. Each of the selected algorithms will be compared and the most suitable will be utilised for the purpose of this project. The IoT system will collect the GPS data from its module and store it in the cloud. Hence, no manual data must be collected. The data must be in real-time and monitored through the cloud.

2. Materials and methods

2.1. Data collection and processing

It was necessary to identify an area in South Africa where potholes are prevalent. St. Francis Bay and Jeffrey's Bay in the Eastern Cape Province were identified for the purpose of this study. Additional data containing pothole images was found online. The latter was produced by [4] who completed a project relating to pothole detection using vision and machine learning methods. The images taken by [4] were taken in the Vaal Triangle area in Gauteng, Stellenbosch and Somerset West in the Western Cape. Initially, it was assumed that these two sets of data would be adequate, however, with the advent of time, it was revealed that more images were required to enhance training for the network. If the dataset is too small, overfitting, which refers to a model that models the training data too well can occur. The detail and the noise in the training data has been mastered to such an extent that the performance of the model on new data is negatively impacted [5]. Thus, it was necessary to produce additional images. These images which were produced by Volkswagen Group South Africa (VWSA) were taken in Uitenhage and surrounding areas in the Eastern Cape Province. These areas were identified as those containing a large number of potholes.

The entire dataset of instances of potholes comprised of 1910 images. The images of the datasets was gathered through a GoPro Hero 3+. This camera was selected because it produces good quality images whilst moving. Thus, it was not necessary to de-blur the images in the pre-processing steps, thereby reducing the processing time. The camera was set at a resolution of 1280 x 720pixels. The goPro used by [4] was set to the time lapse mode at an interval of 0.5 seconds/image and at a resolution of 3680 x 2760pixels.

The collected data was arranged into positive and negative sets. The positive set of data contained images with potholes, while the negative dataset comprised of images without potholes. The data collected by [4] was divided into two scenarios, positive and negative sets as well as divided into a complex and simple scenario. In the simple scenario, it was assumed that the roads were well lit and open, while in the complex scenario, a more real-world scenario is depicted. This scenario includes instances of shadows on the road. This dataset comprised of images which were taken at various times throughout the day. All of the data was separated into the training and test sets.

The images were cropped such that only the road surface was illustrated. The sky and other unnecessary information was cropped out of the image. This had to be done because the computer utilised for training had inadequate GPU memory for training. Reducing the size of the image and eliminating unnecessary data such as the sky and grass resulted in less memory to train the network. The computational time of the algorithms was also reduced as a result of cropping the images. The images were not resized because significant aspects of the image would have been lost thus making it difficult to train the network.

The cropped images containing potholes were annotated by utilising the Image Processing Toolbox in MATLAB. The annotation process involved colouring in the regions in which there were potholes manually. This colouring process was completed by utilising the WACOM Cintiq Pro 13. This compact device is an advanced creative pen display which enables a user to have a direct pen-on-screen connection. Using this device saved time because it would have been extremely difficult and time consuming to colour each pothole using a mouse. Once the masks were created from the semantic segmentation, these were utilised to create the bounding boxes. Since the VWSA dataset was acquired at a later stage, the images within this dataset were annotated using a different graphical annotation tool: labelImg. Figure 1 below shows this tool. Only bounding boxes were created for this dataset. Initially, the results acquired using semantic segmentation and bounding boxes would have been compared, however, at a later stage it was decided that comparisons between various other networks using only bounding boxes would be made.

Figure 1. Labelling of images in labelImg

The overall process to be followed to achieve the desired results is illustrated in Figure 2 below.

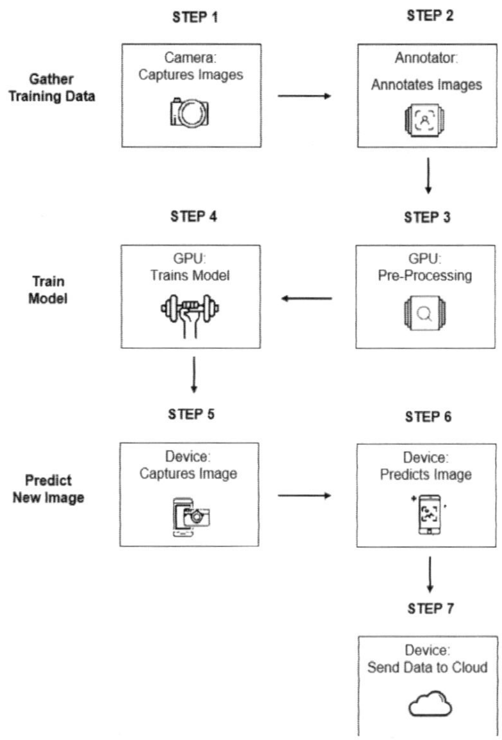

Figure 2. Overall process to be followed to achieve the end goal

2.2. Development and methodology

Various network architectures were utilized for this project. More than one network architecture was utilized to determine which performed the best with the given dataset. The network architectures included SSD (Single Shot Detection) with the Inception v2 backbone, Faster RCNN (Region-based CNN) with Inception v2 backbone and YOLOv3 (You Only Look Once). This section discusses why YOLOv3 was selected as the network used for the application of this project. The overall network architecture is also expounded upon in this section.

The results revealed that although Faster R-CNN was more accurate than SSD, the inference time was extremely slow and unsuitable for real-time detection. Although SSD was fast, the accuracy of the detection was unsatisfactory. These results are shown in Section 3.2. Further research was completed and it was revealed that videos generally shot at a speed of at least 24fps, the Faster R-CNN would likely not be able to keep pace. Since Faster R-CNN is a regional based method which comprises of two phases, proposing regions and processing these, it proves to be somewhat inefficient for real-time detection.

After this comparison was made, it was decided that a third network should be explored to optimise the results. Since single stage-detectors produce a higher inference, further research was conducted on such detectors. Through thorough research, it was revealed that the YOLOv3 network performs better for real-time applications because the detection time per object is less when compared to Faster R-CNN and SSD [6]. Generally, when dealing with objects of large sizes, SSD performs well. SSD utilises upper layers for detection. Consequently, the performance for small objects is not sound. For the application of this project, the size of the potholes could vary significantly, thus resulting in poor accuracy.

YOLO is a deep learning algorithm which utilises convolutional neural networks for the detection of objects. The output of YOLO is generated by applying a 1 x 1 kernel on a feature map. In the case of YOLOv3, 1 x 1 detection kernels is applied on feature maps of three various sizes at three different locations within the network. The shape of the detection kernel is 1 x 1 x (B x (5 + C), where B is the number of bounding boxes a cell on the feature map is able to predict. "5" represents the 4 bounding box attributes and one object confidence. "C" indicates the number of classes.

2.3. Training YOLOv3

The Darknet deep learning framework developed by Joseph Redmon [7] was utilised for this project. The first task is to prepare the dataset. The YOLOv3 annotated files are required to be in a text file format.. The ".txt" file comprises of rows which represent a single bounding box in the image and contains the following information about the bounding box:

<object-class-id> <centre-x> <centre-y> <width> <height>

The first field, object-class-id, is an integer that represents the class of the object. This number ranges from 0 to (number of classes-1). Since this project only comprises of one class, this number is always set at 0. The second and third field, centre-x and centre-y are the x and y co-ordinates of the centre of the bounding box divided by the image width and height. The fourth and fifth field, width and height are the width and height of the bounding box divided by the image width and height.

In YoloV3 trained on the pothole dataset, B = 3 and C = 1, thus the size of the kernel is 1 x 1 x 18.

The second to the fifth entries are all floating-point values that range between the values 0 and 1. The dataset was then split into the training and test sets. 70% of the collected data was utilised for training and 30% was utilised for testing. The process of transfer learning was used in the next step. A pre-trained model containing convolutional weights trained on ImageNet was thus utilised. By utilising these weights, the training time was reduced significantly [8]. The darknet.data, classes.names and darknet-yolov3.cfg files, which is included in the code distribution, requires information relating to the specifications of the object detector and relevant paths. Thus, these files were edited such that the relevant information was provided for each. Various hyper-parameters were configured in the configuration file. Once all the various components for training were set, the training could take place. The training continued till the loss value dropped below a specific threshold [9].

Once the YOLOv3 algorithm was able to accurately detect potholes on the road, it was necessary to upload the GPS co-ordinate of the detected pothole to a cloud server.

The GlobalSat BU-353-S4 receiver was used for this project. This receiver comprises of a highly sensitive, low power consumption chipset in an ultra-compact form factor. It is powered by a SiRF Star IV GPS chipset and provides superior performance in urban canyons and dense foliage. The device is built with a magnet which can stick to the top of a vehicle. Elemental exposure is not a concern because it is able to withstand both freezing and extreme hot temperatures [10].

The location of the detected pothole was required to be sent to a cloud server so that other drivers can be informed of the location of the potholes. The cloud server utilised for this project was Ubidots STEM. Ubidots is a two platform company, which comprises of Ubidots and Ubidots STEM. The standard STEM package is a non-commercial license that allows easier access to students, researchers and hobbyists globally [11]. Ubidots is an Internet of Things (IoT), data analytics and Visualisation Company. Data that is gathered from various sensors can be transformed into useful information, allowing for business decisions and machine-to-machine interactions be made. Educational research is possible through the utilisation of Ubidots. This platform allows for integration of IoT into business and research [11].

Figure 3 below illustrates an overview of the overall system architecture. The image on the road will first be captured via the camera. Once this image is captured, the neural network will run, if a pothole is detected. The GPS location of the detected pothole will be determined via the GPS module. The collected GPS data will be sent to the cloud server, Ubidots. This information will be available for all those who have access to the Ubidots account used to store the data. The relevant parties can then be informed of the potholes on the road and the location thereof.

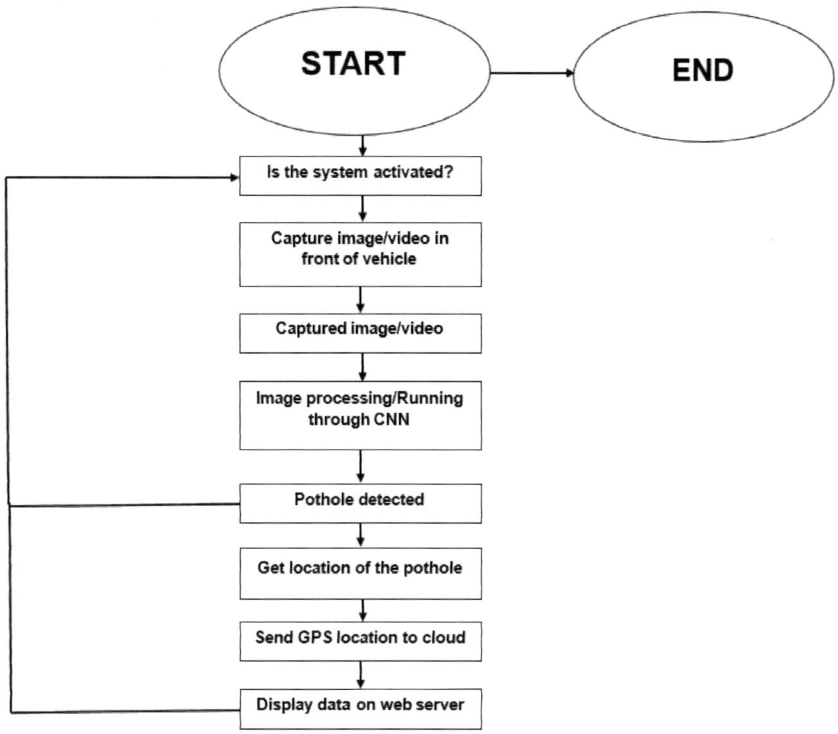

Figure 3. Overall system architecture

3. Results

In this section, the various evaluation metrics to determine which network performed the best is discussed. The three networks are compared and the results gathered from the discussed evaluation metrics is presented. The integrated system performance is discussed in this section too.

3.1. Evaluation-metrics

Mean average precision is a metric utilised to evaluate object detectors. It is the average of the average precision. To comprehend MAP, it is necessary to define the terms 'precision', 'recall' and 'IoU' (Intersection over Union). Precision can be defined as how repeatable a measurement is [12]. It is the percentage of the results that is relevant. An example of precision is how close a second arrow is to the first arrow on a dart board [13].

$$Precision = \frac{TP}{TP + FP} \tag{1}$$

Recall can be defined as the percentage of the total relevant results that is classified correctly by the algorithm [14].

$$Recall = \frac{TP}{TP + FN} \tag{2}$$

If precision is increased then recall will decrease and vice versa.

IoU can can be defined as the ratio of the area of intersection and area of union of the ground truth and predicted bounding boxes. The "ground truth bounding box" is the bounding box and its co-ordinates are provided in the training set [15]. Figure 4 below illustrates that the green box represents the ground truth box while the red box is what the model predicts. It is clear that these two boxes have different co-ordinates. The area of intersection is where the one box overlaps the other and the area of union is the total area covered by both bounding boxes [16].

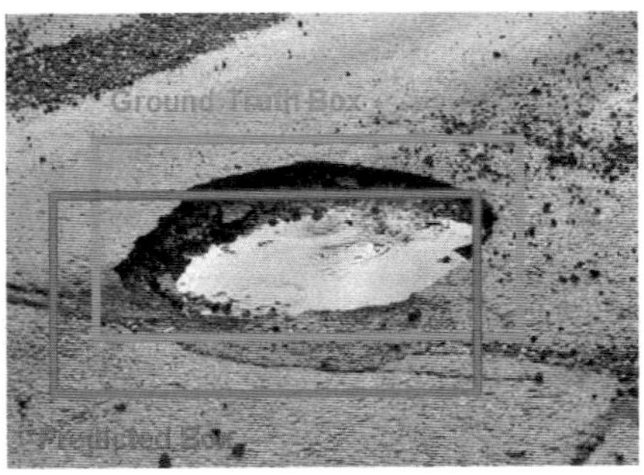

Figure 4. Representation of ground truth box and predicted box and IoU

The confidence score can be defined as the probability that an anchor box contains an object. The confidence is predicted by a classifier. Both the IoU and confidence allows one to determine whether a predicted box is a true positive, false positive or false negative. A threshold value of 0.5 is predefined for the IoU [15].

A detection is considered a true positive (TP) only if the following three conditions are met [17]:

- Confidence score > threshold
- The predicted class matches the class of a ground truth
- The predicted bounding box IoU (e.g. 0.5) is greater than the threshold of the ground-truth

If the above conditions are not met, the prediction is considered a false positive (FP). When the confidence score of a detection that is supposed to detect a ground truth is lower than the threshold, it is considered a false negative (FN) [17]. When the confidence score of a detection that is not supposed to detect anything is lower than the threshold, it is considered a true negative (TN). However, this is not of great significance in object detection [16].

A numerical metric, known as average precision (AP) can be utilised to evaluate the performance of a detector. AP is essentially the precision averaged over all unique recall levels. To reduce the fluctuations in the curve, it is necessary to interpolate the precision at multiple recall levels before actually calculating the metric. The interpolated precision p_{interp}, at a specific recall level *(r)* is defined as the highest precision found for any recall level $r' \geq r$ [18]:

$$P_{interp}(r) = \max_{r' \geq r} p(r') \tag{3}$$

There are two ways to select the levels of recall at which the precision should be interpolated. Traditionally, 11 equally spaced recall levels are selected (i.e., 0.0, 0.1, 0.2,...1.0). A new standard selects all unique recall levels presented by the data. The new method is more advanced to enhance the precision and measure differences between methods with low AP. It is possible to define AP as the area under the interpolated precision-recall curve [18].

$$AP = \sum_{i=1}^{n-1} (r_{i+1} - r_i) p_{interp}(r_{i+1}) \tag{4}$$

where $r_1, r_2, \ldots, r_n$ is the recall levels at which the precision is first interpolated. The calculation of AP only considers one class. Only one class, potholes, was used for this project.

Average recall (AR) is a numerical metric that can also be utilised to compare the object detector performance. AR is essentially the recall averaged over all $IoU \in [0.5, 1.0]$ and can be determined by the following equation [19]:

$$AR = 2 \int_{0.5}^{1} recall(o) do \tag{5}$$

where o is *IoU* and *recall(o)* is the corresponding recall.

3.2. Performance metric results

Table 1 below illustrates the results of the various performance metrics for each of the networks which were trained.

Table 1. Performance metrics results for each network

	Average Precision	Average Recall	F1-Score	mAP @ 0.5 IoU	Detection time per image (average)
Faster R-CNN	0.077	**0.663**	0.137	**0.415**	7.02s
SSD	0.043	0.326	0.076	0.185	4.815s
YOLOv3	0.347	0.32	0.42	0.347	0.836s

The YOLOv3 network achieved the precision value of 0.347 whereas Faster R-CNN and SSD achieved average precision values of 0.077 and 0.043 respectively. However, the recall value of YOLOv3 was lower than that of Faster R-CNN, indicating that while the YOLOv3 network has a higher proportion of positive results in the correctly predicted values, the Faster R-CNN network outperforms it in its ability to correctly predict the positive results. The mAP value of Faster R-CNN was also higher than that of YOLOv3, further proving that Faster R-CNN has a greater level of precision in detecting potholes. It should be noted that, although the size of each detected pothole is not the same, Faster R-CNN performs the best out of all of the networks regardless of the size of the object to be detected. The average detection time per image was the lowest for YOLOv3. Since the project had to be utilized in real-time, the YOLOv3 network was selected as the network on which the other operations would run i.e. acquire the GPS location of the detected pothole and storing it in the cloud. This network was selected because the inference time per image was the lowest.

Video's of the detection of the various networks can be found at the following links:

- Faster R-CNN: https://youtu.be/ZxGJPdVJ8JQ
  SSD: https://youtu.be/8jnu8CKnlKI
- YOLOv3: https://youtu.be/Rgx7thKWPKY
- YOLOv3 real-time detection: https://youtu.be/7QcPcRQIa8o

For the real-time detection using the YOLOv3 network, the vehicle was driven at 60km/h to adhere to the speed limit set in the city. The webcam used for the real-time testing was positioned such that the view of the road was akin to a driver's viewpoint and the maximum area of the road was captured. The webcam was placed inside the vehicle. When driving at a speed of 60km/h and taking into consideration that the detection time per image of the YOLOv3 network was 0.836 seconds, the calculated distance from which potholes can be detected is 13.877m. Determining the distance in real-time was beyond the scope of this project. The detection in real-time can be seen in Figure 5 below. Once the pothole was detected, the GPS location of the detected potholes had to be acquired. The GPS location of the detected pothole was then sent to a cloud server, Ubidots. The GPS data is available for anyone who has access to the Ubidots account created for this project. The longitude and latitude values of each detected pothole is stored on Ubidots. The last detected pothole is illustrated as a pin on a map on Ubidots. This enables one to acquire a visual representation of where the pothole is situated. This visual representation is shown in Figure 6 below.

Figure 5. Screenshot showing the real-time detection

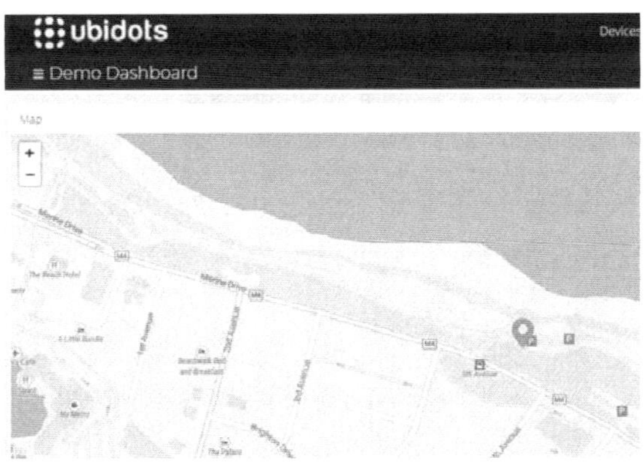

Figure 6. Screenshot of zoomed in map showing the pin of the last detected pothole on Ubidots

3.3. Comparison of various detection methods

Various pothole detection methods exist. Such methods include vision-based, vibration-based, laser based and stereo imaging. Each of these methods have certain advantages and disadvantages. Table 2 below was adapted from [20]. A comparison between the various techniques utilized to detect potholes is illustrated below.

Table 2. Pothole detection implementation methods and technology adapted from [20]

	Vision-based	**Vibration-based**	**Laser-based**	**Stereo-imaging**
Device used	**Camera**	**Accelerometer**	**Laser**	**Cameras**
Technology used	2D Imaging	Force, rotation and orientation	3D reconstruction of the image using light reflection	3D reconstruction using multiple cameras
Response time	High	Low		High
Sensing time	While approaching the pothole	While going through a pothole	While approaching the pothole	
Processing	Complex image	Readings are directly used	Collection of 3D point cloud with their elevations	A complex process of 3D image construction by combining image from different camera perceptions
Cost	High due to expensive parts (lens)	Low	High	
Characterization of pothole	Based on size	Based on vibrations	Based on 3D image constructed	
Detection at night time	Difficult-poor lighting	Can detect	Based on 3D image constructed	
Accuracy	Depends on algorithm used	High	Depends on alignment of cameras and algorithms used	

[21] Utilized a traditional vision-based and machine learning method based on Support Vector Machine (SVM) to detect potholes. In machine learning, SVMs are supervised learning models with associated learning algorithms that analyze data utilized for classification and regression analysis. It was revealed that in more complicated instances such as when mud was present, defects could not be identified correctly [22]. The same was revealed when utilizing deep learning methods. It was found that shadows on the road from lampposts were frequently mistaken for potholes.

In a paper by [23], a deep learning-based algorithm was utilized as a classifier for crack damage detection from concrete images. The CNN algorithm was utilized. It is advantageous to utilize this method because the features are learned automatically and no feature extraction processes need to be conducted. It was revealed that concrete cracks were identified correctly in realistic situations. Utilizing deep learning techniques for the detection of potholes in this instance revealed that potholes could also be correctly identified in real-life situations.

Research revealed that although vision, vibration, stereo-imaging and laser based techniques can be utilized to solve the pothole detection problem, some of these methods have been utilized for more than a decade. SpotholeAI [24] and CarVi [25] are apps that are able to detect potholes. These apps were developed in the last year and both utilize deep learning methods to detect potholes, further proving that such methods are suitable and can be utilized in real-time detection successfully.

4. Discussion

The development and maintenance of roads in developing countries such as South Africa is of great significance. Potholes can result in an unpleasant and potentially dangerous ride. The tyres are likely to get damaged and would require re-alignment. It was thus necessary to create a system whereby driving over a pothole could be minimised. It was possible to determine which method would be the most appropriate to achieve the desired result from research conducted previously. The work completed by [3], [4], [6] and [18] was carefully studied. It was revealed that deep learning methods produce desirable results. It should be noted that although vision, vibration, stereo-imaging and laser based techniques can be utilised to solve the pothole detection problem, some of these methods have been utilised for more than a decade. This research paper presented a comparative evaluation of state-of-the art CNN based object detection models to detect potholes on the road in real-time and from videos. Each of the networks were trained on a custom dataset and the performance was evaluated through the utilisation of various evaluation metrics. The best results for the application of this project was acquired through the YOLOv3 architecture, which worked the best for real-time applications because the inference speed was the fastest of the three evaluated architectures. The Faster R-CNN network proved the most accurate of the three models. SSD performed the worst in terms of accuracy, which could be attributed to the varying sizes of the potholes. Once it was determined that YOLOv3 would be the most suitable architecture for this particular application, the location, i.e. the GPS co-ordinates of the detected pothole had to be determined. The next step was upload these GPS co-ordinates to a cloud server where these could be stored and later illustrated on a map.

5. Conclusions and recommendations

The above-mentioned hypothesis is confirmed and met because the entire system was able to detect a pothole, determine the detected pothole GPS co-ordinates and save these to a cloud server. However, a number of enhancements can be made to the existing solution. Such improvements include increasing the dataset size. Increasing the dataset decreases the chances of overfitting occurring. Overfitting was discussed in Section 2.1. Increasing the dataset size also adds more diversity, i.e. more information is added and thus the fit is improved. A dataset comprising of images of different weather conditions can be included to expand the utilisation of the system. Adding a feature whereby the driver is told the approximate position of the pothole would greatly assist to limit the probability of the driver driving over the pothole. This feature could be integrated into a driver assist system. In future, developments could be made to display a warning of the pothole on a heads-up display.

References

[1] Fan R, Orgunalp U, Hosking B and Liu, I. Pothole Detection Based on Disparity Transformation and Road Surface Modeling. *IEEE Transactions on Image Processing*. **2019**, 897-908.

[2] First Drive: 2018 Mercedes-Benz-S-Class. Available online: https://www.automobilemag.com/news/first-drive-2018-mercedes-benz-s-class/ (accessed on 9 February 2019).

[3] Namala N, Mallesh S and Kumar, A. A Study on Potholes and its Effects on Vehicular Traffic. *IJCR*. **2018,** 1-2.

[4] Nienaber, S, Booysen, M and Kroon, R. Detecting potholes using simple image processing techniques and real-world footage. 34th Southern African Transport Conference (SATC 2015), CSIR International Convention Centre, Pretoria, South Africa, 6-9 July 2015.

[5] Nicholas JA, Herbert Chan, HW and Baker MAB. Machine learning: Applications of artificial intelligence to imaging and diagnosis. *Biophys Rev*. **2019,** 11, 111-118.

[6] Ukhwah E, Yuniarno E.M and Suprapto Y.K. Asphalt Pavement Pothole Detection using Deep Learning method based on YOLO Neural Network. *10.1109/ISITIA.2019.8937176*, 35-40.
[7] Redmon J and Farhadi A. YOLOv3: An Incremental Improvement. *arXiv* **2018**, arXiv:1804.02767.
[8] Tsiakmaki M, Kostopoulos G, Kotsiantis S and Ragos, O. Transfer Learning from Deep Neural Networks for Predicting Student Performace. *Appl. Sci.* **2020,** 10, 2145.
[9] Zhijian H, Fangmin Li, Xidao L and Zuowei, C. A Weekly Supervised Method for Mud Detection in Ores Based on Deep Active *Learning. Mathematical Problems in Engineering.* **2020**
[10] Soundy A.W.R, Packhurst B.J, Brown P, Martin A and Molteno, T.C.A.; Schumayer, D. Comparison of Enhanced Noise Model Performance Based on Analysis of Civilian GPS Data. *Sensors* **2020,** 20, 6050.
[11] Ferrandez-Pastor F.J, Garvia-Chamizo J.M, Nieto-Hidalgo M, Mora-Pascual J and Mora-Martinez, J. Developing Ubiquitous Sensor Network Platform Using Internet of Things: Application in Precision Agriculture. *Sensors.* **2016,** 16, 1141
[12] Xinliang Z, Wanru W, Yunji, Z and Heng, X. An improved YOLOv3 model based on skipping connections and spatial pyramid pooling. *Systems Science and Control Engineering.* **2020**
[13] Goutte C and Gaussier E. A probabilistic interpretation of Precision, Recall and F1-Score, with implication for Evaluation. *Lecture Notes in Computer Science.* **2005,** 325-359.
[14] Jingzhe Ma, Shaobo D, Ye Z, Jing W, Zongmin W, Runzhi, Li, Yongli, Li, Liazhong, Z and Huimin, M. Efficient Deep Learning Architecture for Detection and Recognition of Thyroid Nodules. *Computation Intelligence and Neuroscience.* **2020**, 1-15.
[15] Drid K, Allaoui M, and Kherfi M.L. Object Detector Combination for Increasing Accuracy and Detecting More Overlapping Objects. *Lecture Notes in Computer Science.* **2020**
[16] Li C, Luo B, Hong H, Su X, Wang Y, Li J, Wang C, Zhang J and Wei, L. Object Detection Based on Global-Local Saliency Constraint in Aerial Images. *Remote Sens.* **2020**, 12
[17] Simhambhatla R, Okiah K and Slater, R. Self-Driving Cars: Evaluation of Deep Learning Techniques for Object Detection in Different Driving Conditions. *SMU Data Science Review.* **2019,** 2.
[18] Zhang Z, Ai X, Chan C.K and Danhnoun, N. An efficient algorithm for pothole detection using stereo vision. Proceedings of the IEEE International Conference on Acoustics, Speech and Signal Processing (ICASSP '14), Florence, Italy, 2014, 564-568.
[19] Hosang J, Beneson R, Dollar P and Schiele, B. What makes for effective detection proposals?. *IEEE Transactions on Pattern Analysis and Machine Intelligence.* 2016, 814-830.
[20] Akula A, Bhatia R, Rai R, Gupta V and Aggarwal, N. Convolutional Neural Networks Based Potholes Detection Using Thermal Imaging. *Journal of King Saud University – Computer and Information Sciences*. **2019**, 3.
[21] Lin J, and L Yayu. Potholes Detection Based on SVM in the Pavement Distress Images. *International Symposium on the Distributed Computing Applications to Business, Engineering and Science.* **2010**, 533-547.
[22] Understanding Support Vector Machine Algorithm. Available online: https://www.analyticsvidhya.com/blog/2017/09/understanding-support-vector-machine-example-code/ (accessed on 15 February 2020)
[23] Young-Jin, C, Wooram C and Oral, B. Deep Learning – Based Crack Damage Detection using Convolutional Neural Networks. *Computer-Aided Civil Infrastructure Engineering.* **2017,** 32.
[24] Bhalla, V.,An Artificial Intelligence Assistant to Fix Potholes. *Spothole-AI.*
[25] A windshield mounted cure for the pothole. Available online: https://www.wired.com/story/carvi- pothole-detection/ (accessed on 1 May 2020).

Three-dimensional CFD modelling of urban flood forces on buildings: a case study

G Viccione[1,a] **and C Izzo**[1]

[1] Department of Civil Engineering, University of Salerno, via Giovanni Paolo II 132, 84084, Fisciano, SA, Italy

[a]E-mail: gviccion@unisa.it

Abstract. Italy is among the countries most likely to be exposed to high hydrogeological risk. The recent 2018 Hydrogeological Disruption Report by ISPRA (Higher Institute for Environmental Protection and Research) states that about 90% of Italian municipalities fall within areas with hydrogeological risk, that is where the probability of floods/landslides is high (level P3) or very high (P4) or the hydraulic hazard is of medium level. Floods are the most severe and frequent phenomena affecting landscapes, mainly if they occur in urban contexts. Urban development, if nature-based solutions are not taken into account, determines an increase in the impermeable surface, with the effect of reducing the infiltration and water evaporation, and consequently with an increase in runoff peaks and decrease of the concentration times. In this framework, the work here presented numerically analyses the effect of floods in an urban basin, in term of their three-dimensional interaction with buildings. The site under study is Cervinara, a municipality in the Campania region, Italy, hit by severe floods in 1999. Flow peaks were calculated using the VAPI procedure, a regional methodology based on two-component extreme value (TCEV) distribution. The results show that the level of exposure of the buildings, evaluated by calculating the thrust of the interacting water on the exterior walls, is influenced by the local three-dimensional kinematics.

1. Introduction

The presence of water in different forms is responsible for various processes on Earth, consisting of dynamic interactions between the water itself, in the form of rainfall, evapotranspiration and runouts, landforms and landscapes. The leading actor in the water cycle is the solar energy, spatially and temporally distributed in the atmosphere and Earth surface, producing continuous exchanges of thermal energy. Precipitation is one of the most important and frequent processes between the atmosphere and the inlands and oceans. With respect to the ground, its contribution can be basically divided into two components: one contributes to surface runoffs, the other evaporates or infiltrates, contributing to the deep runoff, the quality of which should be preserved [1]. The distribution of these components depends on the climatic area, soil features, vegetation and anthropic activities [2-3].

Mediterranean regions are characterized by mild and humid winters and hot and dry summers [4]. The Mediterranean Sea has the particularity to be nearly closed, with the evaporation that exceeds the contribution of precipitation and rivers: the Atlantic Ocean supplies the Mediterranean with water through the Strait of Gibraltar. Humidity formation, together with the ongoing increasing temperature trends, promote extreme precipitations [5-7], defined as meteoric events characterized by high intensity and a short duration, capable of triggering paroxysmal events [8-13] with devastating effects on the territory, both in economic terms of damage to infrastructures and buildings, and in terms of loss of human life. Among severe induced phenomena are floods, large masses of water quickly covering commonly dry areas. They may be catastrophic [14-17], caused by rivers embarkment failures, lakes dam breaks, etc. [18-20]. Damages may be even more catastrophic if floods take place in urban environments. With generally increasing impermeable surfaces, urban development alters the

basin hydrological response, reducing infiltration and evaporation [21-26]. Alteration is enhanced due to the fact that urban regions have a higher temperature than rural regions, with buildings facades and other anthropic infrastructure absorbing most of the incident radiation (yielding the so-called urban heat island effect), storing it and then releasing it in the form of thermal radiation overnight [27,28]. Consequently, the outflow and peak volume increases and the concentration time decreases [21, 29]. Nature-based solutions, defined by the European Commission as "solutions that are inspired and supported by nature, which are cost-effective, simultaneously provide environmental, social and economic benefits and help build resilience" [30], consist of solution strategies briefly defined as blu-green infrastructure (BGI), declined for cities specifically as Sustainable Urban drainage Systems (SUDS) for pluvial/river flood risk management [31-36]. Green roofs [37-39], rain gardens [40,41] and permeable pavement [42,43] are among possible solutions to mitigate urban flooding. Current technologies advances, such as light detection and ranging (LiDAR) systems and Unmanned Aerial Vehicles (UAVs) help tackling with flooding phenomena [44] as well as for other geomorphological processes [45].
In this paper, urban flooding is investigated in terms of its three-dimensional (3D) interaction with buildings. The city of Cervinara, Southern Italy, has been chosen as case of study, owning the tragic fast flow slide events occurred in 1999. The procedure, next described in the Material and methods section, is however generally applicable, knowing urban topography and input hydrographs. The literature on the application of porous Shallow Water Equations, two-dimensional (2D) Finite Volume/Difference methods is huge, see for instance [46-49], to name a few. Significant insights on 3D Fluid-Structure Interaction (FSI) analysis can be however noticed, as recently proved by [50-55].

2. Materials and methods

2.1. The urban area under study: Cervinara, Italy

Cervinara (Fig. 1) is located at 284 m. a.s.l., and it is among the largest municipalities in the province of Avellino, Campania Region, Italy, covering nearly 30 km^2. It borders the municipalities of Rotondi, Avella and San Martino Valle Caudina. It is part of the Partenio Regional Park, a protected natural area. The territorial limits fall within the jurisdiction of the Liri-Garigliano and Volturno River Basin Authority. It is bordered to the north by the Caudina Valley and to the south by the Partenio massif. The inhabited area is located in the inner flat part, surrounded by woods and crossed by several torrential water courses.

Figure 1. The area under study: Cervinara Municipality, Avellino Province, Italy.

The area is prone to fast flow slides induced by precipitation due to the presence of steep hill slopes and unsaturated soil conditions. In December 1999 severe flow masses from the near hillslopes caused five victims and several damaged and destroyed buildings [56]. The territory of Cervinara falls within the hydrographic basin of the Volturno river, among the most extended watercourses in Southern Italy. Specifically, the surrounding hydrographic network falls within the catchment area of the Isclero river, a left tributary of the Volturno river. Five main streams cross the municipal area: the Castello stream, the Loffredo stream, the San Gennaro stream, the Remescuso stream, the Conca stream and the Pirozza stream [57, 58].

2.2. Generation of the model in the CAD environment

The first phase involved the generation of the urban model in the CAD (Computer Aided Design) Autodesk AutoCAD ® environment, through the definition in plain of NURBS (Non-Uniform Rational Basis-Splines), commonly used in computer graphics for representing simple to complex curves and surfaces, which can be modified using their control points. NURBS are mathematically defined by a polynomial of degree one less than the order of the curve, e.g. second-order curves are represented by linear polynomials. The urban plan extension was deduced by the CTR (Regional Technical Cartography, 1:5000 scale) sheets (Fig. 2.a) of Campania Region [59], while the ground elevation was obtained from Google Earth. Urban works like buildings, sheds, shelters were then obtained by vertically extruding the NURBS (Fig. 2.b).
The need to first pre-process the urban fabric in a CAD software derives from the fact that the used CFD program Flow-3D [60] needs the STL (Standard Triangulation Language) as input format, specifically for complex geometries, numerically representing the surfaces of 3D bodies as a mosaic of triangles.

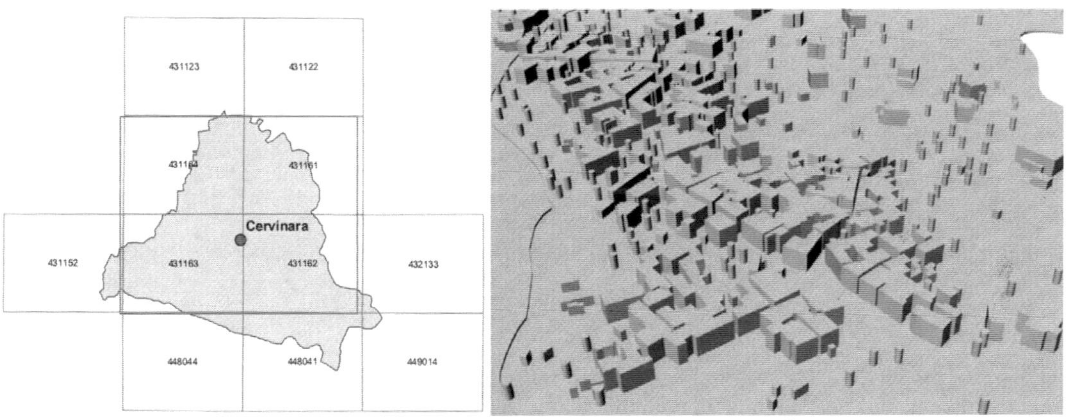

a. b.

Figure 2. a. CTR sheets covering the area under study. b. Geometric model generated in a CAD environment.

2.3. The VAPI procedure

The survey area (Fig.3) covers 13 (first 13 rows in Table 1) of the 17 basins that underlie the municipal area of Cervinara. The northern limit is represented by the railway line that delimits most of the inhabited center. The first 6 basins on the South consist of vegetated mountainous, the rest are nearly flat. For each of the mountain basins (1-6), the maximum average flow rate peaks Q were calculated using the VAPI procedure for a return period T=20 years.
The VAPI procedure [61, 62] is a hierarchical regional frequency analysis for the assessment of the flood peak for different return periods. It is based on the use of the TCEV (Two Component Extreme Value) probability distribution.

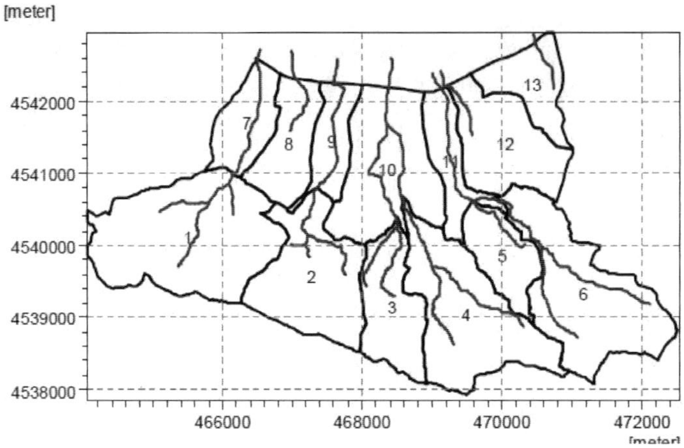

Figure 3. The sub-basins of the study area. Blue lines are the streams conveying runoff downstream.

Table 1. Basin denomination and related morphological properties: maximum elevation H_{max}, mean elevation H_{mea}, minimum elevation H_{min}, Watercourse length L, mean watercourse slope i [57].

#	Basin denomination	Area [km^2]	H_{max} [m a.s.l.]	H_{mea} [m a.s.l.]	H_{min} [m a.s.l.]	L [m]	i [%]
1	Pozzillo	3,734	983	711	473	2166	16,2
2	Conga monte	2,328	1149	720	323	2907	22,6
3	San Gennaro monte (left)	1,447	1252	815	337	2896	27,4
4	San Gennaro monte (right)	2,835	1291	892	314	3886	21,1
5	Castello (left)	1,022	1060	637	334	2382	23,3
6	Castello monte	3,012	1311	792	369	4006	11,5
7	Pirozza (left)	4,740	983	632	264	3902	8,8
8	Pirozza (right)	1,184	624	331	267	2179	5,1
9	Conga	3,145	1149	610	270	4622	8,5
10	San Gennaro	6,761	1291	671	272	5525	8,7
11	Castello	4,941	1311	676	273	6064	7,3
12	Castello (right)	1,815	770	365	271	2888	5,0
13	Fosso Est	1,005	461	301	273	1723	6,0
14	Pirozza – downstream railway	6,674	983	537	254	5242	3,4
15	Conga – downstream railway	4,880	1149	487	254	7442	2,0
16	Castello/San Gennaro - downstream railway	14,508	1311	607	266	7333	3,5
17	Fosso Est – downstream railway	1,804	461	289	265	3154	1,5

The procedure interprets the annual maximum events as a result of the combination of two distinct populations: the first concerns the ordinary maximum events, which are more frequent but less intense, the second, the extraordinary maximum events, which are less frequent but more intense. The probability of occurrence of extreme events is generally very low, but it becomes not negligible in a wider context (regional or national). Extreme phenomena are treated on a regional statistical basis to reduce the uncertainty of estimation in the measurement points and for the assessment of risk even in areas without measurement.

2.4. The CFD solver

Flow-3D® software [60, 63] numerically solves the Reynolds-Averaged Navier-Stokes equations (RANS), consisting of non-linear second order differential equations, ruling the motion of a free surface fluid. RANS are here coupled with the k-ε turbulence model [18, 20, 64], based on the transport equations for kinetic turbulent energy (k) and the dissipation rate (ε). The finite volume approximation is used for spatial discretization. The free surface is modelled with the VOF (Volume of fluid or Volume fraction) method [65]. VOF is a specific algorithm that treats the free surface as a cutting interface moving through a computational grid, and as a means of applying boundary conditions to the surface. Physical space is discretized into computational cells of parallelepiped shape. Sixteen cell meshes were used for spatial discretization (Fig. 4.a). Adaptive cell meshing was adopted to cover areas that need a higher resolution with a finer mesh, e.g. see Fig. 4.b.

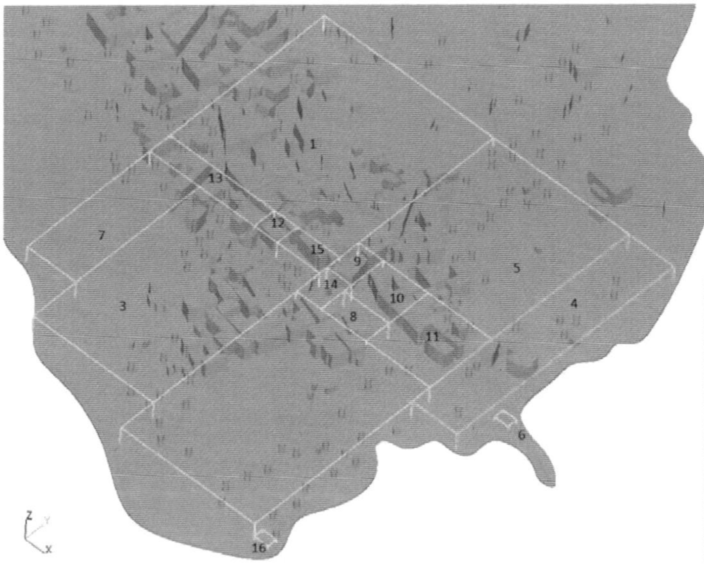

a.

b.

Figure 4. a. Spatial discretization of Cervinara municipality into 16 cell meshes. Cell mesh 6 was adopted for the inflow boundary condition. b. Different spatial resolution to accurately detect flow hydrodynamics.

3. Results

3.1. Data extrapolation

The VAPI procedure returns the highest flow rate peak Q=30m^3/s for basin 6 (Castello basin). Mesh cell 6 (bottom right corner in Fig. 4.a), located at the downstream end of the watershed, was therefore used for the inflow boundary condition. Three simulations were considered, corresponding to the flow rates Q, 2Q, 3Q. To evaluate the flow impact produced by each simulation, three buildings (with upfront red vertical walls in Fig. 5.a) were considered in mesh 8, 9, 12, respectively. Buildings (Fig. 5.b) were chosen according to their spatial location, which was expected along the fluid path trajectory. The instant considered for the data extrapolation of the local interacting pressures p and velocities components v_x and v_y (see global reference system in Fig. 4.a) corresponds to that in which the highest interacting pressure value was detected. Specifically, the following parameters were evaluated for each simulation: pressures p in correspondence with each vertical cell that intercepts the building's red surface and the consequent spatial weighted average pressure $\bar{p}$; the fraction of liquid that fills each vertical cell, to evaluate the whole wet area; the components v_x and v_y along the coordinate x and y directions of the velocity to reconstruct on each vertical cell the velocity component orthogonal to the impacted surface, and ultimately the average velocity $\bar{V}$. The above-mentioned hydrodynamic quantities allow the definition of static and dynamic thrusts components (RHS of Eq.1).

3.2. Thrust calculation

The literature on evaluating the total thrust S on surfaces due to a water impact is huge, see e.g. Pugliese Carratelli et al. [63]. Basically, the impact force can be assumed either to be proportional to the mean pressure $\bar{p}$ or to the square of the mean velocity $\bar{V}$ of the impacting mass. Other models consider a combination of the previous approaches. Therefore, three macro-group models can be identified: hydrostatic, hydrodynamic and mixed models. In this study we referred to the mixed model, in which S is given by the sum of the hydrostatic contribution S_{stat} and the hydrodynamic contribution S_{dyn}:

$$S = S_{stat} + S_{dyn} = \bar{p}A_{wet} + \rho A_{wet}\bar{V}^2 \quad (1)$$

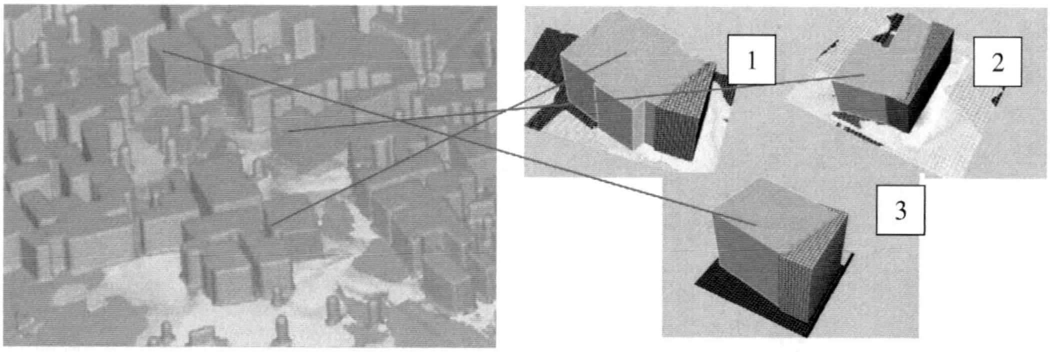

a. b.

Figure 5. a. The buildings (numbered as "1", "2" and "3"), chosen to assess the flood impact (red coloured). b. Mesh definition on the upfront wall surfaces for pressure and velocity data extraction.

where $\bar{p}$ is the average pressure acting on the wet area, ρ is the density of the water fluid (1000 kg/m^3), $\bar{V}$ is the averaged orthogonal component of water velocity respect the impacted wet surface. The results are shown in the Table 2:

Table 2. Static "S_{stat}", dynamic "S_{dyn}", and total "S" thrust numerical calculation. A_{wet} is the wetted interacting area of building for which maximum pressure is exerted.

Water discharge Q at mesh cell 6	Building #	A_{wet} [m^2]	S_{stat} [kN]	S_{dyn} [kN]	S [kN]	S / A_{wet} [kN]
Q = 30 m^3/s	1	4.36	16.22	1.59	17.81	4.08
Q = 30 m^3/s	2	8.35	32.01	67.42	99.43	11.91
Q = 30 m^3/s	3	8.52	70.04	2.79	72.83	8.55
Q = 60 m^3/s	1	35.02	182.86	10.57	193.43	5.52
Q = 60 m^3/s	2	16.71	87.95	81.17	169.12	10.12
Q = 60 m^3/s	3	12.06	62.06	13.16	75.22	6.24
Q = 90 m^3/s	1	20.4	203.42	8.04	211.46	10.37
Q = 90 m^3/s	2	23.28	146.09	74.25	220.344	9.46
Q = 90 m^3/s	3	10.3	49.51	1.95	51.46	5.00

3.3. Discussion

Total thrust S is generally increasing with the water discharge Q except for Building "3". This is basically due to the fact that above Q=60 m^3/s the flooding mass tends to remain on its main course toward Building 3, hence lower flooding volumes deviate to its direction. This evidence is confirmed by the fact that most of the thrust at Building 3 is given by the hydrostatic component S_{stat}, that is S_{dyn}/S_{stat} = 21% and S_{dyn}/S_{stat} = 4%, for Q=60 m^3/s and Q=90 m^3/s, respectively. Besides climatic factors (temperature, rainfall regime) and drainage characteristics (sewer capacity, lamination tanks), the influence of urban topography has been already enlightened in literature, as to be among the most influencing drivers in flooding evolution and related consequences, see for instance [66, 67]. Likewise, interesting is the behaviour of the specific thrust S/A_{wet}: a monotonic trend is exhibited by Building 1, which is the one directly invested by the upcoming flooding. On the contrary, Buildings 2 and 3 exhibit the opposite behaviour. Building 3 exhibits the lowest values of S/A_{wet}, confirming that it is the less exposed among the buildings taken into account.

4. Conclusions

In this paper, a numerical modelling of flood forces on urban buildings is performed and discussed in a case study in Southern Italy. The Flow-3D VOF-based software was used for the solution of the Navier-Stokes ruling equations. The urban fabric was first created in a CAD environment as a collection of 2D splines representing building contours and other urban objects such as tree trunks, then converted as a 3D geometric model, lastly given to Flow-3D as stereolithography input file.
Upstream flow rates peaks, for a twenty-year return period, were calculated with the VAPI procedure, based on the TCEV probabilistic distribution. Obtained maximum flow rate was then given as inflow boundary condition in Flow-3D.
Flow-3D results obtained, consisting of velocity components and pressures at the mesh cells, were post-processed to get the mean hydrostatic pressure $\bar{p}$ and to the square of the mean velocity $\bar{V}$ of the approaching mass near three buildings exterior facades.
The analysis let us drove the following conclusions:

- The nearest Buildings "1" and "2" to the inlet flow condition suffered the highest impacting forces.
- The longer was the free path toward the Building the higher was the dynamic thrust S_{dyn} onto it, as it can be seen in Table 2, comparing S_{dyn} for Buildings "1" and "2".
- Conversely, Building "3" exhibit the lowest S_{dyn} values, proving that upstream Buildings and other urban works act as barriers, making them at a higher hydraulic risk.

This study proves the strong dependence of urban flooding on topography. More reliable results can be obtained considering 3D pressure and velocity fields at the buildings front facades, from which average $\bar{p}$ and $\bar{V}$ values can be obtained for the application of Eq. 1. More in general 3D flooding simulations allow taking into account 3D water deviation effects and building obstruction.

5. References

[1] Stoppiello, M.G., Lofrano, G., Carotenuto, M., Viccione, G., Guarnaccia, C. and Cascini L. A comparative assessment of analytical fate and transport models of organic contaminants in unsaturated soils. Sustainability 2020, 12, 1-24, 2949. doi: 10.3390/su12072949.
[2] Wang, C., Fu, B., Zhang, L., Xu, Z. Soil moisture–plant interactions: an ecohydrological review. Journal of Soils and Sediments 2019, 19(1). doi: 10.1007/s11368-018-2167-0.
[3] Timm, A., Kluge, B., Wessolek, G. Hydrological balance of paved surfaces in moist mid-latitude climate – A review. Landscape and Urban Planning 2018, 175, 80-91. doi: 10.1016/j.landurbplan.2018.03.014.
[4] Britannica, The Editors of Encyclopaedia. "Mediterranean climate". Encyclopedia Britannica, 14 Aug. 2019, https://www.britannica.com/science/Mediterranean-climate. (last accessed on August 8, 2021).
[5] Kelebek, M.B., Batibeniz, F., Önol, B. Exposure assessment of climate extremes over the Europe–mediterranean region. Atmosphere 2021, 12(5), 633. Doi: 10.3390/atmos12050633.

[6] Pelosi, A., Furcolo, P., Rossi, F., Villani, P. The characterization of extraordinary extreme events (EEEs) for the assessment of design rainfall depths with high return periods. Hydrological Processes 2020, 34(11), 2543–2559. doi: 10.1002/hyp.13747.

[7] De Luca, C., Furcolo, P., Rossi, F., Villani, P., Vitolo C. Extreme rainfall in the Mediterranean. In: STAHY Workshop: Advances in Statistical Hydrology, 23–25 May 2010, Taormina, Italy.

[8] Viccione, G., Morriello, P., Immediata, N. Bottom shear stress in channel experiments involving fast-moving viscous free-surface flows. In: 3rd EMCEI Conference, 10-13 June 2021, Sousse, Tunisia. In press.

[9] Viccione, G., Bovolin, V. Simulating triggering and evolution of debris-flows with SPH. In: 5th DFHM Proceedings, June 14-17, 2011. University of Padua, Italy, pp. 523–532. Doi: 10.4408/IJEGE.2011-03.B-058.

[10] Viccione, G., Bovolin, V., Pugliese Carratelli, E. Simulating fluid-structure interaction with SPH. In: AIP Conference Proceedings, 1479 (1), pp. 209–212. Doi: 10.1063/1.4756099.

[11] Rendina, I., Viccione, G., Cascini, L. Kinematics of flow mass movements on inclined surfaces Theoretical and Computational Fluid Dynamics, 2019, 33(2), 107–123. Doi: 10.1007/s00162-019-00486-y.

[12] Cuomo, S., Della Sala, M., Novità, A. Physically based modelling of soil erosion induced by rainfall in small mountain basins. Geomorphology 2015, 243, 106–115. Doi: 10.1016/j.geomorph.2015.04.019.

[13] Papa, M.N., Medina, V., Ciervo, F., Bateman, A. Derivation of critical rainfall thresholds for shallow landslides as a tool for debris flow early warning systems. Hydrology and Earth System Sciences, 2013, 17(10), 4095–4107. Doi: 10.5194/hess-17-4095-2013.

[14] Cornwall, W. Europe's deadly floods leave scientists stunned despite improvements, flood forecasts sometimes failed to flag risks along smaller streams. Science 2021, 373(6553), 372-373. Doi: 10.1126/science.373.6553.372.

[15] Blöschl, G., Hall, J., Viglione, A., et al. Changing climate both increases and decreases European river floods. Nature 2019, 573(7772), 108–111. Doi: 10.1038/s41586-019-1495-6.

[16] Libertino, A., Ganora, D., Claps, P. Evidence for Increasing Rainfall Extremes Remains Elusive at Large Spatial Scales: The Case of Italy. Geophysical Research Letters 2019, 46(13), pp. 7437–7446. Doi: 10.1029/2019GL083371.

[17] Gallus, W.A., Parodi, A., Maugeri, M. Possible impacts of a changing climate on intense Ligurian Sea rainfall events. International Journal of Climatology 2018, 38, pp. e323-e329. Doi: 10.1002/joc.5372.

[18] Kocaman, S., Evangelista, S., Guzel, H., Dal, K., Yilmaz, A., Viccione, G. Experimental and numerical investigation of 3d dam-break wave propagation in an enclosed domain with dry and wet bottom. Applied Sciences (Switzerland) 2021, 11(12), 5638. Doi: 10.3390/app11125638.

[19] Capasso, S. Tagliafierro, B., Guzel, H., Yilmaz, A., Dal, K., Kocaman, S., Viccione, G., Evangelista, S. A Numerical Validation of 3D Experimental Dam-Break Wave Interaction with a Sharp Obstacle Using DualSPHysics. Water (Switzerland) 2021, 13(15), 2133. Doi: 10.3390/w13152133.

[20] Kocaman, S., Güzel, H., Evangelista, S., Ozmen-Cagatay, H., Viccione, G. Experimental and numerical analysis of a dam-break flow through different contraction geometries of the channel. Water (Switzerland), 2020, 12(4), 1124. Doi: 10.3390/w12041124.

[21] Sarchani, S., Awol, F.S., Tsanis, I. Hydrological analysis of extreme rain events in a medium-sized basin. Applied Sciences (Switzerland) 2021, 11(11),4901. Doi: 10.3390/app11114901.

[22] Bian, G., Wang, G., Chen, J., Zhang, J., Song, M. Spatial and seasonal variations of hydrological responses to climate and land-use changes in a highly urbanized basin of Southeastern China. Hydrology Research 2021, 52(2), 506-522. Doi: 10.2166/nh.2021.087.

[23] Wu, J., Wu, Z.-Y., Lin, H.-J., Ji, H.-P., Liu, M. Hydrological response to climate change and human activities: A case study of Taihu Basin, China. Water Science and Engineering 2020, 13(2), 83-94. Doi: 10.1016/j.wse.2020.06.006.

[24] Gao, Y., Chen, J., Luo, H., Wang, H. Prediction of hydrological responses to land use change. Science of the Total Environment 2020, 708, 134998. Doi: 10.1016/j.scitotenv.2019.134998.
[25] Wang, Y., Xie, X., Liang, S., (...), Meng, S., Lu, C. Quantifying the response of potential flooding risk to urban growth in Beijing. Science of the Total Environment 2020, 705, 135868. Doi: 10.1016/j.scitotenv.2019.135868.
[26] Wang, H., Chen, Y. Identifying key hydrological processes in highly urbanized watersheds for flood forecasting with a distributed hydrological model. Water (Switzerland) 2019, 11(8), 1641. Doi: 10.3390/w11081641.
[27] Garuma, G.F. Review of urban surface parameterizations for numerical climate models. Urban Climate 2018, 24, 830-851. Doi: 10.1016/j.uclim.2017.10.006.
[28] Aguilera-Benito, P., Varela-Lujan, S., Piña-Ramirez, C. Thermal behavior in glass houses through the analysis of scale models. Sustainability (Switzerland), 2021, 13(14), 7970. Doi: 10.3390/su13147970.
[29] Fletcher, T.D., Andrieu, H., Hamel, P. Understanding, management and modelling of urban hydrology and its consequences for receiving waters: A state of the art. Advances in Water Resources 2013, 51, 261-279. Doi: 10.1016/j.advwatres.2012.09.001.
[30] European Commission, Nature-Based solutions: State of the Art in EU-funded Projects. Publications Office of the European Union, Luxembourg 2020, https://op.europa.eu/en/publication-detail/-/publication/8bb07125-4518-11eb-b59f-01aa75ed71a1. (last accessed on August 8, 2021).
[31] Costa, S., Peters, R., Martins, R., Postmes, L., Keizer, J.J., Roebeling, P. Effectiveness of nature-based solutions on pluvial flood hazard mitigation: The case study of the city of eindhoven (the netherlands). Resources 2021, 10(3), 24. Doi: 10.3390/resources10030024.
[32] Magliocchetti, M., Adinolfi, V., Viccione, G., Grimaldi, M., Fasolino, I. Small rivers and landscape nbs to mitigate flood risk. Sustainable Mediterranean Construction 2021, (Special Issue 5), 32-36. ISSN 2385-1546.
[33] Hobbie, S.E., Grimm, N.B. Nature-based approaches to managing climate change impacts in cities. Philosophical Transactions of the Royal Society B: Biological Sciences 2020, 375(1794), 20190124. Doi: 10.1098/rstb.2019.0124.
[34] Ourloglou, O., Stefanidis, K., Dimitriou, E. Assessing nature-based and classical engineering solutions for flood-risk reduction in urban streams. Journal of Ecological Engineering 2021, 21(2), 46-56. Doi: 10.12911/22998993/116349.
[35] Schiavo, V., Fasolino, I., Viccione, G. An ecosystem services-based model for the definition of decision support scenarios for urban flood hazard mitigation: a case study. Sustainability (Switzerland). Submitted.
[36] Viccione, G., Nesticò, A., Vernieri, F., Cimmino, M. A Pilot Plant for Energy Harvesting from Falling Water in Drainpipes. Technical and Economic Analysis. Lecture Notes in Computer Science (including subseries Lecture Notes in Artificial Intelligence and Lecture Notes in Bioinformatics) 2019, 11624 LNCS, pp. 233–242. Doi: 10.1007/978-3-030-24311-1_16.
[37] Mobilia, M., Longobardi, A. Event scale modeling of experimental green roofs runoff in a mediterranean environment. Advances in Science, Technology and Innovation 2020, 153–156. Doi: 10.1007/978-3-030-13068-8_37.
[38] Longobardi, A., D'Ambrosio, R., Mobilia, M. Predicting stormwater retention capacity of green roofs: An experimental study of the roles of climate, substrate soil moisture, and drainage layer properties. Sustainability (Switzerland), 2019, 11(24), 6956. Doi: 10.3390/su11246956.
[39] Mobilia, M., Longobardi, A., J. F. Sartor. Including a-priori assessment of actual evapotranspiration for green roof daily scale hydrological modelling. Water (Switzerland) 2017, 9(2), 72. Doi: 10.3390/w9020072.
[40] Bouarafa, S., Lassabatere, L., Lipeme-Kouyi, G., Angulo-Jaramillo, R. Hydrodynamic characterization of Sustainable urban Drainage Systems (SuDS) by using beerkan infiltration experiments. Water (Switzerland), 2019, 11(4), 660. Doi: /10.3390/w11040660.
[41] Perales-Momparler, S., Andrés-Doménech, I., Hernández-Crespo, C., Vallés-Morán, F. Martìn, M., Escuder-Bueno, I. Andreu. J. The role of monitoring sustainable drainage systems for

promoting transition towards regenerative urban built environments: a case study in the Valencian region, Spain. Journal of Cleaner Production, 2017, 163, S113-S124. Doi: 10.1016/j.jclepro.2016.05.153.
[42] Castillo-Rodríguez, J.T., Andrés-Doménech, I., Martín, M., Escuder-Bueno, I., Perales-Momparler, S., Mira-Peidro. J. Quantifying the Impact on Stormwater Management of an Innovative Ceramic Permeable Pavement Solution. Water Resources Management, 2021, 35(4), 1251-1271. Doi: 10.1007/s11269-021-02778-7.
[43] Hernández-Crespo, C., Fernández-Gonzalvo, M., Martín, M., Andrés-Doménech, I. Influence of rainfall intensity and pollution build-up levels on water quality and quantity response of permeable pavements. Science of the Total Environment, 2019, 684, 303-313. 10.1016/j.scitotenv.2019.05.271.
[44] Barrile, V., Bilotta, G., Fotia, A. Analysis of hydraulic risk territories: Comparison between LIDAR and other different techniques for 3D modelling, WSEAS Transactions on Environment and Development, 14, 2018, 45-52.
[45] Bernardo, E., Palamara, R., Boima, R. UAV and Soft Computing Methodology for Monitoring Landslide Areas (Susceptibility to Landslides and Early Warning), WSEAS Transactions on Environment and Development, 17, 2021, 490-501.
[46] Ferrari, A., Viero, D.P. Floodwater pathways in urban areas: A method to compute porosity fields for anisotropic subgrid models in differential form, Journal of Hydrology 2020. 589, 125193. Doi: 10.1016/j.jhydrol.2020.125193.
[47] Varra, G., Pepe, V., Cimorelli, L., Della Morte, R., Cozzolino, L. On integral and differential porosity models for urban flooding simulation. Advances in Water Resources 2020. 136, 103455. Doi: 10.1016/j.advwatres.2019.103455.
[48] Soares-Frazão, S., Zech, Y. Dam-break flow through an idealized city. Journal of Hydraulic Research 2008, 46(5), 648-658. Doi: 10.3826/jhr.2008.3164.
[49] Abderrezzak, K.E.K., Paquier, A., Mignot, E. Modelling flash flood propagation in urban areas using a two-dimensional numerical model. Natural Hazards 2009, 50, 433–460. Doi: 10.1007/s11069-008-9300-0.
[50] Amicarelli, A., Manenti, S., Paggi, M. SPH Modelling of Dam-break Floods, with Damage Assessment to Electrical Substations. International Journal of Computational Fluid Dynamics 2021, 35(1-2), 3-21. Doi: 10.1080/10618562.2020.1811240.
[51] Amicarelli, A., Manenti, S., Albano, R. et al. SPHERA v.9.0.0: A Computational Fluid Dynamics research code, based on the Smoothed Particle Hydrodynamics mesh-less method. Computer Physics Communications 2020, 250, 107157. Doi: 10.1016/j.cpc.2020.107157.
[52] Aureli, F.; Dazzi, S.; Maranzoni, A.; Mignosa, P.; Vacondio, R. Experimental and numerical evaluation of the force due to the impact of a dam-break wave on a structure. Adv. Water Resour. 2015, 76, 29–42. Doi: 10.1016/j.advwatres.2014.11.009.
[53] Teng, J., Jakeman, A.J., Vaze, J., Croke, B.F.W., Dutta, D., Kim, S. Flood inundation modelling: A review of methods, recent advances and uncertainty analysis. Environmental Modelling and Software 2017, 90, 201-216. Doi: 10.1016/j.envsoft.2017.01.006.
[54] Alahmed, S., Zou, Q. Modeling Urban Flood Forces on Buildings Using SPH Method. In: Proceedings of the 20th IAHR APD Congress, 28-31 August, 2016, Colombo, Sri Lanka, 7 pages.
[55] Gómez-Gesteira, M., Dalrymple, R.A. Using a three-dimensional smoothed particle hydrodynamics method for wave impact on a tall structure. Journal of Waterway, Port, Coastal and Ocean Engineering, 2004, 130(2), 63–69. Doi: 10.1061/(ASCE)0733-950X(2004)130:2(63).
[56] Cascini, L., Cuomo, S., De Santis, A. Numerical modelling of the December 1999 Cervinara flow-like mass movements (Southern Italy). In: International Conference on Debris-Flow Hazards Mitigation: Mechanics, Prediction, and Assessment, Proceedings, June 14-17, 2011. University of Padua, Italy, pp. 635-644. Doi: 10.4408/IJEGE.2011-03.B-069.
[57] Autorità di Bacino dei fiumi Liri-Garigliano e Volturno. Relazione generale sugli studi geologici, geomorfologici, idrogeologici e geotecnici. Accordo di Programma tra Autorità di

Bacino dei fiumi Liri-Garigliano e Volturno e Comune di Cervinara, Mitigazione del rischio idrogeologico finalizzato al governo del territorio, 2013, Caserta, Italy. Available online: http://www.ildistrettoidrograficodellappenninomeridionale.it/b.3.1-b.6.1-b.9.1_relazione%20generale%20sugli%20studi%20geologici,%20geomoerfologici,%20idrogeologici%20e%20geotecnici.pdf (last accessed on May 3, 2021).

[58] Cuomo, S., Masi, E.B., Tofani, V. Moscariello, M., Rossi, G., Matano, F. Multiseasonal probabilistic slope stability analysis of a large area of unsaturated pyroclastic soils. Landslides 2021, 18, 1259–1274. Doi: 10.1007/s10346-020-01561-w.

[59] Carta Tecnica Regionale | Geoportale Regione Campania. Available online: https://sit2.regione.campania.it/content/ctr (last accessed on May 3, 2021).

[60] Flow Science, Flow3D User's Manual.

[61] Rossi, F., Villani, P. Regional flood estimation methods, in Coping with Flood, edited by G. Rossi, N. Harmancioglu and V. Yevjevich, pp. 135-170, Kluwer Academic, Dordrecht, NE, 1994.

[62] Rossi, F., Villani, P. Valutazione delle Piene in Campania, CNR-GNDCI, pubbl. n. 1472, Grafica Metelliana & C., Cava de' Tirreni (SA), 1995.

[63] Pugliese Carratelli, E., Viccione, G., Bovolin, V. Free surface flow impact on a vertical wall: a numerical assessment. Theoretical and Computational Fluid Dynamics, 2016, 30(5), 403–414. Doi: 10.1007/s00162-016-0386-9.

[64] Launder, B.E., Spalding, D.B. The numerical computation of turbulent flows. Comput. Methods. Appl. Mech Eng. 1974, 3(2), 269–289. doi: 10.1016/0045-7825(74)90029-2.

[65] Hirt, C.W., Nichols, B.D. Volume of fluid (VOF) method for the dynamics of free boundaries. Journal of Computational Physics 1981, 39(1), 201–225. Doi: 10.1016/0021-9991(81)90145-5.

[66] Xie, L., Zhao, H. Correlation between flood disaster and topography: A case study of Zhaoqing City. Journal of Natural Disasters 2013, 22(6), 240-245. ISSN: 1004-4574.

[67] Albano, R. Investigation on roof segmentation for 3D building reconstruction from aerial LIDAR point clouds. Applied Sciences (Switzerland) 2019, 9(21),4674. Doi: 10.3390/app9214674.

Traffic noise prediction using machine learning and monte carlo data augmentation: a case study on the Patiala city in India

Daljeet Singh[1,a], Priyal Kaler[2], Ishita Lyall[2], Aekamjot Singh[2] and H S Pannu[2]

[1] Mechanical Engineering Department, Thapar Institute of Engineering and Technology, Patiala-147004, India
[2] Computer Science and Engineering Department, Thapar Institute of Engineering and Technology, Patiala-147004, India

[a] E-mail: daljeet.singh@thapar.edu

Abstract. Traffic noise pollution is a serious problem in the modern urban areas especially to design new architecture of smart cities, highways, hospitals, schools for an efficient and healthy environment. To analyse this aspect, we have proposed a machine learning based prediction of sound pressure level on an original dataset collected in Patiala city in India. Vehicular traffic and sound pressure level data was collected on different sites in the city. A total of 502 data samples on the identified sites were obtained for the study. Further this data is augmented using Monte Carlo simulation to 10 times of its initial size and the Artificial Neural Networks (ANN) have been trained and compared with other Machine learning methods for the vehicular traffic noise prediction. The input parameters in the model are traffic volume Q, percentage of heavy vehicles P and the average speed of vehicles V and the output parameter is the equivalent continuous sound pressure level, L_{eq} dB(A). The experimental results show ANN which is trained on the augmented data using Monte Carlo simulations outperforms other advanced methods making it an effective measure for vehicular traffic noise prediction to develop a healthy environment which is free of noise pollution.

1. Introduction

The increased vehicular traffic on the roads is a cause of serious concern as it leads to environmental noise pollution, annoyance, sleep disturbance [1], [2], anxiety to drivers and passengers and various other related health problems [3]. The scientific community and the regulatory agencies the world over are working to address this issue. One aspect related to this problem is the assessment of the present noise levels using different measurement techniques and equipment and prediction of the future traffic noise levels to take adequate preventive measures. Figure 1 depicts the proportion of various noise sources causing annoyance in which the road traffic noise pollution contributes to about 80% of overall noise annoyance according to [4].

Some of the existing traffic noise prediction models to address the issue of traffic noise assessment and plan mitigation measures include FHWA (USA), RLS-90 (Germany), CoRTN (UK), ASJ (Japan) [5] which are country specific and may not be applicable everywhere. There are some methods which are based on machine learning techniques like ANN, GA etc. [6]. The parameters generally considered include traffic volume, percentage of heavy vehicles, average speed of vehicles etc.

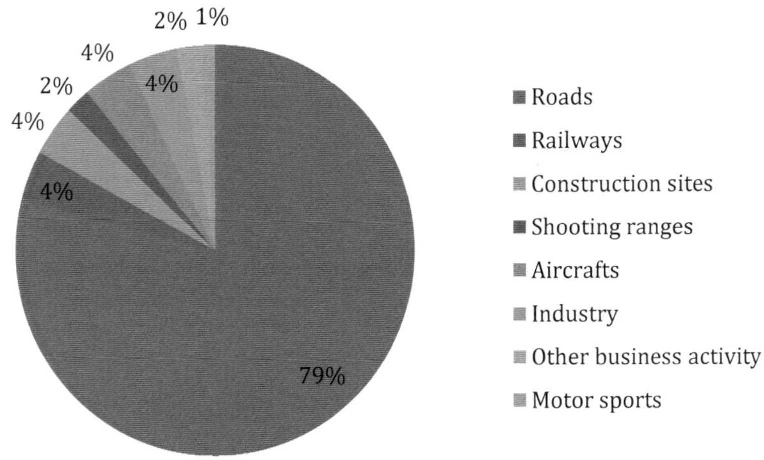

Figure 1. Distribution of noise annoyance created by human beings through various sources [4]. Road traffic noise is responsible for almost 80% of the overall noise annoyance.

A comparison of some of these methods on the criteria of MSE, correlation coefficient and accuracy has been done by Singh et al. [7]–[9]. But the methods considered are limited in number. Also some other aspects like the augmentation of the data sets considered and the effect of the size of the data set needs to be studied. Some of these aspects have been addressed in the present study. A Monte Carlo simulation has been attempted to augment the dataset and see the effects on the results, i.e. the mean square error and the prediction accuracy, obtained using different Machine Learning techniques.

The material and methods have been presented in section 2, results and discussion in section 3 and finally the conclusions drawn from the study are presented in section 4.

2. Materials and Methods

The dataset collected for the study, the measurement locations, equipment and methods used are presented in this section.

2.1 Dataset for the study

The study locations selected were on the highway roads that bring the traffic into Patiala city. The traffic flow observed is in the free-flow condition as there is no congestion, traffic lights, roundabouts etc. on these locations. The input parameters include the traffic volume (Q), percentage of heavy vehicles (P) and the average speed of vehicles (V). The values of these traffic parameters were measured manually using videography. The output parameter considered is the equivalent continuous sound pressure level, L_{eq}. The values of L_{eq} were measured using a class 1 integrating type sound level meter (SLM) confirming with the IEC specifications [10]. The SLM was mounted on a tripod to avoid the human impedance effects and a windscreen was used to cover the microphone. The SLM was kept at a distance of 1m from the edge of the road. A total of 502 datasets were obtained on the three measurement locations during different times of the day and different months throughout the year to account for the variation in the same. A sample dataset is presented in table 1 with a sample of five values for the considered parameters and also the mean and the standard deviation of the entire dataset of 5000 values.

Table 1. Sample dataset where log Q, P and log V are input variables and L_{eq} is the predicted variable in dB(A) units.

S.No.	Leq, dB(A)	log Q	P	log V
1	74.51	3.21	19.49	1.68
2	76.32	3.17	15.18	1.64
3	76.35	3.06	39.12	1.76
4	77.30	3.18	35.62	1.66
5	73.03	3.12	22.50	1.54
...	...	...	...	...
502	75.2	3.1	32.6	1.7
Mean	76.45	3.14	32.74	1.69
S.D.	1.68	0.04	11.68	0.05

In order to achieve the objective of accurately predicting the values of the output parameter of $\mathbf{L_{eq}}$, based on the set of input parameters, the approach of machine learning combined with Monte Carlo simulation was used and is presented in the next section.

2.2 Monte Carlo simulation

The concept of Monte Carlo simulation is used to increase the dataset with random samples rather than using a fixed set of input values for the prediction [11]. The number of experiments using the Monte Carlo method can be increased thousands of times to see more possible outcomes and therefore increase the prediction accuracy as shown in figure 2. It is also observed from figure 2 that as the number of samples increases, the Central Limit Theorem (CLT) states that the distribution becomes normal which is illustrated using the bell curve. The main steps involve studying the probability distribution of the values of the independent parameters or input parameters, generating a range of values for these parameters (using the concept of random sampling) and running the prediction model (machine learning based in the present case) hundreds or thousands of times to get the output results.

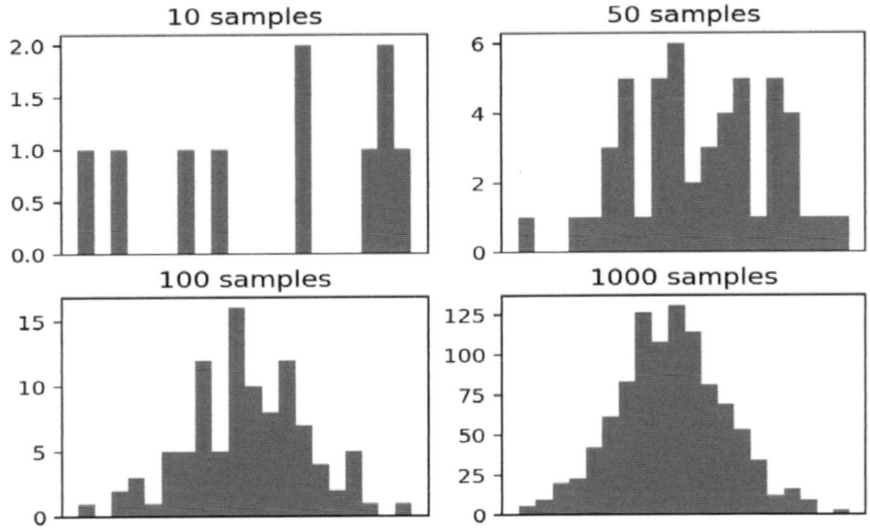

Figure 2. Effect of Monte Carlo simulation as the number of repeated samples increases. The distribution tends to be normal as more samples are extracted.

Monte Carlo simulation is a computational algorithm which randomly extracts samples based upon a given data distribution such as a specified mean and standard deviation as given by the following formulae,

$$\mu_X = \bar{X} = \frac{1}{n}\sum_{i=1}^{n} X_i \tag{1}$$

$$\sigma_X = \sqrt{\frac{1}{n}\left\{\sum_{i=1}^{n} X_i^2 - \frac{1}{n}\left(\sum_{i=1}^{n} X_i\right)^2\right\}} \tag{2}$$

Usually real world problems are deterministic in principle in which random selection is imperative and the original data set is not sufficiently representative of enough population. Monte Carlo repeated random sampling is often used in obtaining a probability distribution, numerical integration and optimization [11]. The simulations are useful where degrees of freedom of input variables are coupled in a complex fashion to generate the output, such as the current problem of noise volume prediction L_{eq}, using traffic volume Q, percentage of heavy vehicles P and their average speed V. Monte Carlo simulation is quite popular in financial modeling, risk management, and complex decision boundary conditions. Central limit theorem (CLT) states that if the sample size is larger than 30 then it approximates to the normal distribution, regardless of the population distribution [12]. Therefore the original data collected was of the size 502 samples can be assumed as an approximation of normal distribution. This relatively smaller dataset of 502 samples has been augmented to 5,000 samples using Monte Carlo simulation. Monte Carlo has been implemented in Excel using NORMINV(.) which takes three inputs (a) random number generator function within a specified range (b) mean and (c) standard deviation of the underlying distribution of smaller dataset. Every time it is used, it creates a new data entry, so we have used this function 5,000 times with the drag function in Excel.

Table 2. Formulas of performance metrics

S.No.	Performance Metric	Formula
1	Root Mean Square Error (RMSE)	$\sqrt{\frac{\sum(x-y)^2}{n}}$
2	Coefficient of determination R^2	$\frac{\text{SS}}{\text{TSOS}} = 1 - \frac{\text{ESS}}{\text{TSOS}}$
3	How good is the fit within the given tolerance *e* (Accuracy)	$\frac{\sum if(\lvert x-y\rvert \leq e)}{n}$

2.3 Results and Discussion

The results obtained by applying the methodology discussed in the previous sections are presented below. Neural Network fitting application in MATLAB R2020a has been used to train a 2-layer feed-forward network with linear output neutrons and sigmoid hidden neutrons. The network is trained using Bayesian Regularization algorithm option with train:validate:test ratio of 70:15:15. Figure 3 shows the outcome of Neural network results for 118 epochs which implies the convergence validation and other parameters using MATLAB fitting application.

Mu refers to the adaption parameter in the underlying optimisation used by ANN during the parameter update calculations. It is defined as momentum update and its value lies in [0,1]. It affects the convergence error and is a control parameter used by ANN during training. It helps to avoid problems such as network getting stuck in the local minima. The details of mu are given in the Levenberg-Marquardt algorithm. 'Gradient' stands for minimum performance gradient. It has a default value of 1e-7. 'gamk' shows the number of parameters used for training, 'ssX' means sum of squares of the parameters, validation checks (val fail) graph at the bottom is about monitoring the possible degradation of the validation failures.

Figure 4 shows the regression plots by the proposed method (Monte Carlo simulation and Neural Networks regression) which is almost 1 for the value of R. Figure 5 shows the frequency histogram of the errors for training and testing target variables (L_{eq}). Figure 6 shows that the mean square error was minimum at 108 epochs.

Various state-of-art machine learning algorithms have been considered for performance comparison such as, GLM: Generalised Linear Model; DT: Decision Tree; RF: Random Forest; ANN: Artificial Neural Network. Finally the proposed regression model using Monte Carlo data augmentation has been considered which has superior performance. Figure 7 and 8 show the performance comparison histograms for R, RMSE [13], [14] & accuracies within 1 dBA and 0.5 dBA. Tables 3-4 show the values of R, RMSE and accuracies in detail corresponding to the figures 7-8. Formulas of all the performance metrics have been described in table 2.

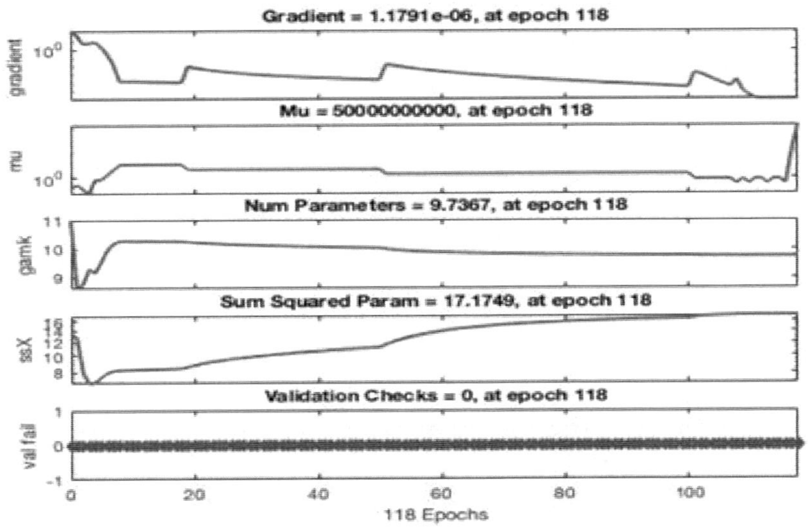

Figure 3. Neural network results for 118 epochs show the convergence validation and other parameters using MATLAB fitting application.

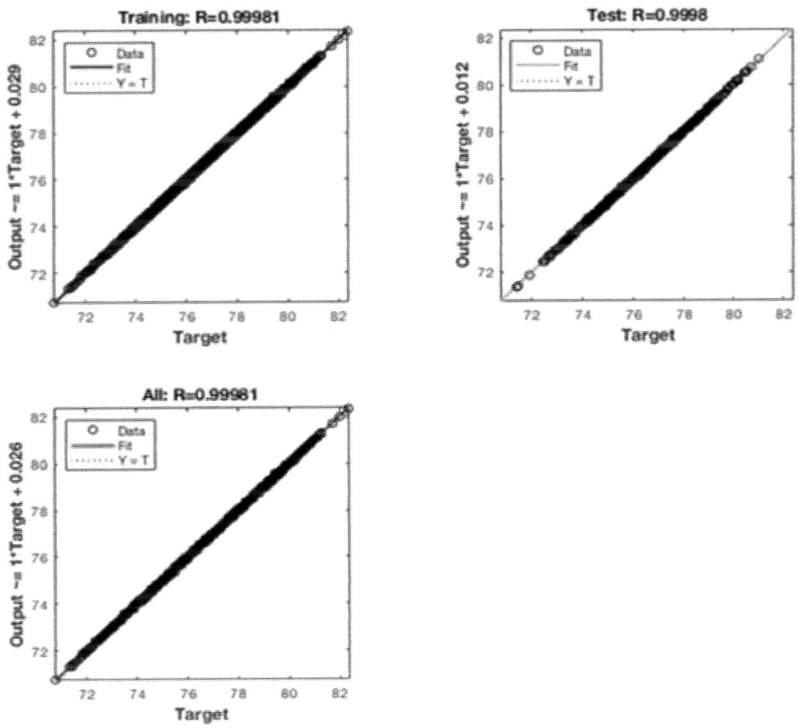

Figure 4. Regression plots for training, testing and all data samples.

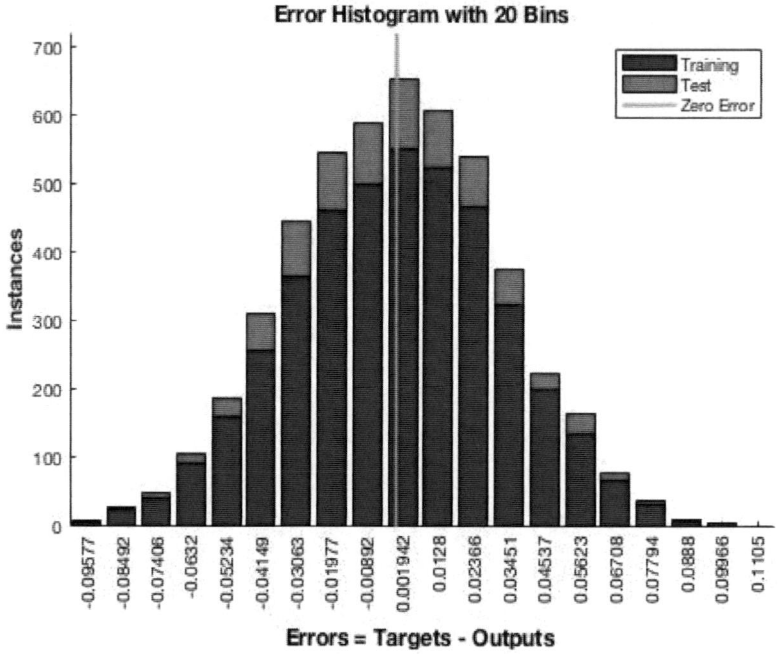

Figure 5. Error histogram for training and testing for 20 bins show the normal distribution and values are negligible due to high accuracy of the trained model.

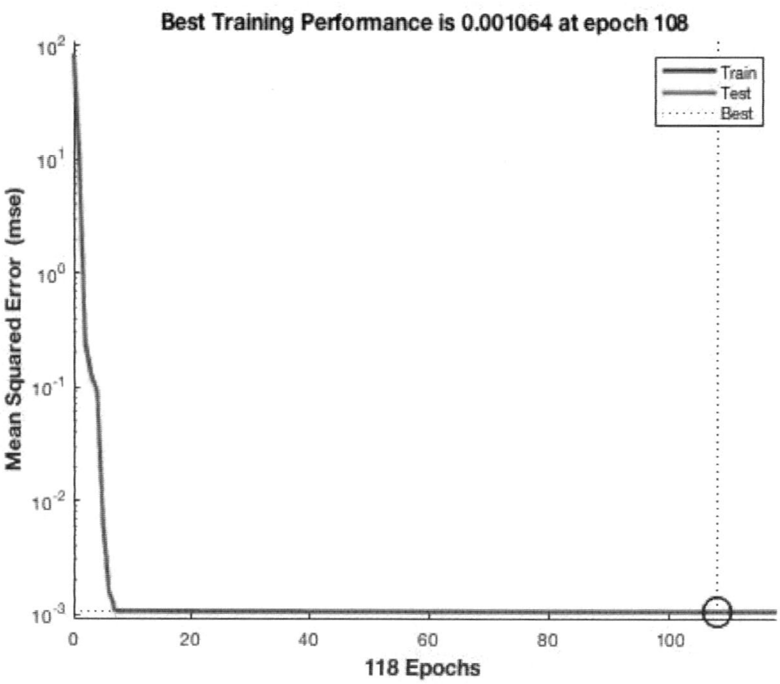

Figure 6. Graph between epochs and mean squared error. Training is best at 108 epochs.

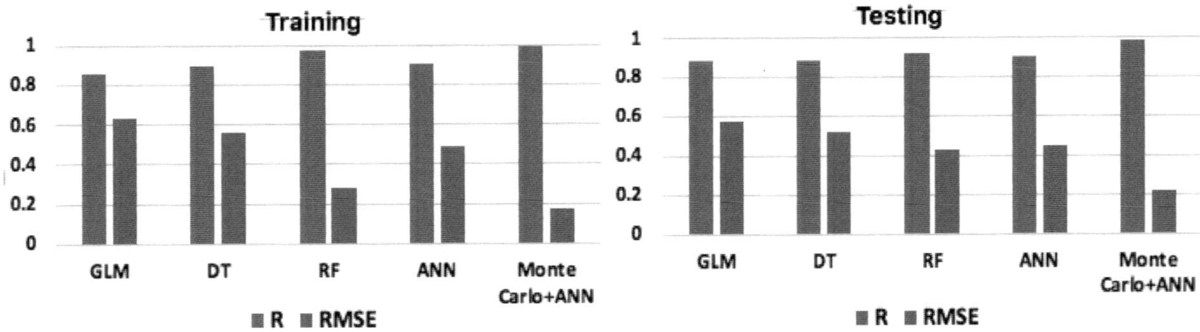

Figure 7. R and RMSE values for training of GLM, Decision Tree, Random Forest, ANN, and proposed method

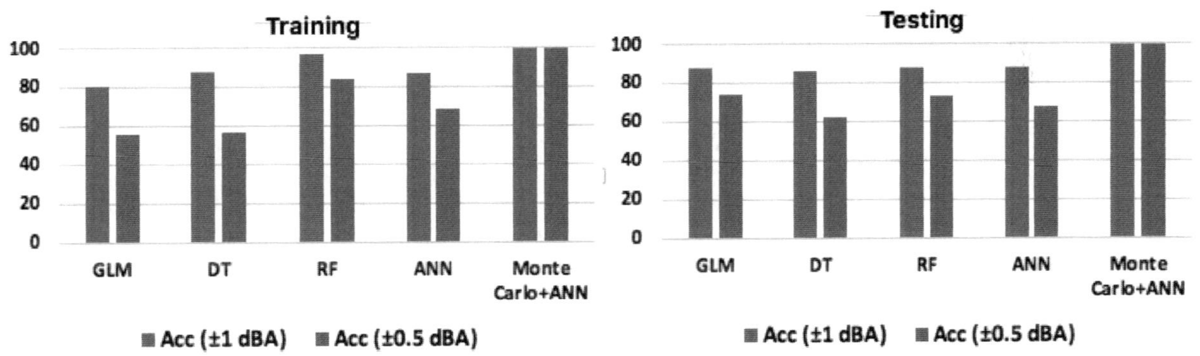

Figure 8. Accuracy values for training and testing of GLM, Decision Tree, Random Forest, ANN, and proposed method

Table 3. Performance comparison during the *training* phase by the underlying techniques

S.No.	Method	R	RMSE	Acc (±1 dBA)	Acc (±0.5 dBA)
1	GLM	0.86	0.63	81	56
2	DT	0.89	0.56	88	57
3	RF	0.97	0.28	97	84
4	ANN	0.9	0.49	87	69
5	Monte Carlo+ANN	0.99	0.18	100	100

Table 4. Performance comparison during the *testing* phase by the underlying techniques

S.No.	Method	R	RMSE	Acc (±1 dBA)	Acc (±0.5 dBA)
1	GLM	0.89	0.58	88	74
2	DT	0.89	0.52	86	62
3	RF	0.92	0.43	88	73
4	ANN	0.91	0.45	88	68
5	Monte Carlo+ANN	0.99	0.22	100	100

3. Conclusions

In the modern world especially in the urban areas, noise pollution is a serious problem which causes health and environmental issues especially around the schools, hospitals, highways and industries. To analyze this problem, we have studied the original dataset collected in the Patiala city including traffic volume, percentage of heavy vehicles, average speed of vehicles and equivalent sound pressure level in dB(A). In this paper, vehicular traffic noise has been predicted using three variables: (a) traffic volume Q, (b) percentage of heavy vehicles P and (c) the average speed of vehicles V. Then the output parameter is predicted which is equivalent continuous sound pressure level, L_{eq} dB(A). The original data of 502 samples were collected from Patiala city in India and then Monte Carlo simulation has been used to augment the data to increase the size to 10 times i.e. 5,000 samples. It has been observed that machine learning algorithms work well for larger dataset as compared to traditional machine learning algorithms with smaller dataset. Thus, Monte Carlo simulation has been observed to justify its advantage for data augmentation. The machine learning algorithms perform better on a larger training dataset since numerous internal parameters (such as neuron weights in case of ANN) needs to be setup accurately for better generalization and accuracy. This dataset enlargement is provided by Monte Carlo simulation in our study. The assumption machine learning algorithms is parameter tuning and data distribution which must be analyzed carefully. Moreover, machine learning depends upon the data distribution, so a collection of techniques must be explored for every data analysis problem. Our future plan is to collect data specific to places and vehicle types while including more input variables for accurate prediction.

References

[1] Öhrström E, Hadzibajramovic E, Holmes M, and Svensson H 2006 "Effects of road traffic noise on sleep: studies on children and adults," *J. Environ. Psychol.* **26** (2) 116–126.

[2] Graziuso G, Mancini S, Francavilla A B, Grimaldi M and Guarnaccia C 2021 "Geo-Crowdsourced Sound Level Data in Support of the Community Facilities Planning. A Methodological Proposal," *Sustainability* **13**(10) 5486.

[3] Babisch W 2003 "Stress hormones in the research on cardiovascular effects of noise," *Noise Heal.* **5**(18) 1.

[4] Amundsen A H and Klæboe R 2005 *A Nordic perspective on noise reduction at the source*, no. 806/2005. Transportøkonomisk institutt.

[5] Steele C, 2001 "A critical review of some traffic noise prediction models," *Appl. Acoust.* **62**(3) 271–287 doi: 10.1016/S0003-682X(00)00030-X.

[6] Kumar P, Nigam S P, and Kumar N 2014 "Vehicular traffic noise modeling using artificial neural network approach," *Transp. Res. Part C Emerg. Technol.* **40** 111-122 doi: 10.1016/j.trc.2014.01.006.

[7] Singh, D, Nigam, S P, Agrawal V P, and Kumar M 2016 "Vehicular traffic noise prediction

using soft computing approach," *J. Environ. Manage.* **183**(1) 59–66 doi: 10.1016/j.jenvman.2016.08.053.
[8] Singh D, Upadhyay R, Pannu H S, and Leray D 2021 "Development of an adaptive neuro fuzzy inference system based vehicular traffic noise prediction model," *J. Ambient Intell. Humaniz. Comput.* **12**(2) 2685–2701.
[9] Singh D, Francavilla A B, Mancini S and Guarnaccia C 2021 "Application of machine learning to include honking effect in vehicular traffic noise prediction," *Appl. Sci.* **11**(13) 6030.
[10] "IEC 61672-1: 2002, International Electrotechnical Commission".
[11] Raychaudhuri S 2008 "Introduction to monte carlo simulation" *Winter simulation conference* 91–100.
[12] Mulekar M S 2006 "Visualizing The Central Limit Theorem," *Alabama J. Math.* 25-36.
[13] Kyong-Il K, Song-Chol K, Song-Chon L and Ryu Chol C 2019 "An Improved Thresholding Method for Wavelet Denoising of Acoustic Signal" *WSEAS Trans. Signal Process.* **15** 65–76.
[14] Vagaská A, Michal P, Bukovský I, Gombár M and Kmec J 2019 "The Application of Neural Networks to Control Technological Process" *WSEAS Trans. Circuits Syst.* **18** 147–153.

Modular enumeration - a new method for solving discrete programming problems

Vitaly O Groppen[1]

[1] North-Caucasian Institute of Mining and Metallurgy (State Technological University), Nikolajev str. 44, Vladikavkaz, 362021, Russia

Abstract. A new approach resulting in a new group of algorithms for finding globally optimal solutions to discrete programming problems with Boolean variables is proposed. It is proved that the effectiveness of this method depends only on the number of variables of the problem to be solved and of the amount of RAM of used computer. The main difference from the existing methods of implicit enumeration is in the ability for some of the modular enumeration procedures to a priori predict, solving any extreme problem with Boolean variables, the number of iterations, the gain in running time in comparison with the traditional enumeration scheme, as well as the amount of RAM used. The experimental results confirm high efficiency of the proposed approaches.

1. Introduction

The efforts associated with the normal tendency to reduce the running time when solving various applied problems can be identified as analytical and computer-based [1] – [4]. The latter dominate in relation to extreme problems described by discrete mathematical models, where search for globally optimal solutions in the general case relies on various enumeration procedures. In this case, as a rule, either complete enumeration methods or implicit enumeration methods developed in the middle of the last century are used, such as dynamic programming, branch-and-bound (B&B) methods, backtracking [5] – [9] and their modifications [10] – [14]. Keeping in mind all the positive features of these methods, it should be recognized nevertheless that all of them have several drawbacks:

- the impossibility to predict a priori neither the number of iterations, nor its upper bound and, as a result, it is impossible to predict the gain in this number from the use of these procedures as compared to the complete enumeration methods;
- operating in a hostile environment, that is in the case when the B&B method, as well as backtracking, makes many mistakes when choosing the direction of search in the branch tree, there may be cases when the number of iterations by these methods exceeds the number of different complete plans, which means that running time of implicit enumeration methods sometimes exceeds the running time of the full enumeration method;
- using dynamic programming the number of analyzed plans can also exceed that of different complete plans - it takes place when, searching a globally optimal solution, it is possible to cut off only a relatively few unpromising groups of variables' vectors due to the specifics of the being solved problem.

Meanwhile, any complete enumeration method makes it possible to predict a priori the number of iterations for globally optimal solutions search to specific problems. Another advantage of complete enumeration is the reduced requirements for the amount of used RAM. As the only drawback of this method is the relatively long runtime, the idea of improving the method of complete enumeration of all combinations of values of variables in relation to discrete optimization problems remains relevant.

The purpose of this work, developing the ideas outlined in [15], is to create a new class of enumeration algorithms, which allow:
a) to reduce the running time of the latter;
b) to adapt the algorithm to the size of the computer's RAM, whereas the goal remains the same - to minimize the computation time;
c) to preserve a possibility of an a priori estimate of the number of iterations in relation to a particular problem to be solved.
The main idea of the algorithms proposed below is in the modular organization of the exhaustive search, which makes it possible to minimize repetitive computations at each iteration by storing their results in RAM and using them as needed.

This paper contains description, analysis, and experimental verification of efficiency of the three new realizations of the complete enumeration method including its modular organization related to searching a globally optimal solution to extreme problems with Boolean variables. Examples and statistics below are based on knapsack problems solving [14].

2. Basic Ideas of Modular Enumeration

The essence of the proposed approach is to implement three stages of finding globally optimal solutions to discrete programming problems. At the first stage, all the variables are grouped into "m" modules and, for each module, all combinations of the values of the variables corresponding to this module are generated and stored in the RAM. At the second stage, for each combination of variables of each module, its part of the objective function and constraints are calculated and stored. The third stage includes two steps:
a) use of the contents of the modules for generation of all the complete plans, corresponding values of the objective function and constraints;
b) their comparison and selection of the best values of the vector of variables.
Below it will be shown how, by modifying these steps, modular enumeration can be adapted to the specificity of various tasks. Further, we assume that these algorithms are solving problems with Boolean variables of the form (see [11] – [13]):

$$\begin{cases} F = \sum_i^n C_i z_i \to \max(min)\,; \\ \sum_i b_{ij} z_i \le a_j;\ j = 1, 2, \dots d; \\ \forall i: z_i = 1{,}0, \end{cases} \tag{1}$$

where $Z = \{z_1, z_2, \dots, z_n\}$ is a vector of Boolean variables, n – their number, whereas $\forall$i,j: C_i, b_{ij}, and a_j are integer constants.

It can be shown that the running time T_1 of the complete enumeration search for problem (1) globally optimal solution can be determined by the following expression:

$$T_1=(d+1)\cdot(n-1)\cdot 2^n\cdot t_0, \tag{2}$$

where t_0 is equal to the sum of the times of addition and multiplication of two integer numbers, whereas the time of two numbers comparison is ignored.
If the modular enumeration is used to solve problem (1), regardless of the number of variables of this problem, the running time of the 3rd stage of this algorithm does not exceed the T_2 value:

$$T_2=(d+1)\cdot[(m-1)\cdot 2^n + \sum_{p=1}^{p=m} n_p\, 2^{n_p}]\cdot t_0; \tag{3}$$

$$k\cdot m\cdot 2^{n/m} \le M, \tag{4}$$

where "m" is the number of modules (m > 1), n_p is the number of variables, belonging to the p-th module, k is the proportionality coefficient, M is the amount of free RAM in the computer used.

This poses three questions:

- What is the optimal number of modules for a specific task?

- What is the gain in running time for finding problem (1) globally optimal solution when applying modular enumeration if compared to the traditional enumeration scheme?

- What should be the strategy of variables distribution between modules?

From (3) it follows that the minimum value of T_2 corresponds to the minimum value of m determined according to (4). Obviously, with a sufficiently large RAM size, the optimal value of m is equal to two. The latter makes it possible to determine the upper bound for the gain η in the time spent on the finding of a globally optimal solution to problem (1) when using modular enumeration in comparison with the traditional enumeration scheme. Considering that the minimum value of m is equal to two and ignoring the second term in square brackets in (3), the ratio of the right-hand sides (2) and (3) results in the following equality:

$$\eta = T_1 / T_2 = n-1. \quad (5)$$

In other words, according to (5), upper bound for the gain in the running time η does not depend on the number of constraints d in (1) and is uniquely determined by the number of variables in this problem. Since when deriving (5), the first two mentioned above stages of searching for system (1) optimal solution by modular enumeration were ignored, associated with them time spent on generating and filling the modules was also ignored, that is why the value of η in (5) can be considered as the upper bound of the gain in computation time. Moreover, considering that, in the first approximation, the time spent by modular enumeration on these two stages can be described as: $T_3 = t_0 \cdot n \cdot 2^{0.5n}$, the following expression is true:

$$\lim_{n\to\infty} T_3/T_2 = 0. \quad (6)$$

Since due to (3) - (4) system optimal number of modules in the modular enumeration algorithms for any amount of free RAM in the computer used can be determined, the problem arises of the optimal distribution of variables between the modules. This problem is solved by the following theorem:

Theorem 1. For the number of variables of the problem (1) "n", which is a multiple of the number of modules "m", the optimal is a uniform distribution of variables in these modules.

Proof: The running time T_1 corresponding to the search by modular enumeration algorithm with the uniform distribution of variables between the modules, i. e. □ $1<p\leq n/m$: $n_p = n/m$ from (3) it follows:

$$T_1 = t_0 \cdot (d-1) \cdot [n \cdot 2^{n/m} + (m-1) \cdot 2^n], \quad (7)$$

We obtain a new, non-uniform distribution of variables between the modules, for which in the uniform distribution we remove g variables ($0 < g < n/m$) from one module and transfer them to the other module. The search time for problem (1) solution in this case is denoted as T_2:

$$T_2 = t_0 \cdot (d-1) \cdot [\,[\frac{n}{m}(m-2) \cdot 2^{n/m} + (\frac{n}{m} - g) \cdot 2^{n/m-g} + (\frac{n}{m} + g) \cdot 2^{n/m+g} + (m-1) \cdot 2^n], \quad (8)$$

The difference T_2 - T_1 is designated bellow as ΔT. After expanding the parentheses and transformations, we get:

$$\Delta T = \cdot t_0 \cdot (d-1) \cdot 2^{n/m} \cdot [\,[(2^{-g} + 2^g - 2) + 2^g \cdot (2^g - 2^{-g})]. \quad (9)$$

It is easy to verify that for any integer values of g in the range [1 - n/m], the right-hand side of (9) is non-negative.

It follows that $T_1 < T_2$. The theorem 1 is proved.

3. Modular Enumeration Algorithms

Modular enumeration is a paradigm that allows you to generate various enumeration algorithms. Three of these algorithms are presented and further explored below.

The main feature of Algorithm 1 presented below is such an organization of the full enumeration process, in which:

a) the conditions of Theorem 1 are satisfied;
b) amount of free RAM in the used computer satisfies (4) thus resulting in m = 2;
c) the number of arithmetic operations is minimal.

Algorithm 1

Step 1. If problem (1) goal function F is maximized, then R= - ∞, otherwise R= + ∞.

Step 2. The entire set of variables Z is divided into m = 2 modules, the number of variables of the first module is equal to the integer part of the ratio | Z | / 2, the number of variables of the second module is equal to | Z | - | Z | / 2, subsets of variables of these modules do not intersect, in other words, for subsets of variables belonging to different modules, the following conditions are true:

$$\begin{cases} \bigcup_{j=1}^{j=2} Z_j = Z; & (10) \\ Z_1 \cap Z_2 = \emptyset; & (11) \\ | Z_2| - |Z_1| \leq 1, & (12) \end{cases}$$

where Z_j is the subset of variables, corresponding to the j-th module (j = 1, 2).
Step 3. i=1.
Step 4. If i = 1, then q is equal to the integer part of the ratio | Z | / 2, otherwise q = | Z| - $|Z_i|$.
Step 5. All the 2^q different states of the variables of the i-th module, as well as the components of the objective function and of constraints of problem (1) corresponding to each such state, are generated, and fixed in the RAM.
Step 6. i=i+1.
Step 7. If i >2 then go to the step 8, otherwise go to the step 4.
Step 8. A new vector of variables is generated by new combination of in-RAM components of variables belonging to different modules. If there is no such a combination, then go to step 12.
Step 9. The new value of F is obtained by substituting the vector of variables obtained at the previous step into system (1).
Step 10. If F is "better" than R and all constraints of system (1) are satisfied, then go to step 11, otherwise, go to step 6.
Step 11. R:= F, the new vector of variables obtained at the 8-th step of the last iteration is stored, go to step 6.
Step 12. The algorithm is completed. The values of R and of the vector of variables stored in memory at the last call of the 11-th step are problem (2) optimal solution.

Example 1. Using the above Algorithm 1, solve the knapsack problem [14] of the form:

$$\begin{cases} 7z_1 + 2z_2 + 4z_3 + 5z_4 \to \max; \\ 2z_1 + 4z_2 + 8z_3 + 3z_4 \leq 12; \\ \forall i\text{: } z_i = 1,0. \end{cases} \quad (13)$$

Determine the effectiveness of the proposed approach in relation to problem (17), provided that m = 2.

1) R = - ∞.
2) Distribution of variables into two modules satisfying system (10) - (12): $Z_1 = \{z_1;z_2\}$; $Z_2 = \{z_3;z_4\}$.

Table 1. All the states of vectors of variables of each of the two modules.

m_1				m_2			
z_1	z_2	ΔF_1	Δb_1	z_3	z_4	ΔF_2	Δb_2
0	0	0	0	0	0	0	0
0	1	2	4	0	1	5	3
1	0	7	2	1	0	4	8
1	1	9	6	1	1	9	11

3) Generation of all states of the vectors of variables of each module and calculation of the corresponding components of the objective function ΔF_i and of the left side of the constraint of the system (17) Δb_i are presented in the Table 1 above.

4) Analysis of all the pairwise combinations of vectors of variables of both modules (repeating 16 times steps 8-11 of Algorithm 1) allows us to obtain a globally optimal problem (13) solution: $R = F_{max} = 14$, $Z_{opt} = \{1,1,0,1\}$.
5) The effectiveness of the proposed approach in solving problem (6) is determined by the gain in running time of Algorithm 1 as compared to the traditional realization of enumeration method solving the same problem. In the latter case, it can be shown that the running time of the search for a knapsack problem globally optimal solution is determined by the expression:

$$T_1=2\cdot(n-1)\cdot 2^n\cdot t_0, \tag{14}$$

where t_0 is the same, as used above in (2).

A similar approach to estimating the search time for the same problem solution by modular enumeration allows us to determine this time as follows:

$$T_2 = t_0\cdot[n\cdot 2^{0.5n} + 2^{n+1}]. \tag{15}$$

The gain in the running time with the Algorithm 1 solving problem (6) is equal to η:

$$\eta = T_1/T_2 \approx 1.33(3). \tag{16}$$

The experimental results presented below in Section 4 confirm the assessment of the Algorithm 1 efficiency given in (5).

Unlike Algorithm 1, Algorithm 2, being also a modular enumeration one, is organized in such a way that not only the number of arithmetic operations at each iteration, but also the number of iterations is minimized during the enumeration process. This is achieved by changing the order of generation of the vectors of variables at step 8: during each iteration at this step a new vector of variables is generated, which consists of two parts corresponding to the best values of goal function components in each module. This strategy allows us not to analyze vectors of variables, which correspond to a priori "bad" values of the objective function, thus giving us an opportunity to get a gain in time exceeding that which is determined by (5).

Since the first seven steps of Algorithms 1 and 2 are the same, the description of Algorithm 2 below begins with the eighth step.

Algorithm 2

Step 8. All vectors of variables of the first module are considered as unlabeled.

Step 9. Among the unlabeled variables' vectors of the first module, the one that corresponds to the best value of the components of objective function belonging to this module, is selected. If all vectors of variables belonging to this module are labeled, then go to step 17.

Step 10. All vectors of variables of the second module are considered as unlabeled.

Step 11. Among the second module unlabeled vectors of variables, the one that corresponds to the best value of the components of objective function belonging to this module, is selected. If all vectors of variables belonging to this module are already labeled, then go to to step 9.

Step 12. Generation of a new vector of variables, the values of which were sequentially obtained at steps 9 and 11 of the last iteration.

Step 13. The vectors of the variables selected in steps 9 and 11 of the last iteration are labeled.

Step 14. If the value of the objective function, calculated with the values of the variables determined at step 12 of the last iteration, is "not better" than the previously found and stored in RAM value, then go to step 9, otherwise go to step 15.

Step 15. If the vector of variables created at step 12 of the last iteration satisfies all the conditions of problem (1), then go to step 16, otherwise, go to step 11.

Step 16. Replacing the value of the objective function stored in RAM with a new one corresponding to the vector of variables generated at step 12 of the last iteration and transit to step 9.

Step 17. The algorithm is completed. The best value of the objective function and corresponding vector of variables are stored in RAM.

 2162 (2022) 012022 doi:10.1088/1742-6596/2162/1/012022

It is easy to verify that, due to steps 9 and 11, the duration of each iteration of Algorithm 2 exceeds the similar parameter of Algorithm 1, however, due to steps 14 - 16, there is a good chance that the number of these iterations will decrease, and these possibility increase with the number of variables. The effectiveness of the above algorithm 2 is demonstrated below in two ways:

a) as applied to the solution of problem (13);

b) as shown by the statistical research presented in the next section.

Example 2. Using the above Algorithm 2, we solve the knapsack problem (13).

The first seven steps of solving problem (13) by algorithms 1 and 2 coincide and lead to the construction of Table 1 (above), that is why they are not presented here. The sequence of problem (13) vectors of variables generated and analysed by Algorithm 2 is shown below in Table 2. Here F denotes the current value of the objective function, G denotes the current value of the left side of the inequality of problem (13), B is the right side of this inequality, and R is the maximum allowable value of the objective function.

Table 2. Vectors of variables analysed by Algorithm 2

№	x_1	x_2	x_3	x_4	F	G	R	Remarks
1	1	1	1	1	18	17	$-\infty$	G>B
2	1	1	0	1	14	9	14	G<B
3	1	0	1	1	16	13	14	G>B
4	1	0	0	1	12	5	14	F<R
5	0	1	1	1	11	15	14	F<R
6	0	0	1	1	9	11	14	F<R

The obtained result coincides with the previously found with Algorithm 1, but the difference lies in the fact that Algorithm 2 needed to analyze only 6 plans, whereas Algorithm 1 needed to analyze 16 plans. The latter does not mean that the search time for a solution to problem (13) by Algorithm 2 is less than that with Algorithm 1: as noted above, the duration of iterations of Algorithm 2 exceeds the duration of iterations of Algorithm 1. A detailed experimental analysis of the efficiency of the modular enumeration methods described above in relation to the knapsack problem is presented below in Section 4.

The third modification of the modular enumeration differs from the algorithms above: that is before the actual enumeration begins, the components of both modules are arranged in the order of deterioration of "their" objective function components. This approach simplifies the procedure of enumeration and, at the same time, gives us chance to reduce its running time by ignoring a priori "bad" vectors of variables. As the first seven steps of Algorithms 1 and 3 coincide, the description of Algorithm 3 below begins with the eighth step.

Algorithm 3

Step 8. All the components of both modules are arranged in the order of deterioration of their components of goal function.

Step 9. i=1

Step 10. j=1.

Step 11. A new, not previously analyzed vector of variables is obtained by combining its i-th component of the first module and the j-th component of the second module.

Step 12. Calculation of a new value of the objective function F, corresponding to the vector obtained at the previous step.

Step 13. If F is “better” than R, then go to the next step, otherwise go to step 16.

Step 14. If the constraints are met, then go to step 15, otherwise go to step 19.

Step 15. R:= F, the new vector of variables obtained at the 11-th step of the last iteration is stored in RAM, go to the next step.

Step 16. If the components of problem (1) full vector of variables corresponding to the second module have values that coincide with the first ordering vector of the elements of this module obtained at step 8, then go to step 21.
Step 17. i=i+1.
Step 18. If $i>2^{|Z1|}$, then go to step 21, otherwise go to step 10.
Step 19. j=j+1.
Step 20. If $j<2^{|Z2|}$, then go to step 11, otherwise go to step 16.
Step 21. The algorithm is over. The best value of the objective function R and corresponding vector of variables are stored in RAM.

Example 3. Using the above Algorithm 3, solve the knapsack problem (13). As in the previous case the first seven steps of solving problem (13) by algorithms 1 and 3 coincide and lead to the construction of Table 1 (above). During step 8 of Algorithm 3 Table 1 is transformed into Table 3 below:

Table 3. All the components of both modules are arranged in the order of deterioration of their goal function components.

m_1				m_2			
z_1	z_2	ΔF_1	Δb_1	z_3	z_4	ΔF_2	Δb_2
1	1	9	6	1	1	9	11
1	0	7	2	0	1	5	3
0	1	2	4	1	0	4	8
0	0	0	0	0	0	0	0

The sequence of problem (13) vectors of variables generated and analysed by Algorithm 3 is shown below in Table 4. In this case, the same designations as in Table 2 are used.

Table 4. Vectors of variables analysed by Algorithm 3

№	x_1	x_2	x_3	x_4	F	G	R	Remarks
1	1	1	1	1	18	17	$-\infty$	G>B
2	1	1	0	1	14	9	14	G<B
3	1	0	1	1	16	13	14	G>B
4	1	0	0	1	12	5	14	F<R
5	0	1	1	1	11	15	14	F<R

It is easy to verify that:

a) the volume of enumeration by Algorithm 3 was the smallest when solving (13) as compared to Algorithms 1 and 2;
b) this version of modular enumeration is close to the "meet in the middle" algorithm [16];
c) the results of solving (13) by all three algorithms above coincide.

4. Experimental Verification

The purpose of the experiments was to test the effectiveness of the above algorithms 1 - 3 in comparison with the traditional organization of exhaustive search in relation to the knapsack problem, the number of variables of which varied in the range from 3 to 20. Each algorithm's criterion of efficiency in relation to knapsack problems of a fixed dimension resulted in an average gain in a problem-solving time. The behaviour of the algorithms was analysed in relation to two types of environments: benevolent and hostile. Benevolent environment below corresponds to the case, when in (1) a_1 is close to the value $\sum_i b_{i1}$, whereas hostile environment is simulated by the opposite condition: a_1 is close to the $\min_i\{b_{i1}\}$.

The computer with the following parameters was used during our experiments:
Processor Celeron N4100 /J4125; RAM 8 Gb.; hard disk 1 Tb.
Operating system Windows 10 – 64.
The order of the experiments was determined by the following procedure:

Algorithm 4

Step 1. n = 3.
Step 2. Using Monte Carlo method we generated all the integer coefficients and constants of problem (1) for a given number of Boolean variables n in the range 1 - 10. Other parameters: d = 1, goal functions are maximized – generated are knapsack problems with the number of variables in the range 3 – 20.
Step 3. The problem obtained in the previous step is sequentially solved by software implementations of algorithms 1, 2, 3 and brute force. For each i-th algorithm (i=1, 2, 3) fixed are the corresponding running times $T_1(n)$, $T_2(n)$, $T_3(n)$, whereas T(n) is the running time of the traditional organization of exhaustive search and the gains in running time $\eta_1(n)= T(n)/T_1(n)$, $\eta_2(n)= T(n)/T_2(n)$, $\eta_3(n)= T(n)/T_3(n)$. Steps 2, 3 are repeated 10 times, after which the average values of the gain in time for each algorithm are recorded in the memory.
Step 4. n = n + 1.
Step 5. If n < 21, then go to step 2; otherwise, go to step 6.
Step 6. The algorithm is completed

Experimental dependences of average values of the gain in time as function of number of variables n for each algorithm - $\eta_1(n)$ (grey squares) for algorithm 1, $\eta_3(n)$ (black oblique crosses) for algorithm 3 and $\eta_2(n)$ (black triangles) for algorithm 2, are presented below in Figure 1 and Figure 2.

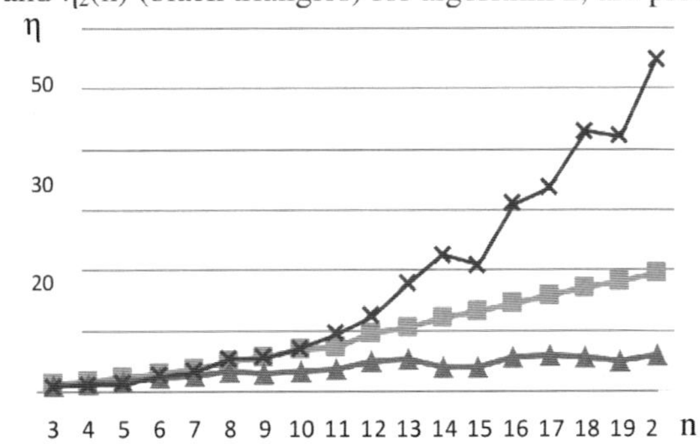

Figure 1. The effectiveness of the above algorithms 1 - 3 in a hostile environment.

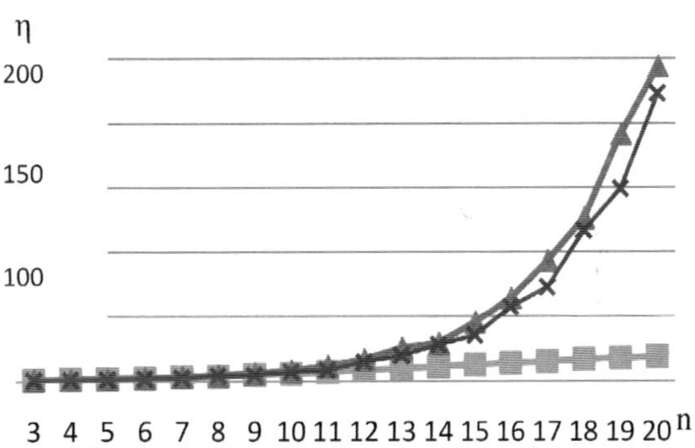

Figure 2. The effectiveness of the above algorithms 1-3 in benevolent environment.

5. Conclusions

1. In contrast to the methods of implicit enumeration, such as B&B, dynamic programming and backtracking, described above approach allows one to create algorithms permitting a priori to evaluate their effectiveness regardless of the specific numerical values of the coefficients and constants of problem (1).
2. The efficiency of these algorithms depends on the environment in which they operate, but their common feature is the increase in efficiency with an increase in the number of variables of the problem being solved.
3. It is easy to make sure that in the case when the number of modules m is equal to two, the third modification of modular enumeration is close to the "meet in the middle" algorithm [16].
4. The gain in the running time presented by (5) is true only for the algorithm 1 above.

References

[1] D. Conte, B. Paternoster, L. Moradi, F. Mohammadi, 2019, Construction of Exponentially Fitted Explicit Peer Methods, International Journal of Circuits, Systems and Signal Processing, Volume 13, pp. 501-506.

[2] T. Atar, T. Mccabe, E. Erdogmus, B. Pulatsu, 2020, Dynamic Analysis of Semi-Circular Dry-Joint Masonry Arches: Small-Scale Experiment and Discrete Element Modeling, WSEAS Transactions on Computers, ISSN / E-ISSN: 1109-2750 / 2224-2872, Volume 19, Art. #19, pp. 137-142.

[3] K. Bibi, T. Feroze, 2020, Discrete Symmetry Transformations of Third Order Ordinary Differential Equations and Applications, WSEAS Transactions on Mathematics, ISSN / E-ISSN: 1109-2769 / 2224-2880, Volume 19, Art. #54, pp. 503-509

[4] J. S. Yeomans, 2019, Water Management using Population-based Simulation-Optimization, WSEAS Transactions on Information Science and Applications, ISSN / E-ISSN: 1790-0832 / 2224-3402, Volume 16, Art. #19, pp. 164-173.

[5] A. H. Land & A. G. Doig. 1960, An automatic method of solving discrete programming problems, Econometrica, Vol. 28, No. 3, pp. 497-520.

[6] C. A. Brown and P. W. Purdom Jr. 1982. An empirical comparison of backtracking algorithms. IEEE PAMI, pp. 309-315.

[7] F. Rossi, P. V. Beek, T. Walsh. 2006. Constraint satisfaction: an emerging paradigm. Handbook of Constraint Programming. Foundations of Artificial Intelligence. Amsterdam: Elsevier. p. 14. ISBN 978-0-444-52726-4.

[8] R. Bellman, 1954. The Theory of Dynamic Programming. The RAND Corporation.

[9] G. Desaulniers, J. Desrosiers, S. Spoorendonk. 2011. Cutting Planes for Branch-and-Price Algorithms. Published online 29 October 2011 in Wiley Online Library (wileyonlinelibrary.com). Wiley Periodicals, Inc.

[10] D. R. Morrison, 2014. New methods for branch-and-bound algorithms. Dissertation submitted for the degree of Doctor of Philosophy in Computer Science in the Graduate College of the University of Illinois at Urbana-Champaign.

[11] V. O. Groppen, and A. A. Berko. 2019. Composite version of B&B algorithm: experimental verification of the efficiency. J. Phys.: Conf. Ser. 1278 012029, pp. 1-8. Available at: https://iopscience.iop.org/issue/1742-6596/1278/1

[12] V.O. Groppen. 2020. Composite Versions of Implicit Search Algorithms for Mobile Computing. Proceedings of the Future Technologies Conference (FTC) 2020, Volume 2, November 5–6, pp. 336 – 348.

[13] V.O. Groppen. 2021, Analysis of the Effectiveness of Composite Versions of Backtracking Algorithms. Conference proceedings, RusAutoConf 2020: Available: Advances in Automation II, Springer, pp 235-244.

[14] L. Caccetta, A. Kulanoot, 2001. Computational Aspects of Hard Knapsack Problems. Nonlinear Analysis, 47: 5547–5558. Available:*doi*:*10.1016/s0362-546x(01)00658-7*

[15] V.O. Groppen. 2021. A New Method Searching Globally Optimal Solutions to Discrete Optimization Problems. Proceedings of the Future Technologies Conference (FTC) 2021, Volume 3, pp 486 – 494.

[16] H. Ellis, S. Sartaj, 1974. Computing partitions with applications to the knapsack problem, *Journal of the Association for Computing Machinery*, **21** (2), pp. 277–292.

Calculation of depressurization coordinate in underground and offshore gas pipelines

G I Kurbatova, V A Klemeshev and V B Filippov
St. Petersburg State University, 7/9 Universitetskaya Emb., St Petersburg 199034, Russia
E-mail: g.kurbatova@spbu.ru, v.klemeshev@spbu.ru, vassiliphilippov@gmail.com

Abstract. The paper introduces a method for calculating the depressurization coordinate of an extended ultra-high pressure gas pipeline from the data measurements of pressure and flow rate at the outlet. The method is based on a mathematical model of a non-isothermal steady single-phase gas mixture flow through a pipeline of constant section on the quasi-linearization method. It allows to calculate the depressurization coordinate for extended underground and offshore ultra-high pressure gas pipelines with high accuracy, while considering varying leakage size and location. Examples of solving problems of practical interest are given. The influence of the flow rate and pressure settings error on the coordinate of depressurization calculating error has been demonstrated.

1. Introduction

An analysis of known accidents in surface and underground gas pipelines leads to the conclusion that most of them are associated with the depressurization of the pipeline. Timely detection and elimination of gas leaks are important tasks. It is especially difficult to locate leaks in underground and offshore pipelines, for which visual observation of the state and the environment is difficult. To date, many approaches have been proposed to solve gas pipeline depressurization site detecting problem. Leak detection methods are usually divided into external and internal [1, 2]. External methods include pipeline inspection, acoustic and optical methods. External methods also include monitoring of the soil adjacent to the gas pipeline, pipeline route satellite surveying, acoustic and optical tests of the gas pipeline state.

Acoustic methods [1, 3, 4] are based on the use of sensors installed along the pipeline. However, the recorded data includes not only the signal of the leak, but also the background signals of the gas pipeline normal operation. One of the problems with acoustic leak detection is separating the signal generated by the leak itself from the background signals. In [4], a method is proposed that allows you to eliminate signal parts that belong to the background noise. The filtered data is processed, and as a result, location of the leak site is determined with high accuracy. The authors of [4] note that this method allows to detect small gas leaks quickly. The effectiveness of the acoustic method depends on the distance between the sensors. For long pipelines, installing and maintaining a large number of sensors can be very costly. Leak detection technique based on acoustic methods alone may be of little value due to the potential for false alarms.

Optical methods [1, 5, 6, 7] are based on fiber optic sensors that can be used to measure various physical and chemical parameters. The optical fiber is placed as close to the pipeline as possible so that it can detect vibrations near the pipeline and other possible changes. When a disturbance occurs, the signal passing through the optical fiber undergoes a change. This change is detected and interpreted in order to determine the type of event that caused the change in the signal. Optical methods can quickly locate and assess the leak, but these methods are expensive.

Thus, using acoustic and optical methods, it is possible to detect the location and size of the gas leak quickly and accurately. However, these methods require expensive equipment to be installed and maintained, resulting in high operating costs. A substantial review of acoustic and optical methods is given, for example, in [1, 8].

Internal (model) methods [9, 10, 11, 12, 13, 14] are based on the gas transportation process computer modeling, i. e., on transportation mathematical models and on computing systems based on these models. Computer modeling has now become a powerful tool in solving many engineering problems in the design and operation of gas pipelines. Computing systems are installed on computers in the dispatch center. They allow real-time monitoring of gas transportation. The disadvantage of model methods is the impossibility of using them to determine the cause of the leak, in addition, model methods are difficult to use when there is more than one leak.

Model methods based on computer simulation are *economical* and *easy to implement*. When using the simulation method, the fact that the change in outlet pressure due to the occurrence of a leak does not occur immediately must be considered. Therefore, in contrast to acoustic and optical methods, the simulation method allows detecting a leak only sometime after its occurrence.

Numerous publications have been dedicated to the gas leakage site location calculation by the modeling method, starting from [11] in the 70s of the last century, and up to the present time, for example, [12].

In this work [12], an extremely simplified mathematical model of gas transportation is used. The model assumes a constant cross-section of the gas pipeline, one-dimensionality, stationary and isothermal processes, constancy of the gas compressibility coefficient, and a negligible value of inertia forces compared to pressure. The system of equations of such a model is easily integrated, which leads to an uncomplicated explicit analytical dependence of the pressure on the coordinate along the pipeline. This model allowed the authors to obtain a simple formula for calculating the coordinates of the gas leakage location based on the values of pressure and flow rate at the inlet and outlet of the gas pipeline. Unfortunately, in most gas pipelines, especially underground and offshore, the processes are essentially non-isothermal and the gas compressibility factor depends both on the changing temperature and on the pressure; therefore, it is possible to directly use the results of [12] in rare cases. However, in terms of methodology, work [12] is of undoubted interest.

A promising direction in the effectiveness study of the various methods for detecting leaks in pipes is presented in the recently published interesting work [15]. It describes in detail a pipeline leak simulator and its study of the effectiveness of acoustic, optical and model methods. The authors conclude that acoustic and optical techniques, combined with simulation techniques, can not only detect, locate and quantify a leak, but also prevent it.

This report [8] made in 2019 at the SPE conference on gas and oil technologies also needs to be noted. In it, along with an overview of papers on gas leak detection using mathematical modeling of processes in pipelines, an overview of papers on new trends associated with the use of measuring probes is given.

In this article, we propose an effective method for calculating the coordinate of the depressurization from the data of pressure and flow rate at the outlet measurements in an extended underground or offshore gas pipeline. This method relates to internal (model) methods. It is based on the Bellman quasi-linearization method [16] and on a mathematical model of a non-isothermal steady natural gas flow through a pipeline, which was thoroughly researched in this book [17]. It possesses several features linked to the choice of the initial approximation and the algorithm for calculating a new iteration in the problem of identifying the coordinate of depressurization which provided its efficiency. The calculation of one problem variant takes no more than a few minutes on an average computer. The proposed method is an effective tool for detecting gas leaks in extended underground and offshore gas pipelines.

Calculation of the gas pipeline depressurization coordinate can be formulated as an inverse problem. A reliable solution of the inverse problem, as it is known, is possible only under the condition that the corresponding direct problem is well studied, namely, an adequate mathematical model of the process has been created and an effective algorithm for calculating the process parameters has been proposed. The model and algorithm for solving the direct problem of the gas mixture flow through an extended offshore gas pipeline at high pressures are given in our book [17]. There is also a solution given to the problem of identifying hard-to-determine model parameters from experimental data, such as the coefficient of hydraulic resistance and the total coefficient of heat transfer. The solution of this problem makes it possible to ensure the adequacy of the gas flow mathematical model while dealing with real problems.

2. Mathematical model of steady gas mixture flow through a gas pipeline of constant circular cross section

Continuity equation:

$$\rho u S = Q^0, \tag{1}$$

Momentum balance:

$$\frac{d}{dz}\left(p + \rho u^2\right) = -\frac{\lambda \rho u|u|}{4R} + \rho g \cos\alpha(z), \tag{2}$$

Heat equation:

$$\rho u c_p \frac{dT}{dz} = \rho u T \left(\frac{\partial V}{\partial T}\right)_p \frac{dp}{dz} + \frac{\lambda \rho u^2 |u|}{4R} + 2\beta \frac{T^* - T}{R}, \tag{3}$$

Equation of state:

$$pV = R_g T Z(p,T), \quad V = 1/\rho. \tag{4}$$

In the system of equations (1)–(4) $Z(p,T)$ is the compressibility factor; V — specific gas volume; R_g — gas constant that depends on the composition of the gas mixture; z, L — coordinate along the axis of the gas pipeline and its length; ρ, p, T, u — density, pressure, temperature and gas velocity, respectively; R, S — inner radius and cross-sectional area of the gas pipeline; g — acceleration of gravity; $\alpha(z)$ — the angle between the direction of gravity and the gas pipeline axis in the zth section; λ — hydraulic resistance coefficient; β — total heat transfer coefficient; c_p — specific heat of the gas mixture at constant pressure coefficient of; Q^0 — gas mass flow; T^* — environment temperature.

In the general case, λ, c_p, β and T^* can be z coordinate functions. Adequacy of the mathematical model (1)–(4) is ensured by the correct choice of following functions, and the compressibility coefficient $Z(p,T)$ dependence from pressure and temperature in the investigated range of pressure, temperature, and gas flow rate changes.

This book [17] describes in detail calculation methods of λ and β, there are also estimates of the influence of their variability on calculated flow characteristics. The specific heat coefficient of the gas mixture c_p (4) can be calculated using one of the known thermodynamic formulas [18] or determined from the corresponding experimental data at the selected equation of state. The system of equations (1)–(4) is supplemented by the boundary condition at the gas pipeline inlet:

$$z = 0, \quad p = p_0, \quad T = T_0. \tag{5}$$

The proposed model covers a wide range of practical tasks. It allows one to take into account the non-isothermal nature of the processes, the complex thermodynamics of the gas mixture at ultra-high pressures, the relief of the route, and the conditions of heat exchange with the environment.

This is how it differs from other works on calculating leak location coordinates based on modeling, for example, from this work [12].

The direct task is to calculate the pressure distributions $p(z)$, temperature $T(z)$ and gas density $\rho(z)$ by the system of equations (1)–(4) under the boundary condition (5). A solution of the system of equations (1)–(4) in a wide range of changes p_0, T_0, Q^0, R, L exists, is unique, and can be obtained numerically, for example, by the Runge – Kutta method.

3. The gas mixture equation of state

It is convenient to use pressure and temperature as independent thermodynamic variables in tasks of gas leak locating. This determines the choice of the equation of state in the form of equation (4). There is some difficulty in the fact that, in a wide range of temperature changes at high pressures, it is not easy to set a single form of the dependence of the compressibility coefficient on p and T for gas mixtures of different compositions. As is known [19, 20], analytical equations of state, such as the Redlich – Kwong, Peng – Robinson, Benedict – Webb – Rubin equations, are more accurate. However, the analytical equations of state are cubic relatively to the specific volume V and to the compressibility factor $Z(p, T)$. This complicates their direct use in model (1)–(3).

When solving real problems, it is most reliable to determine the form of the compressibility coefficient dependence $Z(p, T)$ via experimental data $p - V - T$ obtained for the investigated gas mixture in the investigated range of temperature and pressure changes.

A lot of approximate dependences of the compressibility coefficient $Z(p, T)$ in high-pressure regions have been proposed. For example, in this book [21] the dependence of V.V. Latonov – G.R. Gurevich is proposed, which approximates the well-known graphs of Brown et al. It has the following form:

$$Z_{LG}(p_b, T_b) = (0.4 \lg T_b + 0.73)^{p_b} + 0.1 p_b. \tag{6}$$

Dimensionless temperature T_b and pressure p_b in (6) are equal: $T_b = T/T_r$, $p_b = p/p_r$, p — pressure in MPa, T — gas temperature in K. For pressures not exceeding 30 MPa, authors recommend calculating the pseudocritical parameters p_r, T_r of gas mixtures according to the expressions of Hankinson, Thomas, and Phillips:

$$p_r = 0.006894 \left(709.604 - 58.718 \frac{M}{28.96} \right) \text{ (MPa)},$$

$$T_r = \left(170.491 + 307.44 \frac{M}{28.96} \right) /1.8 \text{ (K)},$$

M (g/mol) — molar mass of a gas mixture.

In this paper, as experimental data for $Z_{\exp} - p - T$, we used calculation data of according to the Redlich – Kwong equation of state for a gas mixture with a predominance of methane, which is typical for many problems of natural gas transportation. (The composition of this mixture of 12 components, critical pressures p_c and mixture temperature T_c, as well as the gas constant R_g are taken from this book [17], their values are given below in the set of parameters (35).) The dependency building algorithm $Z(p, T)$ by experimental data is as follows. To visually determine the nature of $Z_{\exp}$ dependence on pressure and temperature, surface graphs $Z_{\exp}(p, T)$ with a different number of experimental points (from several thousand to a million) are plotted. For the investigated gas mixture in the investigated range of pressure and temperature changes, the analysis of the plotted graphs led to the conclusion that this surface should be approximated by a second-order surface satisfactorily, i. e., the objective function should be of the form:

$$Z(p, T) = a_1 T^2 + a_2 T + a_3 p^2 + a_4 p + a_5 + a_6 T p.$$

Optimal values of $a_1, \ldots, a_6$ coefficients were calculated by the least squares method [22] with a given objective function. There is free source code software for this approach, namely, Gnuplot [23] and LAPACK library [24, 25].

The dependency $Z(\tilde{p}, \tilde{T})$ found via the least squares method algorithm is as follows:

$$Z(\tilde{p}, \tilde{T}) = n_1\tilde{T}^2 + n_2\tilde{T} + n_3\tilde{p}^2 + n_4\tilde{p} + n_5 + n_6\tilde{T}\tilde{p}. \tag{7}$$

Here $\tilde{p}$ and $\tilde{T}$ — dimensionless pressure and temperature related to their critical values p_c and T_c for the given gas mixture, $n_1, \ldots, n_6$ — dimensionless constants, the values of which depend on the composition of the gas mixture.

As is known, the Redlich – Kwong equation in terms of the compressibility factor Z_{RK} [19] is written as:

$$\begin{gathered} Z_{RK}^3 - Z_{RK}^2 + (A - B^2 - B)Z_{RK} - AB = 0, \\ A = \frac{\Omega_a \tilde{p}}{\tilde{T}^{2.5}}, \quad B = \frac{\Omega_b \tilde{p}}{\tilde{T}}, \\ \Omega_a = 0.4274802327, \quad \Omega_b = 0.086640350, \quad \tilde{p} = p/p_c, \quad \tilde{T} = T/T_c. \end{gathered} \tag{8}$$

In Table 1 there is an example of calculating the compressibility coefficients Z (7), Z_{LG} (6), and Z_{RK} (8). The values are present for the given gas mixture at 19 MPa pressure for temperatures of 10, 15, 20, 25, 30 degrees Celsius.

Table 1. The compressibility coefficients values.

T, °C	10	15	20	25	30
Z	0.78969	0.80177	0.81343	0.82466	0.83546
Z_{RK}	0.78888	0.80121	0.81309	0.82448	0.83538
Z_{LG}	0.80146	0.80763	0.81377	0.81987	0.82594

As shown by the calculations, the maximum deviation of compressibility factor Z modulus, calculated by approximation (7), from its exact value Z_{RK} (8) does not exceed 0.00081 in the range of pressure change from 15 to 21 MPa and temperature change from 5 to 35 degrees Celsius.

In additional calculations, the found approximation (7) of the compressibility coefficient was used.

4. The task of locating a gas leak

Let us denote by z_a the dimensional section coordinate of the gas pipeline in which the leakage occurs. In the absence of leakage at the outlet from the gas pipeline in the steady state mode, the flow rate is constant and equal to the flow rate Q^0 at the inlet.

If a leak occurs due to the depressurization of the pipeline, after a while, a new stationary mode is established, in which the pressure and flow rate at the outlet take on new values $P(L)$, $Q(L)$ ($P(L) > P^0(L)$, $Q(L) < Q^0$), in this case, the equalities are:

$$Q = \begin{cases} Q^0, & 0 \le z < z_a, \\ Q(L), & z_a \le z \le L. \end{cases} \tag{9}$$

The difference $(Q^0 - Q(L))$ determines the size of the gas leak. (The values $P(L)$, $P^0(L)$, $Q(L)$, Q^0 are dimensional.)

Let us demonstrate a method for solving the inverse problem of calculating the coordinate z_a for inlet pressures of about 20 MPa, provided that the gas pipeline route is horizontal and values λ, c_p, β and T^* are constant.

Consideration of values λ, c_p, β and T^* dependency from pressure, temperature, and conditions along the route, as well as taking into account the surface pattern of the route, do not introduce any fundamental difficulties. Let us reduce the system of equations (1)–(4), (7) with boundary condition (5) to dimensionless form, setting:

$$\begin{aligned} &\tilde{z} = \frac{z}{l_x}, \quad \tilde{T} = \frac{T}{T_x}, \quad \tilde{p} = \frac{p}{p_x}, \quad \rho_x = \frac{p_x}{R_g T_x Z(p_x, T_x)}, \\ &u_x = \frac{Q^0}{\rho_x S}, \quad \tilde{\rho} = \frac{\rho}{\rho_x}, \quad \tilde{u} = \frac{u}{u_x}, \quad \tilde{T}^* = \frac{T^*}{T_x}. \end{aligned} \tag{10}$$

Let us take independent characteristic values l_x, T_x, p_x as equal: $l_x = L$, $T_x = T_c$, $p_x = p_c$. The resulting dimensionless system of equations is reduced in a known way to a system consisting of two ordinary differential equations, resolved with respect to the derivatives $d\tilde{p}/d\tilde{z}$, $d\tilde{T}/d\tilde{z}$ and to the dimensionless equation of state.

Let us omit the wavy lines at the dimensionless values wherever this does not lead to ambiguity and write the dimensionless model of these processes in the absence of gas leakage.

$$\begin{cases} \rho u = 1, \\ \dfrac{dp}{dz} = \dfrac{f_5 - f_2 f_6}{f_1 - f_2 f_3} = F_1(p, T), \\ \dfrac{dT}{dz} = \dfrac{f_1 f_6 - f_3 f_5}{f_1 - f_2 f_3} = F_2(p, T), \\ \rho = \dfrac{pZ(1,1)}{TZ(p,T)}, \end{cases} \tag{11}$$

$$z = 0, \quad p(0) = \frac{p_0}{p_x}, \quad T(0) = \frac{T_0}{T_x}. \tag{12}$$

Dimensionless functions $f_1, \ldots, f_6$ in system (11) are in the form:

$$\begin{aligned} &f_1(p,T) = 1 + m_1 \frac{T}{p}\left(-\frac{Z}{p} + \frac{\partial Z}{\partial p}\right), \\ &f_2(p,T) = \frac{m_1}{p}\left(Z + T\frac{\partial Z}{\partial T}\right), \\ &f_3(p,T) = -m_3 \frac{T}{p}\left(Z + T\frac{\partial Z}{\partial T}\right), \\ &f_5(p,T) = m_2 \frac{TZ}{p}, \\ &f_6(p,T) = m_5 \left(\frac{TZ}{p}\right)^2 + m_6(T^* - T). \end{aligned} \tag{13}$$

Dimensionless complexes $m_1, \ldots, m_6$ for constant values λ, β, c_p, T^* are equal:

$$\begin{aligned} &m_1 = \frac{(Q^0)^2 R_g T_x}{S^2 p_x^2}, \quad m_2 = -m_1 \frac{\lambda L}{4R}, \quad m_3 = \frac{R_g}{c_p}, \\ &m_5 = \frac{\lambda L (Q^0)^2 R_g^2 T_x}{4RS^2 c_p p_x^2}, \quad m_6 = \frac{2\beta\pi RL}{Q^0 c_p}. \end{aligned} \tag{14}$$

The function $Z(p, T)$ of the dimensionless arguments p and T is written in the form of equation (7). Let us introduce the dimensionless coordinate l of the gas leakage site: $l = z_a/L$.

5. Direct problem in the presence of a gas leak

Let the dimensionless coordinate l of the gas leakage location be known. The gas pipeline is divided into two sections: $z \in [0, l]$ and $z \in [l, 1]$.

In the first section, (11) the system is solved in the region $z \in [0, l]$ with dimensionless complexes (14), calculated for Q^0 with the boundary condition (12). In the second section (11), the system is solved in the region $z \in [l, 1]$, with dimensionless complexes (14), calculated for $Q^0 = Q(L)$, with the boundary condition:

$$z = l : \quad p = p_1(l), \;\; T = T_1(l),$$

in which the values $p_1(l)$, $T_1(l)$ are calculated in the first section.

As a result of solving the direct problem for given values l, p_0, T_0, Q^0 and $Q(L)$, distributions of dimensionless pressure, temperature, and density in the first and second sections are determined.

6. Inverse problem of calculating the dimensionless coordinate l of a gas leak

The inverse problem is solved by the parameter identification method. In this method, it is assumed that the desired parameter is included in the right-hand sides of the differential equations of the system (11). According to [11], let us introduce new dimensionless coordinates y and x, determining them as follows (15):

$$\begin{array}{ll} \textit{in the first section} & \\ y = \dfrac{z}{2l}, & z \in [0,\, l] \rightarrow y \in [0,\, 1/2], \\ \textit{in the second section} & \\ x = 1 - \dfrac{1-z}{2(1-l)}, & z \in [l,\, 1] \rightarrow x \in [1/2,\, 1]. \end{array} \qquad (15)$$

Let us write the differential equations of the dimensionless system (11) in terms of the dimensionless coordinate y for the first section, supplementing them with the condition of the l parameter invariability:

$$\begin{cases} \dfrac{dp_1}{dy} = 2lF_1(p_1, T_1) = \Phi_1(p_1, T_1, l), \\ \dfrac{dT_1}{dy} = 2lF_2(p_1, T_1) = \Phi_2(p_1, T_1, l), \\ \dfrac{dl}{dy} = 0. \end{cases} \qquad (16)$$

Boundary condition:

$$y = 0 : \qquad p_1 = p_0/p_x, \quad T_1 = T_0/T_x. \qquad (17)$$

For the second section, the differential equations of the dimensionless system (11) in terms of the dimensionless coordinate x, supplemented by the condition of the invariability of the parameter l, are in form:

$$\begin{cases} \dfrac{dp_2}{dx} = 2(1-l)\tilde{F}_1(p_2, T_2) = \Psi_1(p_2, T_2, l), \\ \dfrac{dT_2}{dx} = 2(1-l)\tilde{F}_2(p_2, T_2) = \Psi_2(p_2, T_2, l), \\ \dfrac{dl}{dx} = 0. \end{cases} \qquad (18)$$

Boundary condition:

$$x = \frac{1}{2}: \qquad p_2 = p_1(y)\Big|_{y=\frac{1}{2}}, \qquad T_2 = T_1(y)\Big|_{y=\frac{1}{2}}. \tag{19}$$

In formulas (18) $\tilde{F}_1(p_2, T_2)$ and $\tilde{F}_2(p_2, T_2)$ functions are calculated for dimensionless complexes (14), in which the flow rate Q^0 equals $Q(L)$.

In the first and second sections, the l parameter identification problem is solved by the Bellman quasilinearization method [16]. This method in this book [17] solved the task of identifying the λ and β parameters in model (1)–(3) for the given range of problems.

In accordance with Bellman's method, the solution of both system (16) and system (18) is sought iteratively; at each iteration, the solution of systems of nonlinear equations is replaced by the solution of systems of corresponding linearized equations with respect to the previous iteration.

Let us demonstrate the algorithm of the method by the example of solving the system (16) for the first section.

7. Calculation of the dimensionless leak coordinate l at the $(s+1)^{\text{th}}$ iteration

Let us represent system (16) in vector form:

$$\frac{d\mathbf{u}}{dy} = \mathbf{\Phi}(\mathbf{u}), \quad \mathbf{u} = \begin{pmatrix} p_1 \\ T_1 \\ l \end{pmatrix}, \quad \mathbf{\Phi} = \begin{pmatrix} \Phi_1 \\ \Phi_2 \\ 0 \end{pmatrix}. \tag{20}$$

Let the vector of unknowns be found at the st iteration. The vector of unknowns $\mathbf{u}^{s+1}$ at the $(s+1)^{th}$ iteration is determined from the following linearized equation:

$$\frac{d\mathbf{u}^{s+1}}{dy} = \mathbf{\Phi}(\mathbf{u}^s) + J^s(\mathbf{u}^{s+1} - \mathbf{u}^s). \tag{21}$$

Jacobi matrix J^s, which equals:

$$J^s = \begin{pmatrix} \Phi_{1p} & \Phi_{1T} & \Phi_{1l} \\ \Phi_{2p} & \Phi_{2T} & \Phi_{2l} \\ 0 & 0 & 0 \end{pmatrix}, \tag{22}$$

at each point (y) calculated at the values of pressure $p_1^s(y)$, temperature $T_1^s(y)$ and l^s, found at the sth iteration. Elements of the Jacobi matrix are partial derivatives of the functions Φ_1, Φ_2 for p_1, T_1 and l, for example:

$$\Phi_{2T}(y) = \Phi_{2T}(p_1^s(y), T_1^s(y), l^s) = \frac{\partial \Phi_2}{\partial T_1},$$

the designations of the remaining Jacobi matrix (22) elements have a similar meaning.

The simplification of the equation (21) used below is based on the hypothesis of the linear dependence of the quantities p_1^{s+1} and T_1^{s+1} from l^{s+1} parameter in every point (y) of the first section in the accepted range of gas leakage value change:

$$\begin{aligned} p_1^{s+1}(y) &= a_1(y)l^{s+1} + a_3(y), \\ T_1^{s+1}(y) &= a_2(y)l^{s+1} + a_4(y). \end{aligned} \tag{23}$$

The research has confirmed the validity of this hypothesis for both the first and second sections of the gas pipeline. As an illustration of the validity of this hypothesis, Fig. 1 shows the dependence of the pressure in the second section at different points (x) for different values of the parameter l and different flow rates at the outlet and inlet of the gas pipeline (i. e. at different gas leak rates) for the set of parameters presented below (35).

In vector form, linear representations (23) are written in the form:

$$\mathbf{u}^{s+1} = C\mathbf{u}^{s+1} + \mathbf{g}, \quad C = \begin{pmatrix} 0 & 0 & a_1 \\ 0 & 0 & a_2 \\ 0 & 0 & 1 \end{pmatrix}, \quad \mathbf{g} = \begin{pmatrix} a_3 \\ a_4 \\ 0 \end{pmatrix}. \tag{24}$$

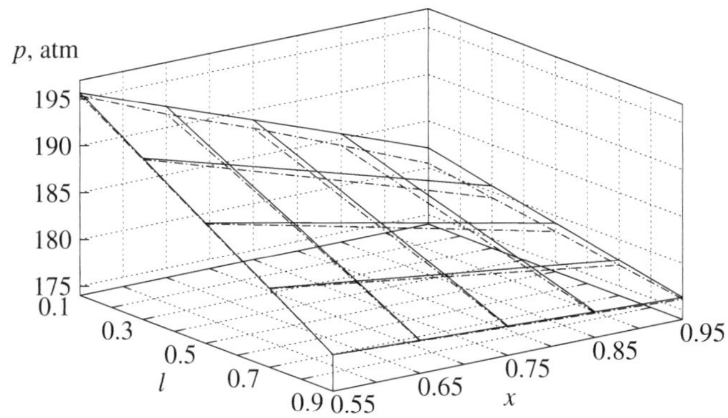

Figure 1. Dependence of pressure on dimensionless l and x at different $Q(L)$; —— $Q(L) = 400\,\mathrm{kg/s}$, — · — $Q(L) = 430\,\mathrm{kg/s}$.

Linearized equation (21) considering representation (24) of the vector $\mathbf{u}^{s+1}$ leads to the following equations for the matrix C and the vector $\mathbf{g}$:

$$\frac{dC}{dy} = J^s C, \tag{25}$$

$$\frac{d\mathbf{g}}{dy} = \mathbf{\Phi}(\mathbf{u}^s) + J^s(\mathbf{g} - \mathbf{u}^s). \tag{26}$$

Matrix equation (25) is equivalent to the system of two ordinary differential equations for the functions $a_1(y)$ and $a_2(y)$. Vector equation (26) is equivalent to the system of two ordinary differential equations for the functions $a_3(y)$ and $a_4(y)$. Boundary conditions for functions $a_1(y)$, $a_2(y)$, $a_3(y)$, $a_4(y)$ follow from the boundary condition (17), which takes place at any iteration, and from the linear representation (23). These boundary conditions are as follows:

$$a_1(0) = 0, \quad a_2(0) = 0, \quad a_3(0) = \frac{p_0}{p_x}, \quad a_4(0) = \frac{T_0}{T_x}. \tag{27}$$

As a result of the numerical solution of the resulting two systems of ordinary differential equations, the functions are calculated $a_1(y)$, $a_2(y)$, $a_3(y)$, $a_4(y)$ in particular, their values at $y = 1/2$.

In the second section, system (18) is solved in a similar way with the boundary condition (19). Linear representation of $p_2^{s+1}(x)$ and $T_2^{s+1}(x)$ functions is used from l^{s+1} in every point (x) of the second section:

$$\begin{aligned} p_2^{s+1}(x) &= b_1(x)\, l^{s+1} + b_3(x), \\ T_2^{s+1}(x) &= b_2(x)\, l^{s+1} + b_4(x). \end{aligned} \tag{28}$$

Functions $b_1(y)$, $b_2(y)$, $b_3(y)$, $b_4(y)$ are calculated under boundary conditions:

$$x = \frac{1}{2}: \qquad b_1 = a_1\left(\frac{1}{2}\right), \quad b_2 = a_2\left(\frac{1}{2}\right), \quad b_3 = a_3\left(\frac{1}{2}\right), \quad b_4 = a_4\left(\frac{1}{2}\right), \tag{29}$$

which follow from conditions (19) valid at any iteration. By calculated $b_1(x)$ and $b_3(x)$ functions the values $b_1(1)$ and $b_3(1)$ at the gas pipeline outlet are defined. From the linear representation (28) for $x = 1$ at the equality at the outlet for the pressure follows:

$$p_2^{s+1}(1) = b_1(1)\, l^{s+1} + b_3(1),$$

which allows to find the l^{s+1} value of the leak site dimensionless coordinate at the $(s+1)^{\text{th}}$ iteration:

$$l^{s+1} = \frac{p_2^{s+1}(1) - b_3(1)}{b_1(1)}. \tag{30}$$

Pressure $p_2^{s+1}(1)$ at every iteration is known and equals: $p_2^{s+1} = P(L)/p_x$, here $P(L)$ — measured pressure at the gas pipeline outlet (dimensional).

Thus, the solution to the problem of calculating the dimensionless coordinate of the gas leak at the $(s+1)^{\text{th}}$ iteration is given by this formula (30). The condition for the iterative process conclusion is the fulfillment of the inequality:

$$|l^{s+1} - l^s| \leq \varepsilon, \tag{31}$$

in which the value ε is determined in result of a computational experiment. To establish the practical convergence of the iterative process, it is sufficient to make sure that with an increase in the iteration number, the value δl^{s+1} decreases, which equals:

$$\delta l^{s+1} = \left| \frac{l^{s+1} - l^s}{l^s} \right|.$$

In contrast to [11], to calculate the pressure and temperature distributions at the $(s+1)^{\text{th}}$ iteration, the proposed method uses the solution of the direct problem with the found coordinate of the leak site l^{s+1}. The solution of the identification problem for the parameter l at the next iterations is carried out similarly to the presented algorithm.

8. Initial approximation

As is known, when solving problems by the iterative method, the choice of the initial approximation is of great importance. In present range of gas leak detection problems, the numerical solution of the direct problem (paragraph 5) at a given location of the gas leak is not difficult, the calculation of one variant takes several seconds on a medium-power personal computer (for example, computers with a 4th Gen Intel Core i5 processor on Haswell architecture).

This allows to numerically form a dependence when making calculations for a specific gas pipeline in the investigated range of pressure and flow rate changes at the outlet:

$$p_2 < p_L < p_1, \quad q_2 < q_L < 1, \tag{32}$$

p_2 — minimum dimensionless outlet pressure at a given dimensionless outlet flow q_L, which is realized at $l \to 1$, p_1 — maximum dimensionless outlet pressure, which is realized at $l \to 0$ (l — dimensionless coordinate of the gas leak, $q_L = Q(L)/Q^0$, q_2 — minimum dimensionless outlet flow rate, which is realized at maximum gas leakage).

For many problems in the investigated range (32) of flow rate changes at the output, the dependence $l(p_L)$ is practically linear. This is the case for the set of parameters (35) of the example given below. Linearity of dependence $l(p_L)$ allows us to propose the following algorithm for calculating the gas leak coordinate l^0 in the initial approximation.

From the solution of the direct problem (paragraph 5) for a given flow rate at the outlet, the values of dimensionless pressures p_1 and p_2 are calculated at $l = 0.05$ and $l = 0.95$ respectively.

If the specified value of the dimensionless pressure p_L at the gas pipeline outlet satisfies the inequality $p_L \leq p_2$, the value l^0 is assigned the value: $l^0 := 0.95$, if p_L satisfies inequality $p_L \geq p_1$, l^0 is assigned the value: $l^0 := 0.05$. For values p_L, which satisfy inequality (32), the value l^0 is calculated by the formula:

$$l^0 = l_1 + \frac{p_1 - p_L}{p_1 - p_2}(l_2 - l_1), \tag{33}$$

in which the values l_1, l_2, p_1, p_2 are:

$$l_1 = 0.05, \quad l_2 = 0.95, \quad p_1 = p_L(0.05), \quad p_2 = p_L(0.95). \tag{34}$$

In equalities (34), all values are dimensionless, value p_L is also dimensionless, equal to the ratio of the dimensional pressure P_L at the outlet to the dimensional characteristic pressure p_x.

To calculate the distributions of dimensionless pressure and temperature in the initial approximation, the direct problem is solved at $l = l^0$ by the system (16) with boundary conditions (17) and by the system (18) with boundary conditions (19).

9. Examples of solving test problems

The following (typical for real natural gas transportation problems) parameters of a steady flow along an extended gas pipeline were set:

$$\begin{array}{llll} R = 0.5\,\text{m}, & L = 300\,\text{km}, & Q^0 = 450\,\text{kg/s}, & \lambda = 0.0087, \\ \beta = 11.0\,\text{W}/(\text{m}^2\text{K}), & p_0 = 200\,\text{atm} & T_0 = 308.15\,\text{K}, & \\ p_c = 4.59776\,\text{MPa}, & T_c = 193.69853\,\text{K}, & T^* = 278.15\,\text{K}, & R_g = 493.58891\,\text{m}^2/(\text{s}^2\,\text{K}). \end{array} \tag{35}$$

(The critical gas mixture parameter values are given in (35) with an accuracy of five digits after the decimal point; in all calculations, a significantly higher accuracy was used.) Parameter values p_c, T_c, R_g in (35) correspond to a gas mixture with a predominance of methane; the composition of the mixture is given in the book [17].

The calculations used the average value of the heat capacity coefficient in the investigated range of pressure and temperature changes. In the examples below, the "experimental" new flow rate value at the outlet $Q(L)$ was set as well as the exact leak site coordinate z_a (that was be further defined). From this data, for this set of parameters (35) and from the solution of the direct problem, the dimensional pressure $P(L)$ atm at the outlet was calculated, which was then taken as an "experimental". The flow rate at the outlet value of the and the leak coordinate of z_a varied in given examples.

For a given flow rate at the outlet, the dimensional pressure values at the gas pipeline outlet at $l_1 = 0.05$ and $l_2 = 0.95$ were calculated from the solution of the direct problem (paragraph 5). At $Q(L) = 400$ kg/s these pressures are:

$$P_1 = 180.684\,\text{atm}, \quad P_2 = 175.865\,\text{atm}. \tag{36}$$

The following designations are adopted:
$z_a^{(s)}$, m — dimensional coordinate of the leak site calculated at the sth iteration; $\delta l^s = |l^s - l^{s-1}|$ — the absolute value of difference between the dimensionless leak coordinates at the sth and $(s-1)^{\text{th}}$ iterations; $\delta_z = |z_a^{(s)} - z_a|$ — the absolute value of the difference between the dimensional leak coordinate at the sth iteration and the given value z_a (m).

Example 1. $Q(L) = 400$ kg/s, $l = 0.01$, $z_a = 3$ km, $P(L) = 180.937$ atm, $P(L) > P_1 \to l^0 := 0.05$; the coordinate of the leak site $z_a^{(3)} = 2986.0$ m, the coordinate of the leak site calculating accuracy $\delta_z = 14.0$ m (table 2).

Example 2. $Q(L) = 400$ kg/s, $l = 0.35$, $z_a = 105$ km, $P(L) = 179.001$ atm, calculating l^0 by formula (33); the coordinate of the leak site $z_a^{(2)} = 104.995$ km, the coordinate of the leak site calculating accuracy $\delta_z = 5.2$ m (table 3).

Example 3. $Q(L) = 400$ kg/s, $l = 0.55$, $z_a = 165$ km, $P(L) = 177.951$ atm, calculating l^0 by formula (33); the coordinate of the leak site $z_a^{(2)} = 164.993$ km, the coordinate of the leak site calculating accuracy $\delta_z = 6.6$ m (table 4).

Example 4. $Q(L) = 400$ kg/s, $l = 0.985$, $z_a = 295.50$ km, $P(L) = 175.682$ atm, $P(L) < P_2 \rightarrow l^0 := 0.95$; the coordinate of the leak site $z_a^{(2)} = 295.5$ km, the coordinate of the leak site calculating accuracy $\delta_z = 8.5$ m (table 5).

Table 2. Calculation for $l = 0.01$.

s	0	1	2	3
l_s	0.05	0.009	0.010	0.010
δl_s	–	0.041	0.0009	$0.6 \cdot 10^{-6}$

Table 3. Calculation for $l = 0.35$.

s	0	1	2
l_s	0.3643	0.34998	0.35000
δl_s	–	0.01431	$0.347 \cdot 10^{-6}$

Table 4. Calculation for $l = 0.55$.

s	0	1	2
l_s	0.560	0.54998	0.55000
δl_s	–	0.01040	$0.434 \cdot 10^{-6}$

Table 5. Calculation for $l = 0.985$.

s	0	1	2
l_s	0.950	0.98503	0.98502
δl_s	–	0.03503	$0.428 \cdot 10^{-7}$

The given examples, as well as numerous calculations for other values of the outlet flow rate and other leak site variants, led to the conclusion that there is a practical convergence of the iterative method. For example, from the calculation data of example 1 it follows that with increasing s value $\delta^{(s)} = \left|\frac{l^s - l^{s-1}}{l^{s-1}}\right|$ decreases as follows:

$$\delta^{(1)} = 0.819, \quad \delta^{(2)} = 0.098, \quad \delta^{(3)} = 0.000061.$$

Influence of the error in setting the pressure $P(L)$ ***at the gas pipeline outlet.*** Let us demonstrate the influence of this error on the coordinate of the leak site calculating accuracy. Assuming that: $Q(L) = 400$ kg/s, $l = 0.35$. From the solution of the direct problem (example 2) it follows: $P(L) = 179.001$ atm.

Let the measured pressure $\tilde{P}(L)$ differ from $P(L)$ by 0.005 atm, namely, $\tilde{P}(L) = 179.006$ atm. Calculation by formula (33) shows that the initial approximation l^0 at this pressure $\tilde{P}(L)$ equals: $l^0 = 0.36336$. The iterative process in this example converges at the third iteration, while $l^3 = 0.34904$, $\delta_z = 278.6$ m.

Influence of the error in setting the flow rate $Q(L)$ ***at the gas pipeline outlet.*** Assuming that flow rate is set with accuracy 1%, let the measured flow rate be: $Q(L) = 396$ kg/s, the dimensionless leak coordinate is: $l = 0.35$. When the flow rate changes, the values P_1 and P_2 (36), namely, $P_1 = 183.458$ atm, $P_2 = 178.215$ atm. Following the above algorithm for the initial approximation choosing, we find that: $l^0 = 0.36406$. The iterative process in this example converges at the third iteration, while $l^3 = 0.35004$, $\delta_z = 13.7$ m.

IOP Publishing
2162 (2022) 012023 doi:10.1088/1742-6596/2162/1/012023

Gas leakage reduction. The numerical experiments have shown that with constant parameters of the numerical solution, a decrease in the leakage value increases the leak site calculating error. For example, with an increase in the gas flow rate at the outlet from 400 kg/s to 425 kg/s, the amount of gas leakage $(Q^0 - Q(L))$ decreases twice at $Q^0 = 450$ kg/s. The leak site coordinates calculating error at $Q(L) = 425$ kg/s equals: $\delta_z = 10.5$ m, which is about twice as much as $\delta_z = 5.2$ m shown in Example 2. This leads to the following conclusion: *To ensure constant accuracy of a gas leak coordinates calculation, it is necessary to use the step along the z coordinate in the numerical solution of the identification problem, which is reduced by about the same factor as the amount of leakage decreases.*

In each real problem the necessary step size Δz in the numerical solution and the accuracy value ε in the iterative process end condition are determined in advance using computational experiments. That ensures the required accuracy of calculating the gas pipeline depressurization coordinate.

The proposed leak detection technique has limitations that are typical for all model methods: it allows calculating the leak site location site only some time after its occurrence, and it does not allow to determine the cause of the leak.

Generalization of the proposed technique for the presence of more than one leak requires additional research.

10. Conclusion

The given examples, as well as solutions of numerous variants of the gas pipeline depressurization coordinate calculation problem for other parameter values, led to the following conclusions.

The proposed technique allows to calculate promptly the depressurization coordinate for extended underground and offshore ($L > 500$ km) ultra-high pressure ($p > 20$ MPa) gas pipelines with an accuracy of tens of meters in a wide leakage size and location range.

The proposed algorithm for choosing the initial approximation ensures a high convergence rate of the iterative process, which allows calculating leakage site location in real time.

The greatest influence on the calculation accuracy is caused by the pressure measurement accuracy.

As shown in this paper, the proposed technique allows calculating the depressurization coordinate even with a decrease in leakage.

Studies have shown that in the case of small leaks, approximately 0.4%, the leak site location can be calculated with an accuracy of up to tens of meters, if the equipment provides high accuracy in measuring the pressure and flow rate at the outlet.

The created calculation programs set is simple to implement, the proposed technique does not require an installation of expensive equipment along the gas pipeline and allows to calculate a leakage site location in extended underground and offshore gas pipelines with high accuracy.

References

[1] Murvay P S and Silea I 2012 A survey on gas leak detection and localization techniques *J. of Loss Prevention in the Process Industries* **25** 966–973
[2] Nicholas E, Carpenter P, Henrie M Hung D Kundert C May 2017 *Proc. of the PSIG Annual Meeting (Atlanta, Georgia)* PSIG-1718
[3] Meng L, Yuxing L, Wuchang W and Juntao F 2012 Experimental study on leak detection and location for gas pipeline based on acoustic method *J. of Loss Prevention in the Process Industries* **25** 90–102
[4] Qingqing X, Laibin Z and Wei L 2013 Acoustic detection technology for gas pipeline leakage *Process Saf. Environ. Prot.*, **91**(4) 253–261
[5] Tanimola F and Hill D 2009 Distributed fibre optic sensors for pipeline protection *J. Nat. Gas Sci. Eng.* **1** 134–143
[6] Rajeev P, Kodikara J, Chiu W K and Kuen T 2013 Distributed optical fibre sensors and their applications in pipeline monitoring *Key Engineering Materials* **558** 424–434

[7] Thodi P, Paulin M, Forester L, Burke J and Lanan G February 2014 *Proc. of the OTC Arctic Technology Conference (Houston, Texas)* OTC-24589-MS
[8] Thiberville C, Wang Y, Waltrich P, Williams W and Kam S I 2019 *SPE Gas & Oil Technology Showcase and Conference (Dubai)* (Society of Petroleum Engineers)
[9] Lapteva T I and Mansurov M N 2006 Detection of leaks with unsteady flow in pipes *Neftegazovoe delo* **2**
[10] Liu A E 2008 Overview: Pipeline Accounting and Leak Detection by Mass Balance, Theory and Hardware Implementation *Quantum Dynamics, inc.* **28** 1–15
[11] Vasilev O F, Bondarev E A, Voevodin A F and Kanibolotskii M A 1978 *Non-Isothermal Gas Flow in Pipes* (Novosibirsk: Nauka) p 128
[12] Obibuike Ubanozie Julian, Ekwueme Stanley Toochukwu, Ohia Nnaemeka Princewill, Igwilo Kevin Chinwuba, Onyejekwe Ifeanyi Michael and Igbojionu Anthony Chemazu 2019 Analytical Model for the Estimation of Leak Location in Natural Gas Pipeline. *Int. J. of Oil, Gas and Coal Engineering* **7**, No. 4, 95–102
[13] Sun L 2012 Mathematical modeling of the flow in a pipeline with a leak *Math. Comput. Simul.*, **82** 2253–2267
[14] Fukushima K, Maeshima R, Kinoshita A, Shiraishi H and Koshijima I 2000 Gas pipeline leak detection system using the online simulation method *Computers & Chemical Engineering* **24** 453–456
[15] Luis Ángel Alarcón-Ramos and Pedro Pablo González Pérez 2021 A Useful Application of a Simulator for Leak Events in Pipelines *Int. J. of Applied Mathematics, Computational Science and Systems Engineering* **3**, 76–84
[16] Bellman R E and Kalaba R E 1965 *Quasilinearization and Non-linear Boundary Value Problems* (New York: American Elsevier Publishing Company Inc) p 218
[17] Kurbatova G I, Ermolaeva N N, Filippov V B and Filippov K B 2020 *Design of Gas Pipelines in the Northern Seas* (Saint-Petersburg: Lan) p 352
[18] Sivukhin D V 2006 *Thermodynamics and Molecular Physics (General Physics Course* vol 2*)* (Moscow: Fizmatlit) p 544
[19] Reid R, Prausnitz J and Sherwood T 1977 *The Properties of Gases and Liquids* (New York: McGraw-Hill) p 548
[20] Modisette J L October 28 2000 *Proc. of the PSIG Annual Meeting (Savannah, Georgia)* PSIG-0008 pp 1–21
[21] Gurevich G R and Brusilovsky A I 1984 *A Reference Manual for Calculating the Phase State and Properties of Gas-Condensate Mixtures* (Moscow: Nedra) p 264
[22] Norman R D, Harry S 1998 *Applied regression analysis* 3rd Edition (New York: Wiley Series in Probability and Statistics) p 736
[23] Gnuplot Available at: http://gnuplot.info/ (Accessed 02 July 2021)
[24] LAPACK Available at: https://www.netlib.org/lapack (Accessed 02 July 2021)
[25] Demmel J W 1997 *Applied numerical linear algebra* (Philadelphia: SIAM) p 416

Biometric Identification based on the Multidimensional Transient Functions of the Human Oculo-Motor System

Vitaliy Pavlenko[1], Tetiana Shamanina[1] and Vladyslav Chori[1]

[1] Odessa National Polytechnic University Shevchenko av. 1, Odessa, 65044, Ukraine

Abstract. Recently, identification systems that use human biometric characteristics in solving the problem of access to information systems are becoming more common. The paper proposes a new method of biometric identification of users of computer systems based on the definition of integral Volterra models of the human oculomotor system (OMS) according to experimental research "input-output" using innovative technology of eye tracking. Developed in the Python IDLE programming environment software to identify OMS. Experimental studies of the OMS of two individuals were performed. Based on the data obtained using the TOBII PRO TX300 eye tracker, the transient functions of the first, second and third orders of the OMS are determined. There is a significant difference in the diagonal intersections of the transition functions of the second and third orders of the two individuals.

1. Introduction

The task of user authentication is quite important in today's world. The most popular method of solving it is input user login and password until now. Password protection alone is not reliable, as it cannot provide the required level of protection without the use of other security mechanisms. Identification systems that use human biometric characteristics to solve the problem of access to information systems are becoming more common [1]-[4].

Eye tracking technology is widely used in research and educational research [5]. A variety of software has been developed and various technologies have been used in research [5]-[7]. The human visual system has unique properties that are easy to identify. Eye movements largely depend on both physical and physiological characteristics, which makes forgery impossible [8], [9]. Computational intelligence approaches are widely adopted in establishing identity based on biometrics [10]-[15]. When tracking eye movements, two positions are suggested for identification. The first is the position of the eye at a specific point on the display. The second is the moment of movement of the eyes when moving the gaze from one point to another. The computer evaluates the data obtained and determines the unique characteristics for each case [16]-[23].

2. The Aim of the Research

Increase the efficiency (reliability) of information protection on the computer through the development of hardware and software identification of human oculo-motor system (OMS) based on nonlinear dynamic model and data of experimental input-output research using innovative eye tracking technology. The Volterra model in the form of multidimensional transient functions (MTF) is used for identification [24].

2.1 The object of research

The process of biometric identification of a computer user on the basis of eye tracking data in dynamics - responses to given test visual stimuli (the process of eye tracking).

2.2 The subject of research

Software tools for constructing the Volterra model – evaluation of multidimensional transient functions of OMS according to the eye-tracking data, determination based on transient functions of informative features and construction of defining rules of optimal classification.

3. Volterra Model and the Method of the Identification OMS

The «input-output» relationship for a nonlinear dynamic system (NDS) with an unknown structure with one input and one output is represented by a discrete Volterra series in the form [25]:

$$y[m] = \sum_{n=1}^{\infty} y_n[m] = \sum_{k_1=0}^{m} w_1[k_1]x[m-k_1] + \sum_{k_1,k_2=0}^{m} w_2[k_1,k_2]x[m-k_1]x[m-k_2] + \\ + \sum_{k_1,k_2,k_3=0}^{m} w_3[k_1,k_2,k_3]x[m-k_1]x[m-k_2]x[m-k_3] + \dots . \quad (1)$$

where $w_1[k_1]$, $w_2[k_1,k_2]$, $w_3[k_1,k_2,k_3]$ are Volterra kernels (discrete weight functions of the 1st, 2nd and 3rd orders); $x[m]$, $y[m]$ are stimulus (input) and response (output) signals (function) of the system; $y_n[m]$ is partial components of the response (convolution of n-th order sequences $w_n[k_1,\dots,k_n]$ and $x[m]$); m is a discrete time variable.

The Volterra series is replaced by a polynomial and is usually limited to the first few terms of the series in practice. In this study we limited ourselves to the first three terms of the series (we chose the degree of the Volterra polynomial model $N = 3$).

The identification problem consists in choosing test signals $x[m]$ and developing an algorithm that allows one to identify the partial components $y_n[m]$, (n=1,2,3) based on the responses $y[m]$ and determine the multidimensional Volterra kernels from them: $w_1[k_1]$, $w_2[k_1,k_2]$, $w_3[k_1,k_2,k_3]$.

Test step signals are used to identify OMS. If the test signal $x[m]=\theta[m]$, where $\theta[m]$ is a unit function (Heaviside function), then the partial components of the response $y_1[m]$, $y_2[m]$, $y_3[m]$ are equal first order transition functions and diagonal sections of second and third order transition functions:

$$y_1[m] = h_1[m] = \sum_{k_1=0}^{m} w_1[m-k_1],$$

$$y_2[m] = h_2[m,m] = \sum_{k_1,k_2=0}^{m} w_2[m-k_1,m-k_2], \quad (2)$$

$$y_3[m] = h_3[m,m,m] = \sum_{k_1,k_2,k_3=0}^{m} w_3[m-k_1,m-k_2,m-k_3].$$

In this case, the responses of the Volterra model of the OMS are calculated based on the expression:

$$\tilde{y}_i[m] = a_i\hat{y}_1[m] + a_i^2\hat{y}_2[m] + a_i^3\hat{y}_3[m], \quad i = \overline{1,N}, \quad (3)$$

where $\hat{y}_1[m] = \hat{h}_1[m]$, $\hat{y}_2[m] = \hat{h}_2[m,m]$, $\hat{y}_3[m] = \hat{h}_3[m,m,m]$ are obtained estimates of the partial components of the model (MTF).

The research uses an approximation identification method [26]. The approximation method of identification in domain time is based on the allocation of the n-th partial component in the OMS response by constructing linear combinations of responses to test signals with different amplitudes [26], [27].

Let at system input test signals are given successively $a_1x[m]$, $a_2x[m]$,…,$a_Nx[m]$ (N is approximation model order, a_1, a_2,…,a_N are different real numbers, which satisfy the term $|a_j|\leq 1$ for $\forall j$=1,2,...,N; $x[m]$ is arbitrary function). Then the linear combination of the OMS responses with the coefficients c_j is amount to the n-th partial component of the OMS response to the input signal $x[m]$. In this case, a methodical error arises in the selection of the n-th partial component, due to the partial components of the OMS response of higher orders $n>N$:

$$\sum_{j=1}^{N} c_j y(a_j x[m]) = y_n(x[m]) + \sum_{j=1}^{N} c_j \sum_{n=N+1}^{\infty} y_n(a_j x[m]), \tag{4}$$

where

$$y_n(x[m]) = y_n[m];$$

$$y(a_j x[m]) = \sum_{n=1}^{\infty} a_j^n \sum_{k1=0}^{m} \dots \sum_{kn=0}^{m} w_n[k_1,\dots,k_n] \prod_{i=1}^{n} x[m-k_i];$$

if c_j is real coefficients such that

$$\mathrm{A}_N \mathrm{c} = \mathrm{b}, \tag{5}$$

where

$$\mathrm{A}_N = \begin{bmatrix} a_1 & a_2 & \dots & a_N \\ a_1^2 & a_2^2 & \dots & a_N^2 \\ \dots & \dots & \dots & \dots \\ a_1^N & a_2^N & \dots & a_N^N \end{bmatrix}, \quad \mathrm{c} = \begin{bmatrix} c_1 \\ c_2 \\ \dots \\ c_N \end{bmatrix}, \quad \mathrm{b} = \begin{bmatrix} b_1 \\ b_2 \\ \dots \\ b_N \end{bmatrix},$$

and $b_l = 1$ at $l = n$ and $b_l = 0$ at $l \neq n$, $\forall l \in \{1,2,\dots,N\}$.

For example, for N=2:

$$\begin{bmatrix} a_1 & a_2 \\ a_1^2 & a_2^2 \end{bmatrix} \cdot \begin{bmatrix} c_1^{(1)} \\ c_2^{(1)} \end{bmatrix} = \begin{bmatrix} 1 \\ 0 \end{bmatrix},$$

$$\begin{bmatrix} a_1 & a_2 \\ a_1^2 & a_2^2 \end{bmatrix} \cdot \begin{bmatrix} c_1^{(2)} \\ c_2^{(2)} \end{bmatrix} = \begin{bmatrix} 0 \\ 1 \end{bmatrix};$$

for N=3:

$$\begin{bmatrix} a_1 & a_2 & a_3 \\ a_1^2 & a_2^2 & a_3^2 \\ a_1^3 & a_2^3 & a_3^3 \end{bmatrix} \cdot \begin{bmatrix} c_1^{(1)} \\ c_2^{(1)} \\ c_3^{(1)} \end{bmatrix} = \begin{bmatrix} 1 \\ 0 \\ 0 \end{bmatrix},$$

$$\begin{bmatrix} a_1 & a_2 & a_3 \\ a_1^2 & a_2^2 & a_3^2 \\ a_1^3 & a_2^3 & a_3^3 \end{bmatrix} \cdot \begin{bmatrix} c_1^{(2)} \\ c_2^{(2)} \\ c_3^{(2)} \end{bmatrix} = \begin{bmatrix} 0 \\ 1 \\ 0 \end{bmatrix},$$

$$\begin{bmatrix} a_1 & a_2 & a_3 \\ a_1^2 & a_2^2 & a_3^2 \\ a_1^3 & a_2^3 & a_3^3 \end{bmatrix} \cdot \begin{bmatrix} c_1^{(3)} \\ c_2^{(3)} \\ c_3^{(3)} \end{bmatrix} = \begin{bmatrix} 0 \\ 0 \\ 1 \end{bmatrix}.$$

The evaluation of transient functions can be set in general as follows:

$$\hat{h}_n^{(N)}[m,\dots,m] = \hat{y}_n[m] = \sum_{j=1}^{N} c_j^{(n)} y(a_j \theta[m]) = \\ = c_1^{(n)} y_{a_1}[m] + c_2^{(n)} y_{a_2}[m] + \dots + c_N^{(n)} y_{a_N}[m], \quad n = 1,2,\dots,N, \tag{6}$$

where $y_{a_j}[m] = y(a_j \theta[m])$ – OMS response to a test signal with an amplitude a_j.

Some cases of recording transient functions for N=3 and $\forall m \geq 0$:

$$\hat{h}_1[m] = c_1^{(1)} y_{a_1}[m] + c_2^{(1)} y_{a_2}[m] + c_3^{(1)} y_{a_3}[m],$$

$$\hat{h}_2[m,m] = c_1^{(2)} y_{a_1}[m] + c_2^{(2)} y_{a_2}[m] + c_3^{(2)} y_{a_3}[m], \quad (7)$$
$$\hat{h}_3[m,m,m] = c_1^{(3)} y_{a_1}[m] + c_2^{(3)} y_{a_2}[m] + c_3^{(3)} y_{a_3}[m].$$

Model responses at N=3 to step test signals with amplitudes a_1, a_2, a_3:

$$\tilde{y}_1[m] = a_1\hat{h}_1^{(3)}[m] + a_1^2\hat{h}_2^{(3)}[m,m] + a_1^3\hat{h}_3^{(3)}[m,m,m],$$
$$\tilde{y}_2[m] = a_2\hat{h}_1^{(3)}[m] + a_2^2\hat{h}_2^{(3)}[m,m] + a_2^3\hat{h}_3^{(3)}[m,m,m], \quad (8)$$
$$\tilde{y}_3[m] = a_3\hat{h}_1^{(3)}[m] + a_3^2\hat{h}_2^{(3)}[m,m] + a_3^3\hat{h}_3^{(3)}[m,m,m],$$

model responses at N=2:

$$\tilde{y}_2[m] = a_1\hat{h}_1^{(2)}[m] + a_1^2\hat{h}_2^{(2)}[m,m],$$
$$\tilde{y}_2[m] = a_2\hat{h}_1^{(2)}[m] + a_2^2\hat{h}_2^{(2)}[m,m]. \quad (9)$$

4. Results

To identify the OMS in the form of MTF according to the data eye tracking program Signal Manager was created to generate test visual stimuli on the computer monitor screen. The obtained physiological features of the OMS, in experiments on eye movement tracking, step signals (bright dots) with different distances a_j (j = 1,2,…,N; N is number of experiments) from the starting position are used. Thus, visual stimuli can be considered as functions $x_j[m] = a_j\theta[m]$, where $\theta[m]$ is a unit Heaviside function. With the help of an eye tracker, the responses of the OMS are recorded, which are used to determine the MTF [28].

In the studies of each respondent, three experiments were performed sequentially for the three amplitudes a_1, a_2, a_3 (N=3) of the test signals in the horizontal direction. The distance between the starting position and the test stimuli is: (1/3)*lx*, (2/3)*lx* and (1.0)*lx*, where *lx* is the length of the monitor screen. Coordinates of the starting position (x=0, y= (1/2)*ly*), where *ly* is the width of the monitor screen. Experimental studies of OMS were conducted using high-tech equipment – eye tracker TOBII PRO TX300 (300 Hz) [29].

In Figure 1 and Figure 2 OMS responses to test visual stimuli with amplitudes a_1, a_2 and a_3. in two individuals are shown that were obtained on different days and at different times of the day.

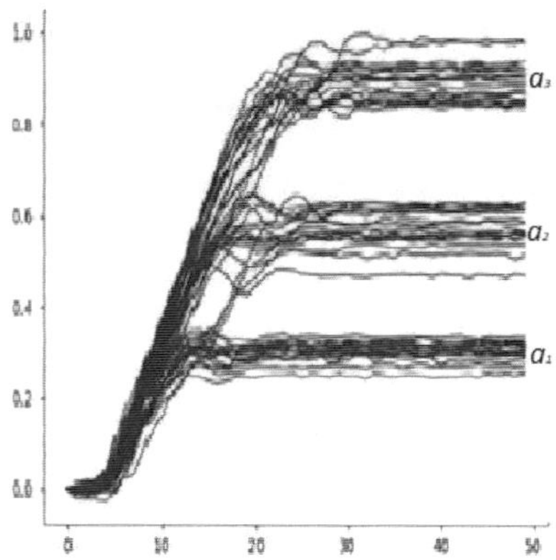

Figure 1. OMS responses to visual stimuli with amplitudes a_1, a_2, a_3 for 1st student

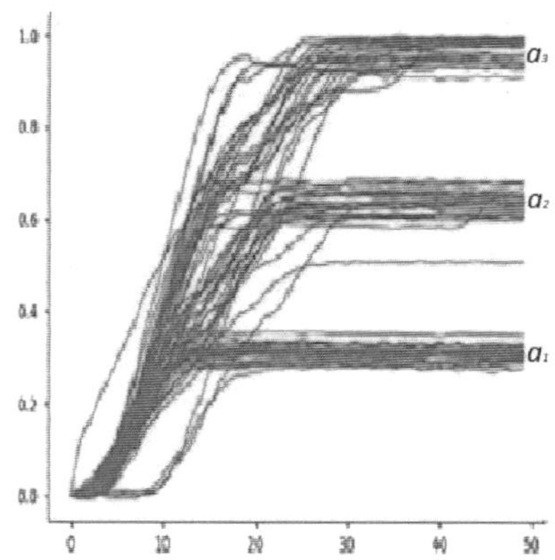

Figure 2. OMS responses to visual stimuli with amplitudes a_1, a_2, a_3 for 2nd student

According to the averaged data of OMS responses to visual stimuli (Figure 3), the transient functions of OMS when using Volterra models of different degree N (N=1,2,3) were determined. Graphs of transient functions for two individuals based on the model at N=1 are presented in Figure 4, at N=2 – in Figure 5 and at N=3 – in Figure 6. As it can be seen from Figure 4 – Figure 6, the obtained transient functions of the 1st order almost coincide for two individuals. However, the diagonal sections of the transient functions of the 2nd (Figure 5) and third (Figure 6) orders in two individuals change significantly in values, therefore, can be effectively used as a source of primary data in building a system of recognition of individuals with application of machine learning.

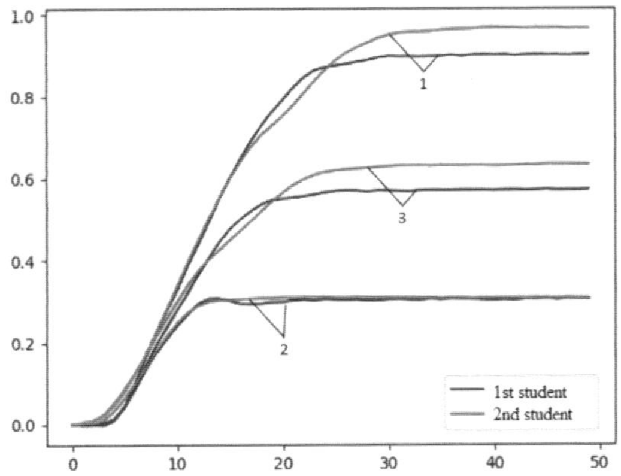

Figure 3. Average responses of OMS of two individuals

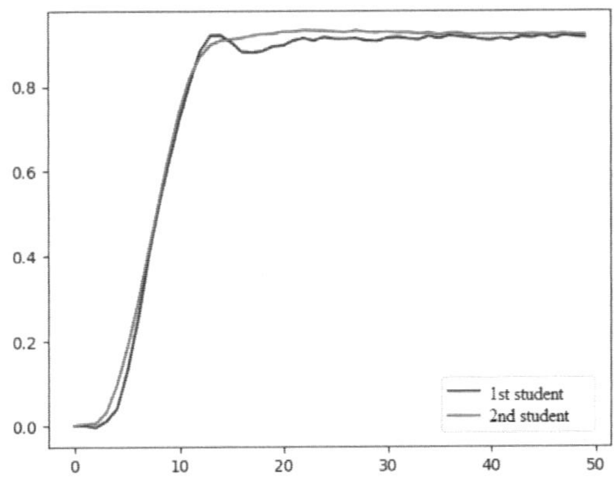

Figure 4. Transient functions 1st orders of two individuals

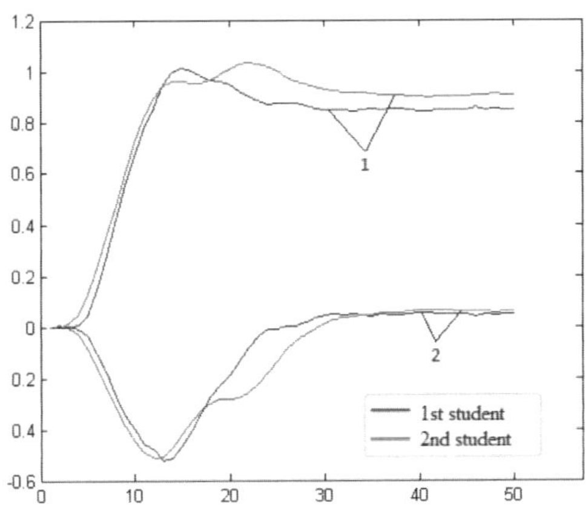

Figure 5. Transient functions: *1* – 1st; *2* – 2nd orders of two individuals

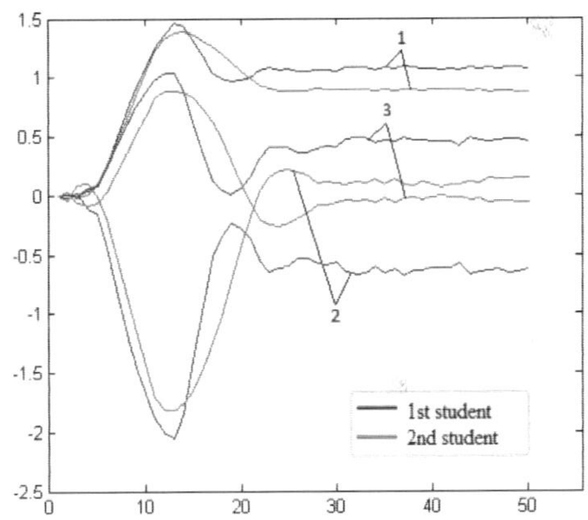

Figure 6. Transient functions: *1* – 1st; *2* – 2nd; *3* – 3rd orders of two individuals

Received responses with the help of calculations on models at $N = 3$ from various amplitudes of test signals. Graphs these are presented in comparison with similar responses OMS for premier and second of the students in Figures 7 and 8, respectively.

4.1 Building a classifier of the individuals

For identity recognition of the individuals based on the OMS nonlinear dynamical model conducted research:

4.1.1 Building a feature space for designing classifier of the individual with using machine learning.

Classifiers construction with using statistical methods of learning the pattern recognition based on the data obtained using eye tracking technology.

The analysis of the reliability of personality recognition in the space of features calculated on the basis MTF consists in forming various combinations of features and evaluating their informativeness based on the classification results on the data sample under study using criterion for the probability of correct recognition (PCR). Thus, all possible pairs of features were investigated by the exhaustive search method. For studies of the informativeness, the features presented in the table (Appendix) are used.

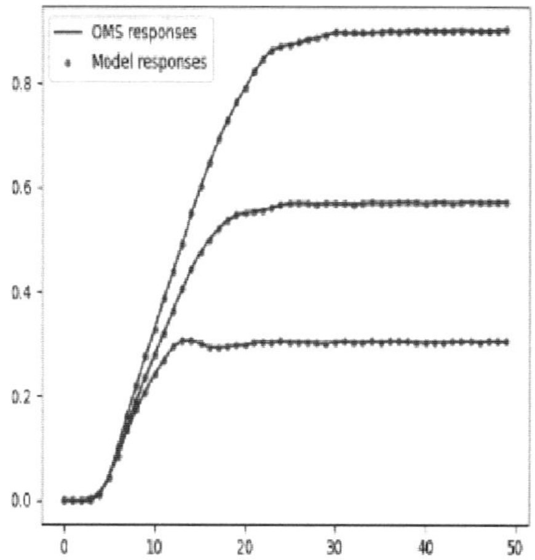

Figure 7. Average responses of the OMS and of the Volterra model for N=3 of first student

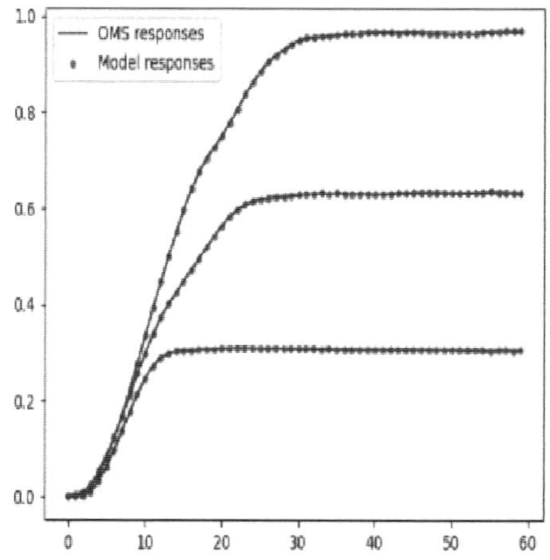

Figure 8. Average responses of the OMS and of the Volterra model for N=3 of second student

To locate features resistant to noise, we conducted the additional research which consisted of adding random 1% and 5% samples from a normal (Gaussian) distribution.

A complete enumeration of features was investigated by the method of statistical solutions for the PCR, from the set of values of which the average value of this indicator was calculated. The pair of features was chosen that maintained the stability of the indicator under perturbation.

The column chart of the maximum recognition reliability values of the best pair of features at different noise levels is presented in Figure 9. For comparison, a couple of features are presented that are less resistant to noise in Figure 10.

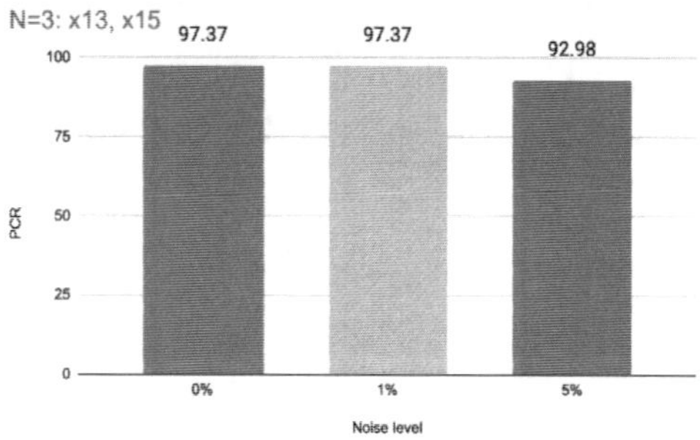

Figure 9. The column chart of the PCR of x_{13}, x_{15} features

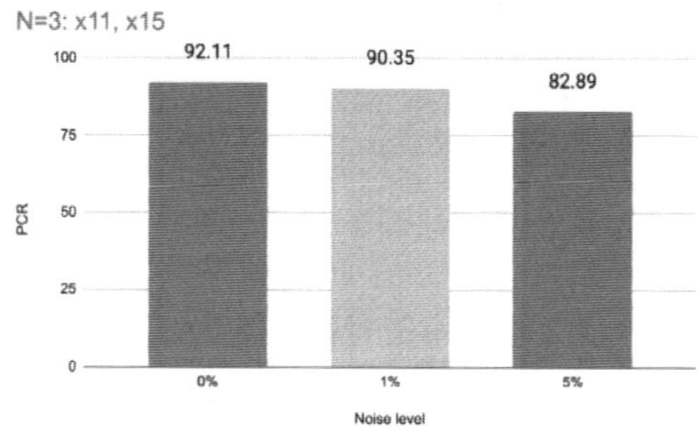

Figure 10. The column chart of the PCR of x_{11}, x_{15} features

Moreover, separately features x_{13} and x_{15} gives the following results:

- x_{13} yields the PCR P=0.8684;
- x_{15} yields the PCR P=0.7368.

So, the best combination of two features with noise resistance that gives high PCR is:

$$x_{13} = \min_{m} h_1'(m) \ \& \ x_{15} = \min_{m} h_3'(m,m,m) \cdot \qquad (10)$$

Bayesian classifier of individuals in two-dimensional features space is provided of the maximum recognition reliability (*P*) at the combination by these features: the PCR is *P*=0.9737.

Analogic results were obtained by means of Support Vector Machine (SVM) [30]. The classifier is built using a 2nd order polynomial kernel:

$$K(x, x') = ((\langle x, x' \rangle + 1)^d, \quad (11)$$

where d is determined by degree of the parameter, d =2.

Scikit-learn library was used (class sklearn.svm.SVC) to apply SVM. The location of the objects of the training sample in the space of features is show in Figure 11.

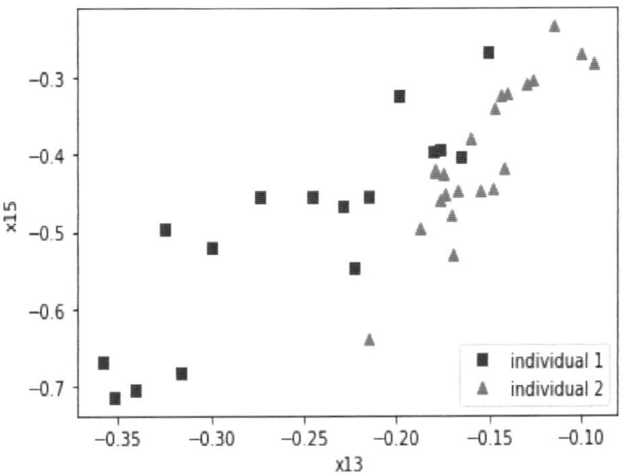

Figure 11. The location of the objects of the training set in the space of features x_{13} and x_{15}

5. Conclusion

Management and differentiation of access to computer systems and their resources is an important aspect of information security. It can be done through user identification. A new method of biometric identification is proposed formed on the determination of integrated Volterra models of the human OMS based upon experimental research "input-output" using innovative eye tracking technology. There is data of two respondents applied in experimental studies. More individuals' data will be involved in future study to expand the research base.

The first, second and third orders transition functions of the OMS were determined based on the data received using the TOBII PRO TX300 eye tracker. Since there is a substantial difference between the diagonal intersections of the second and third order transition functions of two individuals, they can be used for feature space forming to build statistical classifiers for identity recognition.

Promising results were obtained in the problem of identification by behavioral characteristics of a person with a recognition accuracy of more than 97%. A pair of features was selected that are resistant to computational errors, which gives a high result of the probability of correct recognition.

Acknowledgment

The authors are grateful to Prof. M. Milosz for the opportunity to conduct experimental researches at the Laboratory of Motion Analysis and Interface Ergonomics at the Lublin University of Technology (Poland) as well as Dr. M. Dzienkowski for helping to researches with using of the high-tech Eye-Tracker Tobii Pro TX300.

Appendix. Investigated heuristic features

#	Features	Formal definition
1	x_1	$x_1 = \sum_{m=1}^{M} \lvert h_1(m) \rvert$
2	x_2	$x_2 = \sum_{m=1}^{M} \lvert h_2(m,m) \rvert$
3	x_3	$x_3 = \sum_{m=1}^{M} \lvert h_3(m,m,m) \rvert$
4	x_4	$x_4 = \arg\max_m h_1'(m)$
5	x_5	$x_5 = \arg\max_m h_2'(m,m)$
6	x_6	$x_6 = \arg\max_m h_3'(m,m,m)$
7	x_7	$x_7 = \arg\min_m h_1'(m)$
8	x_8	$x_8 = \arg\min_m h_2'(m,m)$
9	x_9	$x_9 = \arg\min_m h_3'(m,m,m)$
10	x_{10}	$x_{10} = \max_m h_1'(m)$
11	x_{11}	$x_{11} = \max_m h_2'(m,m)$
12	x_{12}	$x_{12} = \max_m h_3'(m,m,m)$
13	x_{13}	$x_{13} = \min_m h_1'(m)$
14	x_{14}	$x_{14} = \min_m h_2'(m,m)$
15	x_{15}	$x_{15} = \min_m h_3'(m,m,m)$
16	x_{16}	$x_{16} = \max_m \lvert h_1(m) \rvert$
17	x_{17}	$x_{17} = \max_m \lvert h_2(m,m) \rvert$
18	x_{18}	$x_{18} = \max_m \lvert h_3(m,m,m) \rvert$
19	x_{19}	$x_{19} = \arg\max_m \lvert h_1(m) \rvert$
20	x_{20}	$x_{20} = \arg\max_m \lvert h_2(m,m) \rvert$
21	x_{21}	$x_{21} = \arg\max_m \lvert h_3(m,m,m) \rvert$

References

[1] Resmi K R and Raju G 2020 *International Journal of Computing* **19(4)** 575-82

[1] Cherrat E M, Alaoui R and Bouzahir H 2020 *International Journal of Computing* **19(1)** 11-9

[2] Paliy I, Sachenko A, Kurylyak Y, Boumbarov O and Sokolov S 2009 *Proceedings of the 5th IEEE International Workshop on Intelligent Data Acquisition and Advanced Computing Systems: Technology and Applications*, (IDAACS) pp 425-9

[3] Labati R D, Genovese A, Muñoz E, Piuri V, Scotti F and Sforza G 2016 *International Journal of Computing* **15(1)** 40-9

[4] Friedman L, Rigas I, Abdulin E and Komogortsev O V 2018 *Behav Res Methods* **50(4)** 1374-97

[5] Stuart S, Hickey A, Vitorio R, Welman K, Foo S, Keen D and Godfre 2019 *Physiol Meas* **40(2)**

[6] Lai M-L, Tsai M-J, Yang F-Y, Hsu C-Y, Liu T-C, Lee S W-Y, Lee M-H, Chiou G-L, Liang J-C and Tsai C-C 2013 *Educational Research Review* vol 10 pp 90-115

[7] Brasil A R A, Andrade J O and Komati K S 2020 *International Journal of Computer Applications* vol 176, 24 pp. 1-9
[8] van Renswoude D R, Raijmakers M E J, Koornneef A, Johnson S P, Hunnius S and Visser I 2018 *Behavior Research Methods* **50(2)** 834–52
[9] Wang D, Mulvey F B, Pelz J B and Holmqvist K 2017 *Behavior Research Methods* **49(3)** 947–59
[10] Quaia C and Optican LM 2003 *Strabismus* vol. 11 pp 17-31
[11] Kasprowski P and Ober J 2004 *European Conference on Computer Vision, Prague, Czech Republic* pp. 248-58
[12] Silver D L and Biggs A J 2006 *International Conference on Artificial Intelligence (ICAI)* (Las Vegas, NV, USA) pp 344-8
[13] Cantoni V, Galdi C, Nappi M, Porta M and Riccio D 2015 *Pattern Recognition* 1027-38
[14] Sachenko A, Banasik A and Kapczyński A 2009 *Advances in Soft Computing* 274-9
[15] Yoon H-J, Carmichael T R and Tourassi G 2014 *SPIE Medical Imaging* (San Diego California United States)
[16] Karpov A V and Komogortsev V 2012 *Science Journal of VolSU. Mathematics. Physics* **1 (16)** 42-8
[17] Holland C and Komogortsev O V 2013 *IEEE Transactions on Information Forensics and Security* **8(12)** 2115-26
[18] Komogortsev O V, Holland C D, Jayarathna S and Karpov A 2013 *ACM Transactions on Applied Perception* **10 (4)** 1-18
[19] Komogortsev O V, Holland C D, Karpov A and Price L R 2015 *ACM Transactions on Applied Perception* 1-14
[20] Rigas I, Economou G and Fotopoulos S 2012 *Pattern Recognition Letters* 786-92.
[21] Lohr D J, Friedman L and Komogortsev O V 2019 *Evaluating the Data Quality of Eye Tracking Signals from a Virtual Reality System: Case Study using SMI's Eye-Tracking HTC* v1
[22] Griffith H K, Katrychuk D and Komogortsev O V 2019 *IEEE/CVF International Conference on Computer Vision Workshop (ICCVW)*
[23] Hooge I T C, Holleman G A, Haukes N C and Hessels R S *Behavior Research Methods* 2018 1–10
[24] Pavlenko V, Salata D, Dombrovskyi M and Maksymenko Y 2017 *Mathematical Methods and Computational Techniques in Science and Engineering: AIP Conf. Proc. MMCTSE, UK, Cambridge* (Melville, New York) vol 1872 pp 110-117
[25] Doyle F J, Pearson R K, and Ogunnaike B A 2002 *Identification and control using Volterra models* (Germany: Springer Publ) 314
[26] Pavlenko V D and Pavlenko S V 2018 *Applied Aspects of Information Technology* **01(01)** 9-29.
[27] Pavlenko V, Pavlenko S and Speranskyy V 2014 *Advanced Data Acquisition and Intelligent Data Processing* (River Publishers) pp 233-270
[28] Pavlenko V, Ivanov I, and Kravchenko E 2017 *Proceedings of the 9th IEEE Int. Conf. on Intelligent Data Acquisition and Advanced Computing Systems: Technology and Applications (IDAACS'2017)* vol 2 pp. 645-50
[29] Pavlenko V, Milosz M and Dzienkowski M 2020 *Journal of Physics: Conference Series* (Rome Italy IOP Publishing) vol 1603 pp 1-8
[30] Vapnik V 2010 *The Nature of Statistical Learning Theory* (Springer-Verlag New York Inc)

Fast and accurate control of gates for quantum hypercomputation based on coherent domains of water

Luigi Maxmilian Caligiuri
Foundation of Physics Research Center (FoPRC), Cosenza, ITALY
E-mail: luigimaxmilian.caligiuri@foprc.org; max.caligiuri@gmail.com

Abstract. Quantum electrodynamics (QED) predicts that, under suitable conditions for temperature and density, liquid water can be described as a two-phases system composed by a coherent and a non-coherent phase. The first one is characterized by the formation of macroscopic regions, called "coherent domains" (CDs), in which all the molecules are phase - correlated among them and with a self-trapped electromagnetic field that leaks out from the CDs in the form of an evanescent tail. This feature, along with the possibility for the CDs to have an excited energy spectrum, allow them to interact each other through quantum tunneling of virtual photons. We have already shown how to exploit this dynamics to realize quantum hypercomputation by using a set of universal quantum gates made of interacting water CDs. In this paper we discuss how to accurately and fast control the macroscopic quantum states of such water coherent domains, by using electromagnetic potentials to modify the phase of their wavefunctions. Finally, a possible experimental set-up designed to this aim, based on the electric version of the Aharonov-Bohm effect, is discussed.

1. Introduction

As known, quantum computation manipulates quantum information represented by quantum bits (qubits) to perform logical operations. Besides the theoretical interest behind the study of quantum computation, one of its most fascinating feature is its alleged "supremacy" with respect classical computation making use of bits and its promises in respect to its technological applications in emerging and strategical fields as, for example, computer science, cryptography and cybersecurity, simulation of very complex physical systems, quantum communication, finance, and so on. So far, many systems and arrangements has been proposed as candidates to realize "physical" qubits, namely the physical implementation of the logical quantum bits of information.

Generically, any quantum two-levels system could be considered to this aim and many proposals has been advanced among which, for example, trapped ions, superconductors, quantum dots, optical cavities, molecular spins and so on [1-4].

Nevertheless, due to the very special features required for a physical system to allow a good implementation of a logical qubit [5], all the quantum computational architectures considered so far are characterized by specific strengths as well as weaknesses [2-4]. Apart from the requirement to ensure the long-term stability of the system against environmental decoherence, one of the most challenging issues of the actual quantum computing systems concerns the precise and fast setting up of the quantum states involved in computation and their control.

As pointed out by Di Vincenzo [5], the actual realization of any quantum computing scheme must satisfy, in particular, the following requirements:

- the physical realization of well-defined qubits;
- the initialization of the quantum system in a well-defined initial state, such, for example, the qubit $|0\rangle$;
- the set-up and "programming" of a set of universal quantum gates;

As already shown in previous papers [4,6-12], the coupling between excited states of "coherent domains" (CDs) of water, due to the exchange of virtual photons through tunneling effect, can be exploited to construct a set of universal quantum gates for quantum computation, including, in particular, the three one-qubit T-gate, S-gate and Hadamard gate as well as the two-qubits CNOT gate [4,10].

More specifically, such coupling is due to the interaction between the excited coherent quantum states of water enclosed in tubular structures made of metamaterials (MTMs).

In fact, according to the theory of QED coherence in condensed matter [13], liquid water is a two – phases system, composed by a coherent and a non-coherent fraction of molecules. The former is composed by arrays of CDs, namely quantum macroscopic spatial domains in which all the molecules oscillate in phase with each other and with a self-trapped electromagnetic field at a common frequency ω_{coh}. On the contrary, the non-coherent phase is made by an ensemble of uncorrelated molecules (vapor phase) filling the interstices between coherent domains.

A fundamental consequence of such coherent dynamics is the emergence of a phase-locking between the matter and e.m. quantum fields so that the phase relationship between them keeps a constant value inside CDs. Moreover, the coherent e.m. field spreads across the CD boundaries in the form of evanescent field whose tail can overlap the analogous evanescent fields produced by the other CDs in the surroundings [4,6-13].

In the case of liquid water, it has been shown [14] the coherent oscillations of matter and e.m. field involve the excited state $5d$ of the water molecule at $12.06\,eV$, that is just below the ionization threshold, with a weight of 0.13 so that every water CD is a source of a plasma of quasi-free electrons (i.e. 0.13 electrons per water molecule, meaning about more than 10^4 quasi-free electrons per CD).

The presence of such plasma of electrons makes it possible for the CDs to have excited coherent states produced by electron coherent "vortices" [14] that can be easily excited by external, even weak, energy source so originating metastable coherent excited states. Due to the coherence, the energy is absorbed by the CD as a whole. Moreover, the coherent vortices, being coherent, are characterized by a very long lifetime (they cannot be excited nor decay thermally) and have a quantized magnetic moment that allows them to align to an external static magnetic field.

The particularly long lifetime of unperturbed excited coherent states is due to the energy gap $\Delta E < 0$ that characterizes the energy of the coherent state with respect the non-coherent one. The situation is quite similar to that occurring in the superconductor state in which to break a Cooper's pair (so reverting the superconductor to a usual conductor) it is needed to supply energy to the system.

The energy of an excited coherent vortex, associated to a single CD, is given by [14]

$$E_n = \frac{\vec{L}_n^2}{2K} - g\vec{L}_n \cdot \vec{B} \tag{1}$$

where $\vec{L}_n$ is the angular momentum, K is the momentum of inertia of the CD, $g = e/2m$ the gyromagnetic ratio and $\vec{B}$ the external magnetic field. From equation (1) we see energy depends on the quantized angular moment and, then, is related to the rotational frequency of the vortices.

The Hamiltonian of a system composed by N interacting excited CDs can be written, within our model [4,9,10], as:

$$H = \sum_{i=1}^{N} \varepsilon_i |i\rangle \langle i| + \sum_{i \neq j}^{N} \Gamma_{ij} |i\rangle \langle j| \quad (2)$$

where $|i\rangle$ is the excited states of the i-th coherent tubular CD, ε_i its relative excitation energy and Γ_{ij} the "strength" of the tunneling interaction between the i-th and j-th CDs. The physical realization of quantum computation algorithms implies to map logical qubits to physical states of the considered quantum system, namely the "physical" qubits, and then to implement a suitable set of quantum gates to perform logical operations.

The possibility to exploit the interacting water CDs to perform practical quantum computation strongly depends on the ability to set and control the physical parameters and the quantum states appearing in the Hamiltonian (1). In particular, in order to use the above model of quantum hypercomputation to realize an actual computer architecture, we need a method to suitable change the quantum state of water CDs able to ensure:

- to "read" the quantum state associated to a given water CD;
- the precise control of the quantum states, namely the possibility to change the state vector describing water coherent domain from an initial state $|\Psi_0\rangle$ to to a desired new quantum state $|\Psi_1\rangle$;
- that the above operations will not put the quantum state out of coherence.

In the following we'll show how to achieve these goals within the same framework of QED quantum coherence in condensed matter by considering the active role of e.m. potentials on the coherent states of liquid water.

2. Quantum coherent states of water as eigenstates of quantum phase operator

One of the most important results emerging from the above outlined framework is that the quantum state describing a water CD is a coherent state characterized by a well-defined value of matter and e.m. field common frequency of oscillation This fact has a number of very important mathematical and physical consequences as we'll briefly outline below.

First of all, a coherent domain in condensed matter can be described by a macroscopic wavefunction given by

$$\Psi(\vec{x}, t) = \Psi_0(\vec{x}, t)\, e^{i\Theta(\vec{x}, t)} \quad (3)$$

where $\Psi_0(\vec{x}, t)$ is the wavefunction amplitude and $\Theta(\vec{x}, t)$ its phase. In the case of water, the phase $\Theta(\vec{x}, t)$ also includes the rotational frequency of coherent vortices so that the wavefunction (3) is considered to give a complete description of the energy spectrum of a CD.

From a mathematical viewpoint, coherent states can be considered as eigenvectors of a suitable quantum phase operator [14,15] which is fundamental for a quantum mechanical description of such states since the information about their physical structure is just contained in the phase.

In [15] a detailed treatment concerning the definition and the properties of quantum phase operator has been carried out. We recall that in quantum optics, a coherent state $|z\rangle$, with $z = |z|\, e^{i\phi}$ $(\phi \in R)$, of quantized electromagnetic field, also called a "Glauber state", is an eigenstate of the annihilation operator $\hat{a}$, namely

$$\hat{a}\,|z\rangle = z\,|z\rangle \tag{4}$$

and can be expanded, in terms of the occupation states $\{|n\rangle\}$ of Fock space, as [15]

$$|z\rangle = \exp\left(-\frac{1}{2}|z|^2\right)\sum_{n=0}^{\infty}\frac{z^n}{\sqrt{n!}}\,|n\rangle \tag{5}$$

whose mathematical properties are well-known.

Such coherent states involve a linear combination of an infinite number of eigenvectors of the number operator $\hat{N} = \hat{a}^\dagger\hat{a}$ and consequently an infinite number of quanta. As discussed in [15], an effective quantum phase operator should act on a coherent state by changing its phase, so we consider the so-called "phase shifting" operator, defined as

$$\hat{P}\left(\phi_0\right) \equiv \exp\left(-i\hat{N}\phi_0\right) \tag{6}$$

whose effect on a coherent state of the type (5) is just to generate a phase shift ϕ_0 namely

$$\hat{P}\left(\phi_0\right)|z\rangle = \left|e^{-i\phi_0}z\right\rangle \tag{7}$$

The operator $\hat{P}$ is a function of $\hat{N}$ only, (i.e. the number operator itself generates a phase shift of the quantum state) so the phase is a quantity canonically conjugated to the number of quanta, namely, in the phase representation [15]

$$N = -\frac{\partial}{\partial\phi} \tag{8}$$

and we can define a quantum phase operator $\hat{\Theta}$, satisfying the commutation relation

$$\left[\hat{N}, \hat{\Theta}\right] = -i \tag{9}$$

that leads to the following uncertainty relation

$$\left(\Delta\hat{\Theta}\right)\left(\Delta\hat{N}\right) \geq \frac{1}{2} \tag{10}$$

For a very high number of quanta ($N \to \infty$), as occurs in the coherent states, relation (10) implies the value of phase is always well-defined since $\Delta\Theta \to 0$. This characterizing feature of coherent states demonstrates the consistency of a quantum "phase" operator of which coherent states are eigenvectors and that could be used to define macroscopic quantum state associated to water CDs and to their time-evolution. The latter, in fact, is described by the unitary time-evolution operator

$$U\left(\tau\right) = \exp\left(-i\hat{H}\tau\right) \tag{11}$$

where H is the Hamiltonian operator applied for a time interval τ. As discussed above, we need to control such Hamiltonian in a precise way in order to act on specific qubits and to implement the desired quantum gates [4]. More specifically, within our model of quantum computation, this means to control the physical state of the water CDs and their dynamical evolution that, in turn, implies the ability to set, "measure" and modify the quantum phase $\Theta\left(\vec{x}, t\right)$ in equation (3) through the action of a suitable Hamiltonian. The Schrodinger equation implies the quantum state $|\psi\left(t_2\right)\rangle$ at a given time t_2 is obtained from the state at a previous time t_1 through the equation

$$|\psi(t_2)\rangle = U(t_1,t_2)\,|\psi(t_2)\rangle \tag{12}$$

where $U(t_1,t_2)$ is the unitary time-evolution operator (11) given by

$$U(t_1,t_2) \equiv \exp\left[-i\frac{\hat{H}}{\hbar}(t_2-t_1)\right] \tag{13}$$

By comparing equations (13) and (6) we argue [15] the time-evolution operator is formally identical to the phase-shift operator, provided that we assume

$$\hat{H} = \hbar\omega_0\hat{N} \tag{14}$$

with $\omega_0 \equiv \phi_0/\tau$ and $\tau \equiv t_2 - t_1$ so we can write

$$\hat{P}(\tau) \equiv \exp\left(-i\frac{\hat{H}}{\hbar}\tau\right) \tag{15}$$

and, finally

$$\exp\left(-i\omega_0\tau\hat{N}\right)|\phi\rangle = \exp\left(-i\frac{\hat{H}}{\hbar}\tau\right)|\phi(t_0)\rangle = |\phi+\phi_0\rangle = |\phi(t_0+\tau)\rangle \tag{16}$$

Equation (16) clearly means the time evolution of a coherent domain coincides with a change of the value of its overall phase.

This is perfectly consistent with the picture of a water CD as a macroscopic quantum object whose components have lost their individuality and share among each other only their common oscillation frequency ω_{coh} . If we assume the Hamiltonian in (15) fully describes the interaction between a CD and its surrounding environment, equation (16) tells us that such interaction results in phase shift due, as we'll show in the following, to the action of the electromagnetic vector and scalar potentials.

3. The physical meaning of macroscopic quantum coherent state and the role of phase

The coherent states of water so far considered are similar to those defined in quantum optics where they describe, for example, the states associated to coherent e.m. radiation field of a laser beam. More precisely, coherent states have properties similar to those characterizing "classical" coherent light and, in fact, they can be considered as the most closely quantum mechanical approximation of these classical states. Consequently, as laser light can be represented, under certain conditions, as idealized classical coherent light, in the same way, laser light can be quantum-mechanically described by quantum coherent states.

The connection between coherent states and light appears also natural since, as well as different states of light are distinguished by intensity and phase, a coherent state, being defined by a complex number, is determined by its amplitude and phase.

Analogously to the case of quantum optics, such coherent states represent the most closely quantum mechanical approximation of classical states in the limit where the number of quanta tends to infinite.

In order to better understand the physical meaning of the macroscopic quantum wavefunction associated to coherent states of water we recall, from the above discussion, that the concept of coherent state has been firstly introduced to correctly get macroscopic radiation waves from quantum electrodynamics, basing on the idea that these macroscopic radiations, whose overall behavior is characterized by classical - like properties expressed in terms of average field values, may be formed by the condensation of a large number of photons.

As we have also seen, these states, being characterized by a well-defined value of quantum phase, are not eigenstates of number operator $\hat{N}$, since they are characterized by large uncertainty in particle numbers. This is just what happens in condensed matter when the conditions for the occurrence of a coherent quantum macroscopic state are verified, namely when a large ensemble of interacting matter and e.m. quanta (considered as elementary quantum oscillators) perform a runaway towards the more stable coherent state leading to the formation of CDs [6,9,13]. These states correspond to the evolution of the system of interacting matter and e.m. field around classical paths, whose stationary solution show a space-structure composed by an array of CDs, macroscopic quantum structures resulting by the condensation of many quanta (belonging to the matter and e.m. fields coherently interacting each other).

It has been shown [13] the true vacuum state of a coherent systems in condensed matter is that in which a very high number of elementary quantum components oscillate in phase (named "coherent ground state" or CGS).

Formally, we consider a state representing a macroscopic quantum field which contains a very high number of quanta with a specific phase, whose vacuum (the CGS in our case) is not the state $|0\rangle$ of the Fock space, defined by $\hat{a}\,|0\rangle = 0$, but a new vacuum in which many bosons (the quasiparticles composing the system) are condensed.

In this case, if we indicate as $|0\rangle\rangle$ the vacuum associated to $\hat{a}$, namely

$$\hat{a}\,|0\rangle\rangle = 0 \tag{17}$$

and consider the Bogoliubov transformation

$$\hat{a} \to \hat{\alpha}\,(z) + z \tag{18}$$

where z is a c-number, the vacuum state $|0\,(z)\rangle$ for $\hat{\alpha}\,(z)$, satisfying

$$\hat{\alpha}\,(z)\,|0\,(z)\rangle = 0 \tag{19}$$

is the coherent state of $\hat{a}$ with eigenvalue $-z$ [15] that is

$$\hat{a}\,|0\,(z)\rangle = -z\,|0\,(z)\rangle \tag{20}$$

and it can be obtained from $|0\rangle\rangle$ by the transformation

$$|0\,(z)\rangle = \exp\left(-\frac{1}{2}|z|^2\right)\exp\left(-z\hat{a}^\dagger\right)\exp\left(-z^*\hat{a}\right)|0\rangle\rangle \tag{21}$$

Equation (21) shows the vacuum state $|0\,(z)\rangle$ of $\hat{\alpha}$ is a superposition of states characterized by many $\hat{a}$- particles and can be considered as a condensation of them. From the above discussion, we argue the coherent state of matter and e.m. fields, including a very high number of quanta oscillating in phase and giving rise to the CDs arrays in condensed matter, is conceptually analogous to the coherent state $|z\rangle$ above considered, namely the vacuum state with particle condensation. This result further confirms that the quantum state describing coherent domains of water must be an eigenstate of the phase operator $\hat{\Theta}$ since it is characterized by a well-defined common oscillation frequency of matter and e.m. field.

More interestingly, such coherent states resemble, for example, the wavefunctions representing superfluid and superconductor physical systems.

As well as the macroscopic model of superconductivity is based on the hypothesis that there is a macroscopic wave function $\Psi\,(\vec{x}, t)$ which describes the whole ensemble of superconducting electrons, the theory of QED coherence in matter predicts a macroscopic wavefunction associated to the coherent domains which describes the collective coherent behavior of matter (water molecules in our case) and e.m. field whose quanta can be considered as a condensate of quasiparticles that can be observed at a macroscopic scale.

The assumption of a macroscopic quantum wavefunction describing the whole ensemble of matter-e.m. interacting fields has further consequences, apart from the role played by phase, concerning, first of all, the meaning of $|\Psi(\vec{x},t)|^2$. For a single particle, this quantity represents the density probability to find the particle at point $\vec{x}$ at time t and give rises to the normalization condition $\int |\Psi|^2 dV = 1$. Analogously, we assume the macroscopic wavefunction $\Psi(\vec{x},t)$ representing the whole ensemble of coherently interacting matter+e.m. fields quanta, satisfies the normalization condition

$$\int \Psi^*(\vec{x},t)\Psi(\vec{x},t)\,dV = N \tag{22}$$

$$|\Psi(\vec{x},t)|^2 = n(\vec{x},t) \tag{23}$$

where N is the total number of particle in the ensemble and $n(\vec{x},t)$ its local density. In particular, when considering the equation (23), there must be a sufficiently large density of quanta in order to make the definition of local density meaningful.

Finally, the consideration of a macroscopic wavefunction describing the whole coherent ensemble (as the water CD in our case) has the advantage to allow the quantum dynamical characterization of the system irrespective of the detailed microscopic underlying dynamics.

4. Fast and precise control of coherent quantum state of water through e.m. potentials

4.1. The e.m. potentials in quantum mechanics

It is known that classical electrodynamics can be expressed either in terms of the "physical" electric and magnetic fields $\vec{E}(\vec{x},t)$ and $\vec{B}(\vec{x},t)$ either in terms of electromagnetic potentials ϕ and $\vec{A}$, respectively defined as the scalar and vector e.m. potentials being the components of the four-vector $A_\mu = \left(A_0, \vec{A}\right)$ with $A_0 = \phi$. Differently from classical mechanics, in quantum theory, which is based on the Hamiltonian or Lagrangian formulation, the local electromagnetic fields disappear from the equation of motion leaving the place to scalar and vector potentials.

It is known from the general theory that the equation governing the motion of a particle of charge q and mass m in a region in which there is a e.m. field described by the four-potential A_μ, is the Schrodinger equation:

$$i\hbar\frac{\partial}{\partial t}\Psi = \left[\frac{1}{2m}\left(-i\hbar\vec{\nabla} - \frac{q}{c}\vec{A}\right)^2 + q\phi\right]\Psi \tag{24}$$

where the Hamiltonian of the system is

$$H = \frac{1}{2m}\left(-i\hbar\vec{\nabla} - \frac{q}{c}\vec{A}\right)^2 + q\phi \tag{25}$$

The e.m. potentials are defined so that the electric and magnetic fields expressed as

$$\vec{E} = -\frac{\partial\vec{A}}{\partial t} - \vec{\nabla}\phi \qquad \vec{B} = \nabla\times\vec{A} \tag{26}$$

are invariant under gauge transformation of potentials

$$\vec{A}\rightarrow\vec{A}' = \vec{A} + \vec{\nabla}\Lambda \qquad \phi\rightarrow\phi' = \phi - \frac{\partial}{\partial t}\Lambda \tag{27}$$

Journal of Physics: Conference Series **2162** (2022) 012025 doi:10.1088/1742-6596/2162/1/012025

where Λ is some scalar function and the Schrodinger equation (24) assumes the invariant form

$$i\hbar\frac{\partial}{\partial t}\Psi' = \left[\frac{1}{2m}\left(-i\hbar\vec{\nabla} - \frac{q}{c}\vec{A}'\right)^2 + q\phi'\right]\Psi' \quad (28)$$

provided that the wavefunction transforms as

$$\Psi'(\vec{x},t) = \Psi(\vec{x},t)\exp\left[i\frac{q}{\hbar c}\Lambda(\vec{x},t)\right] \quad (29)$$

In particular it can be shown [16] that the solution of the Schrodinger equation (24) can be written, in a covariant form, as

$$\Psi(\vec{x},t) = \Psi_0(\vec{x},t)\, e^{i\frac{q}{\hbar}\left(\int_{\vec{x}_0}^{\vec{x}}\vec{A}(\vec{x}')d\vec{x}' - \int_{t_0}^{t}\phi(t')dt'\right)} \quad (30)$$

provided that

$$\frac{\partial\vec{A}}{\partial t} = \nabla\phi = 0 \quad (31)$$

where the points $(\vec{x}_0, t_0)$ and $(\vec{x}, t)$ define the path of the particles in the space-time region in which the e.m. field is present and

$$\Psi_0(\vec{x},t)$$

is the solution of the (24) when $\phi = \vec{A} = 0$, namely in the absence of e.m. field.

Equation (30) shows that when a charged particle enters in a region in which a e.m. field is present it "accumulates" a phase given by

$$\Delta\Theta = \frac{q}{\hbar}\left(\int_{\vec{x}_0}^{\vec{x}}\vec{A}(\vec{x}')\,d\vec{x}' - \int_{t_0}^{t_1}\phi(t')\,dt'\right) \quad (32)$$

If we consider a situation in which only the vector potential $\vec{A}(\vec{x})$ or the scalar potential $\phi(t)$ is present, we can respectively write

$$\Delta\Theta_{\vec{A}} = \frac{q}{\hbar}\int_{\vec{x}_0}^{\vec{x}}\vec{A}(\vec{x}')\,d\vec{x}' \quad (33)$$

and

$$\Delta\Theta_{\phi} = -\frac{q}{\hbar}\int_{t_0}^{t_1}\phi(t')\,dt' \quad (34)$$

From equation (33) we obtain, if $\nabla\times\vec{A} = 0$

$$\nabla\Theta = \frac{q}{\hbar}\vec{A} \quad (35)$$

and, from equation (34)

$$\frac{d\Theta}{dt} = -\frac{q}{\hbar}\phi \quad (36)$$

4.2. The role of e.m. potentials in the dynamical evolution of a coherent system

As we have shown, when a charged particle enters a region in which a e.m. field is present it acquires a phase given by equation (32). If Ψ describes a non - coherent state of a system, then its phase-shift due to the e.m. potentials has no effect on its dynamical evolution and can be even eliminated through a global phase transformation. On the other hand, the presence of potentials (or their modification through the gauge Λ) affects only the phase of wavefunction and not its amplitude that depends by the "physical" fields (the amplitude, in fact, remains unaffected by gauge transformation).

The case is totally different if the wavefunction Ψ describes a macroscopic coherent quantum system in which the phase plays an essential role.

As we have seen, in the coherent quantum state of water, the wavefunction Ψ (both of particles and e.m. field) acquires a macroscopic meaning and a wave-like behavior, a situation quite similar to that occurring in laser or in superconductivity where the entire collection of microscopic objects (respectively photons and super-electrons) is described in terms of an electromagnetic field with a well-defined phase.

A simple picture of what happens in this case would be a stationary situation in which $|\Psi(\vec{x},t)|^2$ is approximatively constant and the Hamiltonian H of the system can be replaced by the energy E of the particles. In this case the Schrodinger equation for the wavefunction (3) can be written as

$$-\hbar\frac{\partial\Theta}{\partial t} = -E \tag{37}$$

that confirms the quantum features of the particles reduce to the phase of wavefunction. For a non-coherent system, the particles cannot have the same energy and the time-evolution of the phase of their respective wavefunctions differs for all the particles. This means the values of phase are, in general, uniformly distributed and they don't contribute to the value of observable macroscopic quantities describing the system as a whole.

On the contrary, in a coherent system, as that composed by the water CDs, all the elementary components of the system are characterized by the same quantum state, just described by the macroscopic wavefunction (3), and by the same energy value carachterizing the CGS [9,13]. As a result, as already discussed above, all the elementary constituents of the system (water molecules and its associated coherent e.m. field in our case) form a phase-locked state in which they are all strongly-correlated. In these circumstances, the phases are not randomly distributed and sum up to a well-defined value affecting the macroscopic variables as described by the macroscopic wavefunction (3).

Another way to see this is to remember that coherent systems are generally characterized by non-linear dynamics (that originates the quantum phase transition to the non-coherent state to CGS) [9,13] in which the Hamiltonian (25) contains self-interaction terms as $V = V(\Psi)$.

As already recognized [10,14], in such complex systems, the phase acquires an important physical meaning (so ceasing to be a pure mathematical entity as in classical electrodynamics) able to drive its dynamics since the electromagnetic potentials of the different components of the system determine their phase evolution that, in turn, drives the system evolution.

As regards as the microscopic physical mechanism underlying the action of e.m. potentials on the coherent evolution of the system, we remember the water CDs can be considered as an enormous reservoir of quasi-free coherent electrons that are involved in the CD oscillation. This means the e.m. potentials could act on each of them according to the Hamiltonian (25) modifying their frequency of oscillation and, consequently, the overall phase of the CD.

Furthermore, due to the coherence of the excited states of water CD, the coherent quasi-free electrons are not be able to dissipate their energy thermally. This is conceptually similar to what happens in a superconductor where we have supercurrents composed by Copper pairs described by coherent wavefunctions and for which the e.m. potentials are able to modify their quantum behavior at a macroscopic scale.

To be able to modify the macroscopic quantum phase of a coherent state means, within our model, to have the skill to control the quantum state of water CDs which are fully characterized, being macroscopically coherent, by their quantum phases. The precise tuning of this phase would allow us exactly controlling the quantum state of the system and its excited states (each of them is associated to a well-defined rotational frequency of coherent electron vortices), i.e. the physical parameters driving the evanescent tunneling interaction between CDs and then the quantum gates based on them [4,10].

A space-time modification of would determine a substantial change in the physical state and evolution of the water CD so that, more generally, the interaction among two or more CDs or between a CD and the surrounding "environment" can be considered as mediated by the action of the phase shift operator.

Then, by properly manipulate the space-time values of e.m. potentials, it could be possible to "program" and control the quantum gates using water CDs as physical substrate.

In this way we can easily satisfy the requirement needed for quantum computation about the control of states, by setting a correspondence between a given value of phase and the initial quantum state of computation, for example

$$|\Phi_0\rangle \to |0\rangle \tag{38}$$

and recalling that, since any coherent state corresponds to an eigenstate of quantum phase operator with a well-defined value of phase, we can always state a bijective correspondence between such value and the corresponding quantum state of CD, namely

$$\Theta_i \to |\Theta_i\rangle \tag{39}$$

Finally, a further very intriguing feature of phase tuning through e.m. potentials is its long - range as well as superfast propagation [6,10,14], since the interaction that determines the change of phase, in this case, is driven by a phase velocity and is able to propagate even faster than light without violating any fundamental principle since it involves only a local supply of energy arising from quantum vacuum fluctuations.

5. A possible experimental set-up to control quantum phase of water CDs

As we have seen, water CDs are macroscopic quantum objects described by a the wavefunction (3). In [4,10] we have considered tubular structures whose walls are composed by metamaterials (MTMs) filled by coherent water so that the inner volume of such tubes could be considered as unique tube-shaped CD. We have also shown [4,6,10] how the interaction between such tubular CDs could be used to realize universal quantum gates for attaining a quantum hypercomputation systems, provided that the state of each of these CDs (i.e. their quantum phase) could be controlled with precision.

In the following we'll discuss a possible conceptual experimental set-up able to achieve the control of quantum phase of water coherent domains. To this aim we'll exploit the relationship between e.m. potentials and phase of the water CDs and, in particular, the equation (34), to show how to set-up a given phase difference between two water CDs.

The theoretical treatment we'll based on is in somewhat similar to that considered in the study of the electric Aharonov-Bohm effect (see [17] for a good treatment). In our configuration we assume $\vec{A} = 0$ and $\vec{\nabla}\phi = 0$ locally and consider two close water CDs (each enclosed in a MTM-made container), placed on either side of a large plane capacitor as shown in figure 1.

A voltage difference $\Delta V \equiv V$ is induced between the two CDs in the regions in which $\vec{E} = 0$. If we indicate with $V_1(t)$ the potential at the left face of the capacitor and with $V_2(t)$ the

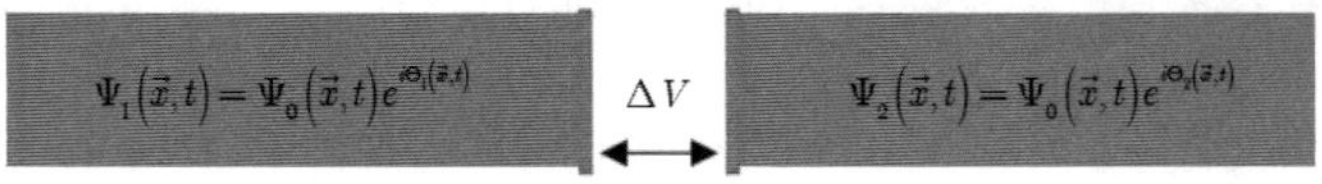

Figure 1. Two water coherent domains subject to an electric potential difference V. Positive potential is on he left, negative potential on the right. A charged capacitor is placed between the two CDs, generating the electric field directed from the left to the right.

potential at the right face we can write the phase difference between the two CDs, by using the equation (34) on both the sides of the capacitor and taking the difference between the two terms

$$\delta\Theta = \frac{e}{\hbar}\int_{t_1}^{t_2} V(t)\,dt \tag{40}$$

where $V(t) = V_2(t) - V_1(t)$, $\Delta t = t_2 - t_1$ is the time interval during which the potentials are active on both the faces of the capacitor and, in our case, $q = -e$. Equation (40) shows that by suitably adjusting the potential V as well as the time interval Δt we could, in principle, obtain a desired phase shift between the wavefunctions of the two water CDs. In this way it could be possible, by considering a "reference" isolated CD (namely whose phase doesn't change with time), to set the value of phase of any other water CD through equation (40).

As regards as the last point, the phase difference so "stored" in the CDs can be measured by exploiting the Josephson effect that my occur between the two involved CDs. In fact, as we have discussed above, the energy of excited levels of CDs is "stored" in the quasi-free electrons coherent vortices who have an associated quantized angular momentum given (in absence of external magnetic field), through equation (1), by

$$L_n = \frac{K}{{}^{-}\!h}E_n \tag{41}$$

which creates a quantized magnetic flux Φ confined inside CDs, namely the necessary condition for the appearance of a supercurrent along the tubular CDs.

In order to obtain an observable supercurrent, a physical connection between the CDs should be realized by using, for example [18], superconducting wires originating from the two CDs connected through an insulating Josephson junction J, located far from the capacitor as represented in figure 2.

Measuring such a supercurrent would give an indirect measure of the phase shift $\delta\Theta$ by using the first Josephson equation

$$I = I_c \sin(\delta\Theta) \tag{42}$$

where I is the supercurrent and I_c is the "critical" current that depends on the configuration of the junction J [18].

Despite its conceptual simplicity, this measurement process could produce several systematic effects able to affect the measurements results as, for example, the interference of the electric field of the capacitor on the superconducting wires or the non-vanishing electrical fields produced by the thermal noise inside the metamaterial structures enclosing water CDs.

Furthermore, particular care must be taken in placing water CDs in a region from which the electric field is entirely excluded a condition that could be satisfied, for example, by putting tubular CDs inside suitable Faraday cages.

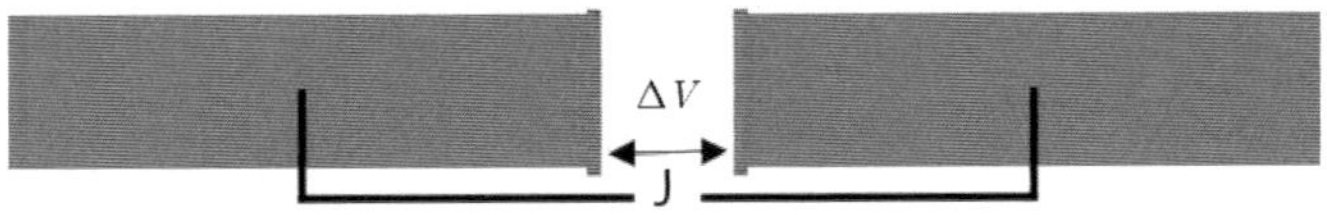

Figure 2. Superconducting wires connecting the two water CDs in the suggested experimental set-up.

6. Conclusions and outlook

According to the theory of QED coherence in condensed matter, liquid water could be described as a two-phases systems in which one phase is "coherent" and characterized by the formation of macroscopic regions, called "coherent domains" (CDs), in which all the water molecules are phase - correlated among them and with a self-trapped electromagnetic field.

The CDs admit excited energy states due to the formation of coherent electron vortices that allow them to interact each other through quantum tunneling of virtual photons. We have already shown how to exploit this novel kind of quantum interaction to realize a set of universal quantum gates able to perform quantum hypercomputation, provided we are able to control and "program" the macroscopic quantum state of CDs to realize the required quantum gates and the logical operations.

Such quantum states are characterized by a well-defined value of phase, being eigenstates of quantum phase operator. So, manipulating these phases would allow us to set and control the associated quantum states of water CDs.

In this paper, basing on the physical significance of electromagnetic potential in quantum mechanics, we have discussed how to use e.m. scalar vector to modify the phase of water CDs. A conceptual experimental setup able to perform such phase control is finally discussed.

Despite its theoretical foundation, this scheme needs to be implemented to consider, for example, the design of a suitable method to measure the phase difference between the involved water CDs able to avoid the main issues previously described as well as the possibility to use the e.m. vector potential, apart the scalar potential, to generate the required phase field. All these issues will be carefully analyzed in forthcoming publications.

We think our proposal, although introductory at all, could represent a further strategical step towards the actual realization of quantum hypercomputer system using water molecules as physical and computational substrate.

7. References

[1] Resch S and Karpuzcu U R Quantum Computing: An Overview Across the System Stack 2019 **Preprint** arXiv:1905.07240
[2] Hidary J D 2019 *Quantum Computing: An Applied Approach*, in *Plastics* (Cham, Switzerland: Springer Nature) 47–59
[3] Nielsen M A and Chuang I L 2016 *Quantum Computation and Quantum Information (10th Anniversary Edition)* (Cambridge, UK: Cambridge University Press) 277-82
[4] Caligiuri L M 2021 Quantum (hyper)computation through universal quantum gates in water coherent domains *Proc. of the 5th International Conference on Applied Physics, Simulation and Computing* (Salerno) pp 199–208
[5] Di Vincenzo D P 2000 The Physical Implementation of Quantum Computation *Fortschr. Phys.* **48** 9-11 771-83
[6] Caligiuri L M and Musha T 2016 *The Superluminal Universe: from Quantum Vacuum to Brain Mechanism and beyond* (New York, USA: NOVA Science Publisher)
[7] Caligiuri L M and Musha T 2019 Quantum hyper-computing by means of evanescent photons *J. Phys.: Conf. Series* **1251** 012010
[8] Caligiuri L M and Musha T 2019 Accelerated Quantum Computation by means of Evanescent Photons and its

Prospects for Optical Quantum Hypercomputers and Artificial Intelligence *Proc. of the 2019 International Conference on Engineering, Science, and Industrial Applications* (Tokyo) pp 199–208

[9] Caligiuri L M 2020 *Frontiers in Quantum Computing* ed L.M. Caligiuri (New York, USA: NOVA Science Publisher) pp 1–37

[10] Caligiuri L M 2020 *Frontiers in Quantum Computing* ed L.M. Caligiuri (New York, USA: NOVA Science Publisher) pp 57–102

[11] Caligiuri L M 2021 Accelerated quantum computation based on quantum coherent dynamics of evanescent photons in liquid water *Intelligent Computing* ed K. Arai (Cham, Switzerland: Springer Nature) vol 1 pp 169-87

[12] Caligiuri L M and Musha T 2016 Superluminal Photons Tunneling through Brain Microtubules Modelled as Metamaterials and Quantum Computation *Advanced Engineering Materials and Modeling* eds A. Tiwari, N. Arul Murugan, and R. Ahula (New York, NJ, USA: Wiley Scrivener Publishing LLC) pp 291-333

[13] Preparata G 1995 *QED Coherence in Matter* (Singapore, London, New York: World Scientific Publishing Co Pte Ltd) 195-217

[14] Del Giudice E and Tedeschi A 2009 Water and Autocatalysis in Living Matter *Electromagnetic Biology and Medicine* **28** 1 46-52

[15] Caligiuri L M 2020 *Frontiers in Quantum Computing* ed L.M. Caligiuri (New York, USA: NOVA Science Publisher) pp 39–56

[16] Jing J, Zhang Y F, Wang K, Long Z W and Dong S H 2017 On the time-dependent Aharonov-Bohm effect *Physics Letters B* **774** 87-90

[17] Peshkin M and Tonomura A 1989 *The Aharonov-Bohm effect* (Berlin, Germany: Springer-Verlag)

[18] Bachlechner T C and Kleban M 2019 Testing the electric Aharonov-Bohm effect with superconductors *arXiv*: 1909.11668v1

AUTHOR INDEX

Acri, G 1, 40
Anfuso, C 40
Bajare, D 154
Belmusto, G 1, 8
Bianco, Francesco 116
Bliznakov, Zhivko 135
Bliznakova, Kristina 127, 135
Bretón, J G Cerón 96
Bretón, R M Cerón 96
Caligiuri, Luigi Maxmilian 16, 258
Caridi, F 1, 8, 40
Carlotta, Bogena 173
Chori, Vladysla 249
Crupi, V 1, 8
Del Pizzo, Lara Ginevra 116
Dukov, Nikolay 127, 135
Encheva, Elica 127
Faggio, G 1
Filippov, V B 235
Francavilla, A B 57, 67, 78, 145
Francavilla, A 86
Fuentes, M L Espinosa 96
Gajjar, K 194
Graziuso, G 57, 78, 86, 145
Greco, C 40
Grillo, R 1
Groppen, Vitaly O 226
Grunwald, Joschka 162
Guarnaccia, C 57, 67, 78, 86, 96, 145
Guida, M 78
Gurgone, S 40
Guzmán, A A Espinosa 96
Hovsepyan, Sirarpi 173
Izzo, C 205
Kaler, Priyal 216
Klemeshev, V A 235
Kurbatova, G I 235
Lara, E Ramírez 96
Licitra, Gaetano 116
Lima, C 49
Longobardi, A 67
Lozada, S E Carranco 96
Lyall, Ishita 216
Majolino, D 1, 8
Mancini, S 57, 67, 78, 86, 145
Martinelli, E 49
Mercorelli, P 194
Mercorelli, Paolo 162, 173
Messina, G 1
Nolasco, V G Bautista 96
Paladini, G 1, 8, 40
Pannu, H S 216
Pavlenko, Vitaliy 249
Pepe, M 49
Praticò, Filippo Giammaria 116
Pundiene, I 154
Schiaffino, Gloria 116
Severino, R C Lara 96
Shamanina, Tetiana 249
Silvestri, Stefano 116
Singh, Aekamjot 216
Singh, Daljeet 216
Spingos, I 185
Testagrossa, B 40
Vallianatos, F 185
Van Niekerk, T 194
Venuti, V 1, 8, 40
Vermiglio, G 40
Viccione, G 67, 205
Vilnitis, M 154
Vitola, L 154
Wilm, Thomas 194